FEB 14 1972

AF676819

LIBRARY
WARNER - LAMBERT RESEARCH INSTITUTE
MORRIS PLAINS, NEW JERSEY

Workshop Meeting
on the
Chemistry of Gonadotrophins
and
the Development of
the Ovary in Infancy.

Participants of the two Workshop Conferences at Hartlebury Castle, which has been the home of the Bishops of Worcester since the ninth century. The original castle was built as a fortress against the Vikings.

Workshop Meeting on the Chemistry of Gonadotrophins,
Birmingham, England, 1969.

Gonadotrophins and Ovarian Development

Proceedings of the two Workshop Meetings on the Chemistry of the Human Gonadotrophins and on the Development of the Ovary in Infancy, Birmingham, 1969.

Edited by

W. R. Butt
B.Sc(Lond), Ph.D.,
D.Sc(Birm), F.R.I.C.

A. C. Crooke
M.A., M.D.(Cantab),
F.R.C.P. (Lond).

and

Margaret Ryle
M.A.(Oxon), Ph.D.(Cantab).

E & S LIVINGSTONE
EDINBURGH & LONDON
1970

The Williams & Wilkins Company
Baltimore

© Longman Group Limited, 1970

All rights reserved. No part of this publication may be reproduced, stored in a retrieval system, or transmitted in any form or by any means electronic, mechanical, photocopying, recording or otherwise, without the prior permission of the publishers (E. &. S. Livingstone, Teviot Place, Edinburgh)

ISBN O 443 00719 5

612.62
W892

Printed in Great Britain

Preface

Since the first meeting of the Gonadotrophin Club in Geneva in 1953, there has been a number of other gatherings, usually of small numbers of people, to discuss the gonadotrophins in all their aspects. At the Vista Hermosa Workshop on Gonadotrophins in Mexico in 1968, Professor Lunenfeld suggested that the subject had now grown to the point where it would be useful to hold a meeting confined to one aspect only of this growing subject, the chemistry of gonadotrophins, and that this meeting should be held in Birmingham. The Birmingham group were honoured by this proposal and set to work to organize it. They soon found, however, that it was necessary to confine the meeting to human gonadotrophins and to restrict the immunology of gonadotrophins to that area in which immunological properties threw light on structure. Otherwise the meeting would have outgrown its resources.

When arrangements were well under way Dr Hannah Peters wrote from Copenhagen to say that her group would like to hold a meeting on the Development of the Ovary immediately after the meeting on the Chemistry of Gonadotrophins since a number of participants would be interested in attending both. The local arrangements for this meeting were therefore also made by the Birmingham group.

The meetings were made possible by financial support from Messrs. G.D. Searle & Co. Ltd., Organon Laboratories Ltd., and the Nordisk Insulinfond, Copenhagen.

Individual participants wish to thank the Editors and Publishers of journals and books for permission to reproduce previously published figures.

The editors are most grateful to Mollie Ball, Janet Stacey, Moira Wardill, Kevin Morrison and Donald Weetman for their willing cooperation with secretarial and other assistance during the meeting.

We are most grateful to E. & S. Livingstone, for agreeing to publish these proceedings and for their willing cooperation.

W. R. Butt
A. C. Crooke
Margaret Ryle

1970

List of Participants

Workshop Meeting on the Chemistry of Gonadotrophins

Mr. J. Adamson	G. D. Searle and Co. Ltd., Lane End Road, High Wycombe, Bucks.
Prof. O. P. Bahl	Department of Biochemistry, State University of New York and Buffalo, Health Sciences Center, G-56 Capen Hall, Buffalo, N.Y. 14214, U.S.A.
Dr. D. R. Bangham	Division of Biological Research, National Institute for Medical Research, Mill Hill, London, N.W.7.
Prof. S. A. Barker	Department of Chemistry, University of Birmingham.
Prof.G. Bettendorf	Universitats-Frauenklinik und -Polilinik, 2 Hamburg 20, Martinstrasse, 52, Germany.
Dr. W. R. Butt	Department of Clinical Endocrinology, Birmingham and Midland Hospital for Women, Birmingham 11.
Prof. R. E. Canfield	College of Physicians and Surgeons of Columbia University, 630 West 168th Street, New York, N.Y. 10032 U.S.A.
Dr. M. F. Chaplin	Department of Chemistry, University of Birmingham 15.
Dr. A. C. Crooke,	Department of Clinical Endocrinology, Birmingham and Midland Hospital for Women, Birmingham 11.
Dr. P. Donini and Miss Donini	Laboratori di Ricarca, Istituto Farmacologico, Serono, Sede Centrale, Via Casilina 125, Rome.
Dr. A. Eshkol	Institute of Endocrinology, Tel Hashomer Government Hospital, Israel.
Dr. P. Franchimont	Agrégé, Hôpital Universitaire de Bavière, Institut de Médecine, Département de Clinique et de Pathologie Médicales, Liège, Belgium.

Dr. M. Gallagher	G. D. Searle and Co. Ltd., Lane End Road, High Wycombe, Bucks.
Dr. Grasslin	Universtats-Fraienklinik und -Poliklinik, 2 Hamburg, 20 Martinstrasse, 52, Germany.
Dr. C. J. Gray	Department of Chemistry University of Birmingham.
Dr. R. O. Greep	Harvard Medical School, Laboratory for Human Reproduction and Reproductive Biology, Boston, Massachusetts 02115, U.S.A.
Dr A. S. Hartree	Department of Biochemistry, University of Cambridge.
Dr. W. M. Hunter	M.R.C. Clinical Endocrinology Research Unit, 2 Forrest Road, Edinburgh EH1 2QW.
Dr. P. Jaquet	Hôpital Universitaire de Bavière, Institut de Médecine, Départment de Clinique et de Pathologie Médicales, Liège, Belgium.
Dr. M. Jutisz	Collège de France, Laboratoire de Physiologie Cellulaire, 11 Place Marcelin Berthelot, Paris V^{e}.
Dr. J. F. Kennedy	Department of Chemistry, University of Birmingham.
Dr. T. Klacansky	First Clinic of Obstetrics and Gynaecology, Bratislava, Zochova 5, Czechoslovakia.
Dr. J. A. Loraine	M.R.C. Clinical Endocrinology Research Unit, 2 Forrest Road, Edinburgh EH1 2QW.
Dr. S. S. Lynch	Department of Clinical Endocrinology, Birmingham and Midland Hospital for Women, Birmingham 11.
Prof. B. Lunenfeld	Director of Institute of Endocrinology, Tel Hashomer Government Hospital, Israel.
Dr. F. Naftolin	Department of Human Anatomy, University of Oxford.
Prof. W. D. Odell	Department of Medicine, Harbor General Hospital, 1000 West Carson Street, Torrance, California 90509, U.S.A.
Dr. R. F. Palmer	G. D. Searle and Co. Ltd., Lane End Road, High Wycombe, Bucks.
Dr. W. D. Peckham	University of Pittsburgh School of Medicine, Department of Physiology, Pittsburgh Pa. 15213. U.S.A.

Dr. S. Raiti	Institute of Child Health, 30 Guildford Street, London, W.C.1.
Dr. P. Rathnam	The New York Hospital-Cornel Medical Center, Department of Medicine, 525 East 68th Street, New York, N.Y. 10012, U.S.A.
Dr. P. Roos	Biokemiska Institutionen, Uppsala Universitet, Uppsala, Sweden.
Dr. M. Ryle	Department of Clinical Endocrinology, Birmingham and Midland Hospital for Women, Birmingham 11.
Prof. B. B. Saxena	The New York Hospital-Cornell Medical Center, 525 East 68th Street, New York, N.Y. 10021, U.S.A.
Dr. A. Schuurs	N.V. Organon, Kloosterstraat 6, Oss, Holland.
Prof. M. Stacey	Department of Chemistry, University of Birmingham.
Dr. M.P. Stack-Dunne	Division of Biological Standards, National Institute for Medical Research, Mill Hill, London, N.W.7.
Dr. V. C. Stevens	The Ohio State University, Department of Obstetrics and Gynecology, 410 West 10th Avenue, Columbus, Ohio 43210, U.S.A.
Dr. M. Theoleyre	Collège de France, Laboratoire de Physiologie Cellulaire, 11 Place Marcelin Berthelot, Paris V^{e}.
Dr. H. van Hell	N.V. Organon, Kloosterstraat 6, Oss, Holland.

Workshop Meeting on the Development of the Ovary in Infancy

Dr. S. Ben-Or	Department of Physiology, Hadassah Medical School, Hebrew University, Jerusalem, Israel.
Dr. H. Brandau	Universitäts-Frauenklinik, 23 Kiel, Germany.
Miss A.G.S. Byskov	The Finsen Laboratory, Copenhagen, Denmark.
Dr. A. C. Crooke	Department of Clinical Endocrinology, Birmingham and Midland Hospital for Women, Birmingham 11.

Dr. R. Deanesly	School of Agriculture, Cambridge University.
Dr. B. T. Donovan	Department of Physiology, Institute of Psychiatry, De Crespigny Park, London, SE5.
Dr A. Eshkol	Institute of Endocrinology, Tel Hashomer Government Hospital, Israel.
Dr. L. Franchi	Department of Anatomy, Birmingham University.
Dr. B. Gondos	Department of Pathology, Harbor General Hospital, Torrance, California, U.S.A.
Dr. R. O. Greep	Laboratory for Human Reproduction and Reproductive Biology, Harvard Medical School, Boston, Massachusetts, U.S.A.
Dr. M. Jutisz	Collège de France, Laboratoire de Physiologie Cellulaire, Paris V^e^, France.
Dr. C. W. Lloyd	Worcester Foundation for Experimental Biology, Boston, Massachusetts, U.S.A.
Dr. B. Louwerens	N.V. Organon, Oss, Holland.
Prof. B. Lunenfeld	Institute of Endocrinology, Tel Hashomer Government Hospital, Israel.
Prof. G. Mayer	Laboratoire D'Histologie et Embryologie, Faculté de Médecine, Bordeaux, France.
Prof. W. D. Odell	Department of Medicine and Physiology, Harbor General Hospital, Torrance, California, U.S.A.
Dr. T. Pedersen	The Finsen Laboratory, Copenhagen, Denmark.
Dr. H. Peters	The Finsen Laboratory, Copenhagen, Denmark.
Dr. J. Presl	Institute for the Care of Mother and Child, Prague 4, Czechoslovakia.
Dr. M. Ryle	Department of Clinical Endocrinology, Birmingham and Midland Hospital for Women Birmingham 11.
Dr. H. E. Stegner	Universitats-Frauenklinik, Hamburg 20, Germany.
Dr. V. C. Stevens	Department of Obstetrics and Gynaecology, The Ohio State University, Columbus, Ohio, U.S.A.

Dr. U. D. Sutaria — Department of Clinical Endocrinology, Birmingham and Midland Hospital for Women, Birmingham 11.

Dr. J. Weisz — Worcester Foundation for Experimental Biology, Shrewsbury, Massachusetts, U.S.A.

Contents

Chemistry of the Human Gonadotrophins — Page

Some Observations concerning the Physicochemical, Biological and Immunological Properties of Human Pituitary Luteinizing Hormone. *by W.D. Peckham.* 1

Purification of Human Pituitary FSH and LH controlled by Disc Electrophoresis and Carbohydrate Analysis. *by D. Grässlin, Y. Yaoi and G. Bettendorf.* 15

Comparison of the Characteristics of Human Pituitary and Urinary FSH and LH during purification. *by V.C. Stevens, D.G. Anderson and J.E. Powell.* 22

Purification and Partial Chemico-Physical Characterization of Human FSH from Menopausal Urine. *by P. Donini, D. Puzzuoli, I. D'Alessio, G. Bergesi and S. Donini.* 39

Studies of the state of Human Urinary FSH in a crude extract on its Purification. *by M. Théoleyre and M. Jutisz.* 57

Further studies on the Purification and Characterization of Human Chorionic Gonadotrophin. *by H. Van Hell and A. Schuurs.* 70

Chemical studies on a Pituitary FSH Preparation. *by M.F. Chaplin C.J. Gray and J.F. Kennedy.* 77

The action of a Neuraminic Acid-Free Derivative of FSH on mouse ovaries cultured *in vitro. by M. Ryle, M.F. Chaplin, C.J. Gray and J.F. Kennedy.* 98

Papain digestion of FSH from Human Pituitary Glands. *by P. Rathnam and B.B. Saxena.* 107

Amino Acid composition of some Purified Urinary LH Preparations. *by D.G. Anderson, P. Donini and V.C. Stevens.* 117

Purification and Properties of Human Pituitary Luteinizing Hormone. *by M. Braikevitch and A.S. Hartree.* 131

Studies on the Chemistry of Interstitial Cell-Stimulating Hormone. *by H. Papkoff and C.H. Li.* 138

On the Isolation and Characterization of Subunits of Human Pituitary Luteinizing Hormone. *by L.E. Reichert, Jr., D.N. Ward, G.D. Niswender and A.R. Midgley, Jr.* 149

Amino and Carboxy-Terminal sequences of Human Chorionic Gonadotrophine. *by O.P. Bahl.* 155

Studies of the Chemistry of Human Chorionic Gonadotrophin. *by R.E. Canfield, G.M. Agosto and J.J. Bell.* 161

The effects of Chemical Modifications on the Biological and Radioimmunological Activity of Pituitary Follicle Stimulating Hormone. *by W.R. Butt, S.S. Lynch, M.F. Chaplin, C.J. Gray and J.F. Kennedy.* 171

Evidence for the presence of fragments of FSH in Plasma. *by W.M. Hunter and H.K. Amin.* 177

The Cross Reaction between Human Chorionic Gonadotrophin and Hypophyseal Luteinizing Hormone. *by P. Franchimont.* 181

Production rates of Follicle Stimulating Hormone in Adult Human Males. *by S. Raiti, R.M. Blizzard, R. Penny and C.J. Migeon.* 185

Summary of Discussion. 199

Development of the Ovary in Infancy

Page

Introduction. *by Hannah Peters.* 222

Part I. Normal Development

Kinetics of Follicle growth in the ovary of the immature mouse. *by T. Pedersen.* 223

The development of the ferret ovary. *by R. Deanesly.* 226

Development of the stroma in the immature mouse ovary. *by A.G. Byskov, T. Pedersen and H. Peters.* 228

Electron microscope studies of the interstitial tissue in the immature mouse ovary. *by H.-E. Stegner.* 232

Germ Cell Relationships in the developing rabbit ovary. *by B. Gondos.* 239

Part II. Gonadotrophins and Development

Ovarian development in infant mice. Dependence on Gonadotrophic Hormones. *by A. Eshkol, B. Lunenfeld and H. Peters.* 249

Ultrastructural ovarian changes after deprivation of Endogenous Gonadotrophins and the effect of FSH and HMG in the deprived immature animals. *by H.-E. Stegner, B. Lunenfeld and A. Eshkol.* 259

Development of the ovary under different experimental conditions. *by S. Ben-Or.* 266

Some further observations on the *in vitro* growth of infantile mouse ovaries in response to Gonadotrophins. *by M. Ryle.* 272

Effect of Gonadotrophins on Protein Synthesis in the ovary of the mouse. *by S. Ben-Or and R. Broza.* 282

Changes in the rate of Nucleic Acid and Protein Synthesis in juvenile mouse ovaries after FSH Stimulation. *by A. Eshkol, B. Hardy and C. Pariente-Coriat.* 297

Histochemical Localization of Enzyme activities in normal and Gonadotrophin Stimulated infant mouse ovaries. *by H. Brandau.* 307

Cell population Kinetics of the ovary of the immature mouse after FSH Stimulation. *by T. Pedersen.* 312

An *in vitro* test for changes in Sensitivity to FSH of infantile mouse ovaries. *by M. Ryle.* 325

Reactions of the Postnatal ovary after Transplantation into Castrated normal and Sterilized rats. *by G. Mayer and A.J. Duluc.* 332

Part III. Interactions between the Hypothalamus, the Hypophysis and the Ovary

Pituitary Gonadotrophins and circulating LH in immature rats. A comparison between normal females and males and females treated with Testosterone in Neonatal life. *by J. Weisz and M. Ferin.* 339

The development of the mouse ovary after Testesterone Propionate injection on Day 5. *by H. Peters, I.N. Sørensen, A. Byskov, T. Pedersen and T. Krarup.* 351

Ultrastructural changes in the immature mouse ovary after injection of Androgen in the Neonatal period. *by H.E. Stegner and M. Poelman.* 362

Radioactive Oestrogen uptake by the Brain, Pituitary, Uterus and Skeletal Muscle of the rat during Postnatal development. *by J. Presl.* 368

Studies of Hypothalmic-Pituitary-Gonadal Interrelations in Prepubertal cattle. *by W.D. Odell.* 371

Index 386

Some Observations Concerning the Physicochemical, Biological and Immunological Properties of Human Pituitary Luteinizing Hormone

William D. Peckham

SUMMARY

A preparation rich in luteinizing hormone activity was obtained by ion exchange and exclusion chromatography of an ethanolic ammonium acetate extract of human pituitary glands. Disc electrophoresis of this material at an alkaline pH revealed the presence of three major protein components. Electrophoresis in a sucrose density gradient at pH 8.6 accomplished a separation of these three components, each of which possessed luteinizing hormone activity. Although each isolated component appeared to be homogeneous by disc electrophoresis at an alkaline pH and by analytical exclusion chromatography, all three exhibited small but significant thyrotrophic activity. After radioiodination, the three components showed similar binding and displacement characteristics with anti-HCG.

In another study, fresh frozen human pituitary glands were extracted with aqueous buffer, and the extract was treated with diisopropyl fluorophosphate (DFP). When the crude extract was chromatographed on Sephadex G-100 and the effluent fractions were assayed for HLH by radioimmunoassay, the expected peak associated with biologically active HLH appeared. In addition, however, there emerged later in the chromatogram a second even larger immunoreactive peak which contained protein with negligible activity in the OAAD assay, thus indicating the existence of immunologically active but biologically inert 'HLH' whose identity remains unclear. The biologically active HLH from this chromatogram, after further purification by ion exchange and exclusion chromatography, was shown to contain the same three electrophoretic components which had been previously isolated. An attempt to isolate these components by density gradient electrophoresis at pH 4.1 was only partly successful.

INTRODUCTION

When these investigations were first undertaken in 1964, there was little information, with the exception of the report by Squire *et al.* (1962), concerning the degree of heterogeneity of the best existing preparations of human LH. In attempting to isolate a well-characterized human LH and demonstrate its homogeneity by rigorous criteria, we have found three forms of the hormone, differing in electrophoretic mobility, but all possessing high biological activity.

MATERIALS AND METHODS

The crude gonadotrophin mixture which served as the major source of starting material for the isolation of human LH was provided by Dr A. F. Parlow and was obtained from 700 acetone-preserved human pituitaries by extraction with ethanolic ammonium acetate according to the procedure of Steelman *et al.* (1958).

All preparative procedures were carried out in a cold room at temperatures of 2 to 5°C using buffers containing 0.01 per cent toluene. Many of the analytical and preparative methods employed here have been described in detail in a previous paper dealing with the isolation of primate growth hormones (Peckham, 1967).

LH potencies were determined by the ovarian ascorbic acid depletion method (Parlow, 1961) and FSH potencies by the ovarian HCG-augmentation method (Steelman and Pohley, 1953). TSH potencies were assessed by the method of McKenzie (1958), with modifications as described by McKenzie and Williamson (1966). Both LH and FSH potencies are expressed as units per mg of protein, a unit being defined as the activity of one mg of the appropriate NIH ovine gonadotrophin standard (either NIH-LH-S1 or NIH-FSH-S1), while TSH potencies are expressed as USP units per milligram of protein. Unless otherwise noted, all dosages for bioassays were based on protein determination by the method of Lowry *et al.* (1951) with bovine serum albumin as standard, using hormone solutions which had been equilibrated by gel filtration with 0.02 M-phosphate, pH 6.8, in 0.2 M-NaCl.

Radioimmunoassays for HLH were either by the double antibody procedure of Midgley (1966) or by the charcoal method described by Neill *et al.* (1967).

RESULTS

The crude gonadotrophin, which weighed 560 mg and contained 135 u of LH and 3000 u of FSH, was extracted with 17 ml of 0.05

M-phosphate buffer, pH 6.8, and twice with 9 ml of 0.01 M-ammonium acetate buffer, pH 6.1. The combined extracts contained 210 mg of protein and virtually all of the LH and FSH activities. This protein was equilibrated with 0.01 M-ammonium acetate, pH 6.1, and chromatographed on a column (2.5 x 43 cm) of CM-cellulose as described by Steelman *et al.* (1959). Material which passed through the column unretarded accounted for nearly all of the FSH activity and only about 5 per cent of the total LH activity, while that which was eluted by 1.0 M-ammonium acetate contained about 60 per cent of the total LH and less than 3 per cent of the total FSH. Thus, in contrast to the report by Hartree (1966), a good separation of LH and FSH was accomplished by this procedure.

A solution (5 ml) containing 40 mg of protein (LH potency = 1.3 u./mg) from the CM-cellulose chromatography was applied to a column (2.0 x 72 cm) of superfine Sephadex G-100 which had been equilibrated with 0.02 M-phosphate, pH 6.8, containing 0.2 M-NaCl, and the column was eluted with the same buffer at a rate of 3.4 ml per hour. The elution pattern is shown in Fig. 1. Fractions 27 to 34,

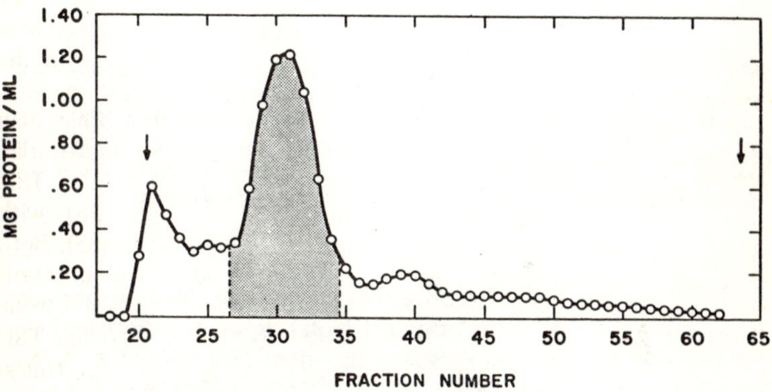

Figure 1. Elution pattern obtained from the chromatography of crude human LH (after separation from FSH by CM-cellulose chromatography) on a column (2.0 x 72 cm) of superfine Sephadex G-100. The volume of each fraction is 3.4 ml. Vertical arrows at left and right mark the void volume and the total solvent volume, respectively, of the column. The shaded area represents those fractions which contained LH activity.

which were found to contain the LH activity, were combined. These fractions yielded 21 mg of protein, and showed a two-fold enhancement (to 2.6 u./mg) in LH potency over that of the material prior to exclusion chromatography. A portion (14 mg) of this preparation was equilibrated with 0.05 M-phosphate buffer, pH 6.8, to yield 3 ml of solution. This was applied to a column (1.0 x 20 cm) of DEAE-Sephadex A-50 which had also been equilibrated with 0.05

M-phosphate, pH6.8, and the column was eluted with the same buffer at a rate of 25 ml per hour. The elution pattern for this chromatogram is shown in Fig. 2. Most of the LH activity was found in fractions 5

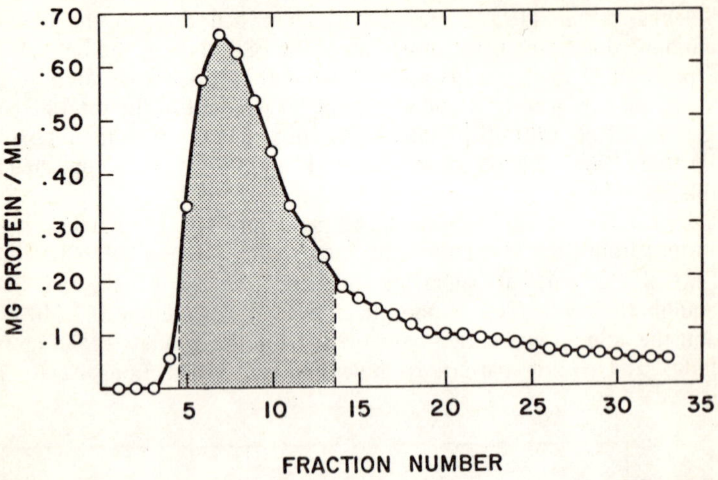

Figure 2. Elution pattern obtained from the chromatography of partially purified human LH (after Sephadex G-100 chromatography) on a column (1.0 x 20 cm) of DEAE-Sephadex A-50 under conditions of equilibrium with 0.05 M phosphate, pH 6.8. The volume of each fraction is 1.8 ml. Shaded area represents those fractions which contained most of the LH activity. Approximately half of the protein applied remained bound to the column under these conditions.

to 13. These combined fractions contained a total of 7 mg of protein with approximately twice the LH potency of the protein before DEAE-Sephadex chromatography; the disc electrophoretic patterns before and after DEAE-Sephadex chromatography are in accord with a two-fold increase in specific activity. Although the potency of this preparation had increased to about 5 u./mg, pronounced heterogeneity was apparent when the material was examined by disc electrophoresis (Fig. 3A). It was therefore subjected to electrophoresis in a sucrose density gradient column containing 0.025 M-Tris-HC1, pH 8.6, as the buffer. A potential of 600 volts was applied for a period of 30 hours; during this time the current fell from 4.5 to 3.5 mA. Quantitative separation of the three major components was not achieved, but the degree of resolution which was produced was sufficient to allow the isolation of several hundred micrograms of each (Fig. 4). These have been designated HLH-I, HLH-II and HLH-III, in order of their increasing electrophoretic mobility in 7.5 per cent polyacrylamide gel at pH 9.5 (Fig. 3).

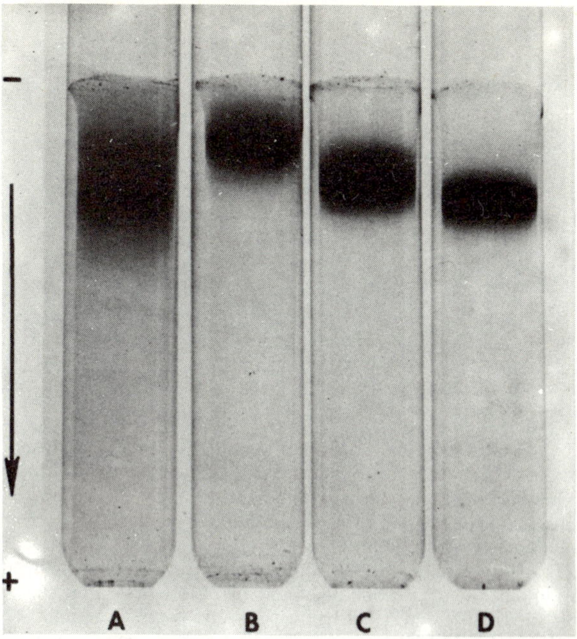

Figure 3. Disc electrophoresis at pH 9.5 (7.5 per cent polyacrylamide, 5 mA per tube) of human LH preparations. A, combined fractions 5 to 13 from DEAE-Sephadex chromatography of human LH (Fig. 2); B, HLH-I; C, HLH-II; and D, HLH-III. The three individual components (B, C and D) were obtained from density gradient electrophoresis (Fig. 4) of the mixture (A).

The LH potencies of the three components isolated by density gradient electrophoresis were: HLH-I, 7.6 u./mg (95 per cent confidence limits 5.3 to 10.7); HLH-II, 10.7 u./mg (95 per cent confidence limits 8.4 to 13.6); and HLH-III, 8.5 u./mg (95 per cent confidence limits 5.2 to 13.7). Each value represents the mean of four replicate potency estimates. The TSH potencies of these preparations were: HLH-I, 0.20 USP units per mg (95 per cent confidence limits 0.13 to 0.36); and HLH-III, 0.20 USP units per mg (95 per cent confidence limits 0.12 to 0.40). While a precise TSH potency estimate was not obtained for HLH-II, a preliminary assay indicated definite TSH activity (approximately 0.4 USP units per mg).

A small quantity of each of the biologically active electrophoretic components was chromatographed on a column (1.1 x 87 cm) of superfine Sephadex G-100 which had been equilibrated with 0.02 M-phosphate, pH 6.8, in 0.2M NaC1. Elution patterns for these chromatograms are shown in Fig. 5. In each case, all of the protein appeared as a single, sharp, symmetrical peak, with very similar distribution coefficients (Porath, 1962): HLH-I, $K_D = 0.218$; HLH-II, $K_D = 0.215$; and HLH-III, $K_D = 0.200$.

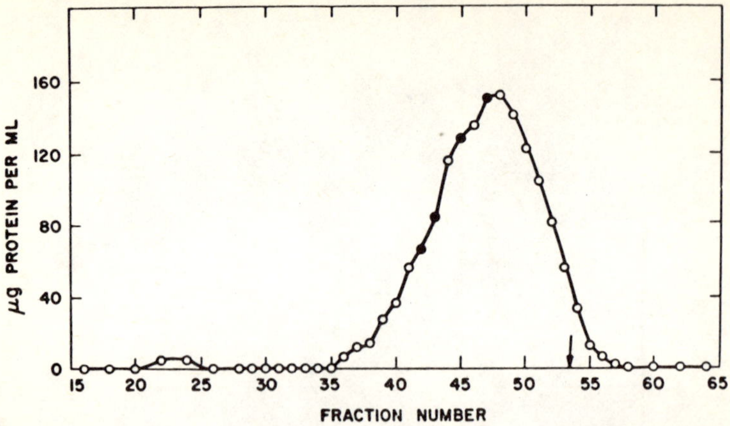

Figure 4. Column electrophoresis in a sucrose density gradient, at an *alkaline* pH, of HLH (6.2 mg) obtained from DEAE-Sephadex chromatography (Fig. 2). Running time was 30 hr with a current of 3.5 to 4.5 mA in 0.025 M Tris-HC1, pH 8.6. The volume of each fraction is 4.0 ml. Position of the sample zone at beginning of electrophoresis is indicated by the vertical arrow; the anode is at the left. Closed circles represent those fractions which were homogeneous by disc electrophoresis at pH 9.5 and possessed high LH activity; fractions 47 and 45 correspond to HLH-I and HLH-II, respectively, while fractions 42 and 43 comprise HLH-III. Disc electrophoresis of fractions 48 through 55 in the descending limb of the protein peak revealed no HLH-II or -III, and only fractions 48 and 49 showed a band whose mobility was similar to HLH-I. The biological activity of protein from the combined fractions in the descending limb was quite low and probably can be accounted for by its content of HLH-I.

Several micrograms of HLH-I, -II and -III were iodinated by ^{131}I by the method of Greenwood *et al.* (1963). The per cent of each radioiodinated hormone which was bound to anti-HCG under the conditions routinely employed in the double antibody radio-immunoassay for HLH (Midgley, 1966) was the same. Also, displacement of each of the three radioiodinated hormones by varying quantities of unlabelled HLH produced three dose response curves which were virtually identical.

In order to ascertain whether the appearance of three HLH components was related to the action of proteolytic enzymes in the pituitary extracts, 100 fresh frozen human pituitary glands were extracted twice with 300 ml of 0.03 M-phosphate buffer, pH 6.3, and the combined extracts (600 ml) were promptly treated with 5 ml of isopropyl alcohol containing 0.1 ml of DFP. After stirring the solution overnight, 300 g of ammonium sulphate were added with stirring. The resulting precipitate was then removed and extracted with 65 ml of 0.05 M-phosphate, pH 6.8, in 0.5 M-NaC1. This solution, containing 31 mg of protein per ml, was chromatographed on a column (5 x 115 cm) of Sephadex G-100 previously equilibrated with 0.05 M-phosphate, pH 6.8, in 0.5 M-NaCl. After protein analysis of the effluent fractions,

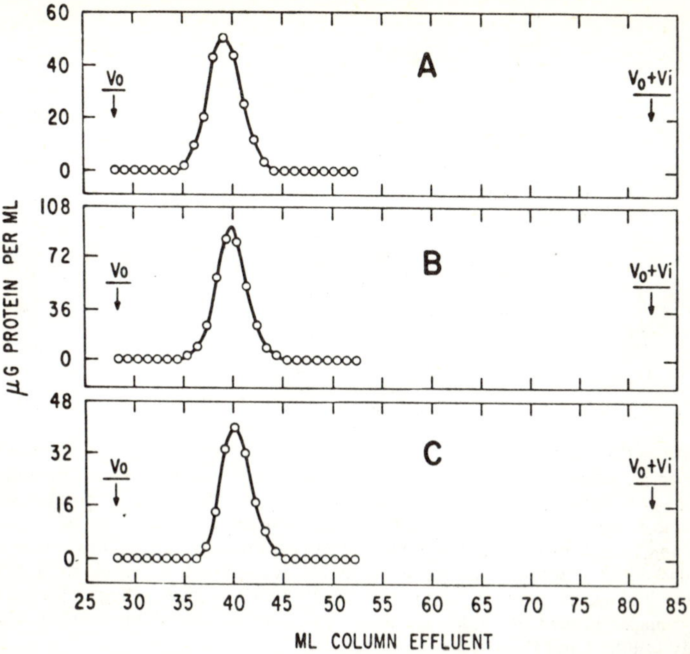

Figure 5. Elution patterns obtained from analytical exclusion chromatography of human LH preparations on a column (1.1 x 87 cm) of superfine Sephadex G-100. A, HLH-III (0.21 mg); B, HLH-II (0.35 mg); and C, HLH-I (0.16 mg). The void volume and total solvent volume of the column for each run are indicated by vertical arrows labeled V_o and $V_o + V_i$, respectively.

many of the same fractions were radioimmunoassayed for HLH by the charcoal method (Neill *et al.* 1967) with anti-HCG as the antibody in order to locate the LH with precision. Elution patterns for this chromatogram in respect to concentrations of both protein and immunoassayable LH are shown in Fig. 6. The position of the first peak detected by radioimmunoassay was consistent with previous experience, based on bioassay, regarding the behaviour of HLH on Sephadex G-100. Protein from the first peak (Figure 6, fractions 17 to 27) was further purified by sequential chromatography on CM-cellulose, DEAE-Sephadex and Sephadex G-100, all as described above, to yield 5 mg of protein with an LH potency of about 3 u./mg. Although the potency of this preparation was somewhat low (probably due to the wide cuts taken at each stage of purification in order to minimize losses) compared to that of the 3-component HLH (Fig. 3A) which had been isolated by a similar procedure, disc electrophoresis revealed the presence of three bands corresponding to HLH-I, -II and -III (as well as several extraneous bands not seen in Fig. 3A). In addition, the relative intensities of these bands were similar to those of

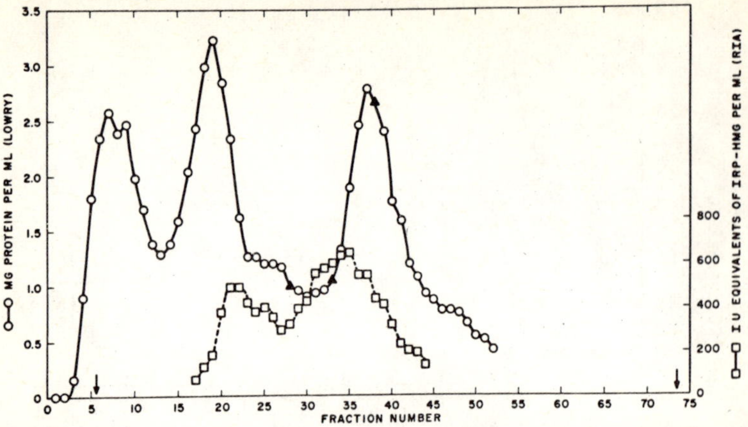

Figure 6. Chromatography of DFP-treated pituitary extract (2.0 g of protein in 64 ml) on a column (5 x 115 cm) of Sephadex G-100. The volume of each fraction is 23 ml. Vertical arrows at left and right mark the void volume and the total solvent volume, respectively, of the column. The concentration of protein in each fraction is indicated by circles connected by solid lines, while squares connected by dashed lines represent the concentrations of radioimmunoassayable HLH; most of the biologically active LH appears in fractions 17 to 27. Fractions represented by triangles instead of circles are those which were selected for more detailed study (Tables I and II).

the 3-component HLH previously isolated. This experiment therefore fails to provide evidence that the existence of three HLH components is the result of the action of proteolytic enzymes present in pituitary extracts, at least of proteolytic enzymes which are inhibited by DFP.

When the purified HLH preparation (3 u./mg) obtained from DFP-treated pituitary extracts was subjected to electrophoresis at 600 volts in a sucrose density gradient column containing 0.025 M-potassium acetate buffer, pH 4.1, the separation illustrated in Fig. 7 (upper curve) resulted. In order to concentrate the very dilute protein solutions, enough ammonium sulphate was added to each fraction to make the final concentration 3 M, and the precipitated protein was collected and dissolved in a much smaller volume of buffer. The lower curve in Figure 7 shows the amount of protein per ml in each fraction which was precipitated in this way. It was now possible to examine each fraction by disc electrophoresis, and, on the basis of the disc electrophoretic patterns in the acidic system of Reisfeld *et al.* (1962) with a running pH of 3.8, individual fractions 23 to 29 and 35 to 40 were combined to form pools A and B, respectively (Figure 7). The LH potency of A was found to be 7.5 u./mg. (4 assays, 95 per cent confidence limits 5.5 to 10.3), while that of B was 11.5 u./mg (3 assays, 95 per cent confidence limits 6.2 to 21.1). Although both A and B

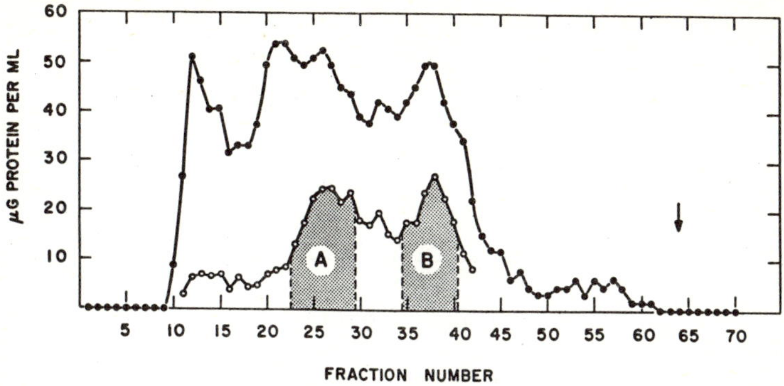

Figure 7. Column electrophoresis in a sucrose density gradient, at an *acid* pH, of HLH (5 mg; 3 u./mg) derived from fractions 17 to 27, Fig. 6, after further purification by chromatography on CM-cellulose, DEAE-Sephadex and Sephadex G-100. Running time was 31 hr. with a current of 4 to 5 mA in 0.025 M potassium acetate buffer, pH 4.1. The volume of each fraction is 3.4 ml. Position of the sample zone at beginning of electrophoresis is indicated by the vertical arrow; the cathode is at the left. The *upper* curve (closed circles) represents the concentration of protein in each fraction, while the *lower* curve (open circles) represents the μg of protein per ml of each fraction which precipitated at 3 M ammonium sulphate. Shaded areas labeled A and B represent pooled precipitated protein from fractions 23 to 29 and 35 to 40, respectively.

appeared to be homogeneous in the Reisfeld system and by analytical exclusion chromatography, disc electrophoresis at pH 9.5 showed that two components corresponding to HLH-II and -III were present in B, and that most of the protein in A did not migrate out of the sample gel. It was later found that most of the component corresponding to HLH-I was present in fractions 30 to 34 along with some HLH-II and -III; these fractions also possessed high LH activity.

In addition to the first peak corresponding to biologically active HLH, radioimmunoassay of effluent fractions from the exclusion chromatography of DFP-treated pituitary extracts also revealed (Fig. 6) a second large peak emerging at a position where biologically active HLH had not been found before. Three fractions from this second peak were therefore selected for bioassay and for confirmatory radioimmunoassay in a somewhat different system (double antibody method with anti-HLH instead of anti-HCG). The same fractions were also analyzed in a double antibody radioimmunoassay using anti-HCG with radioiodinated HLH-I, -II or -III as trace. Results of the various assays are summarized in Table I & II. From these data and from the ratios shown in Table II it is clear that there is in human pituitary extracts a substance (or substances) other than HLH which competes with or in some other way decreases the binding of ^{131}I-HLH to antibody and thus interferes with the radioimmunoassay of HLH. It also appears that this phenomenon does not depend on the use of a particular trace or antibody, or on the method employed for separating bound and free trace.

TABLE I

Luteinizing hormone potencies of three human pituitary fractions as determined by bioassay (OAAD) and radioimmunoassay (RIA)

Assay	#28 (Fig. 6)	#33 (Fig. 6)	#38 (Fig. 6)
OAAD	102*	26	<0.7
RIA–1**	325	599	164
RIA–2***	315	390	159
RIA–3****	564	284–630[1]	83–146[1]

* All potencies are expressed as i.u. (or i.u. equivalents for RIA) of IRP–HMG per mg. of protein.

** Charcoal–HLH–antiHCG system.

*** Double antibody–HLH–antiHLH system.

**** Double antibody–HLH–antiHCG system; potency estimates did not differ regardless of whether radioiodinated HLH–I, –II or –III was used as trace.

1 Log-dose response curve was not parallel with that of standard; numbers rep- present what the potency estimates *would* be at the upper and lower limits of linearity *if* parallelism existed.

TABLE II

Ratios of immunopotency (RIA) to biopotency (OAAD) for three human pituitary fractions

Ratio	#28 (Fig. 6)	#33 (Fig. 6)	#38 (Fig. 6)
LH by RIA–1 / LH by OAAD	3.2	23	>234
LH by RIA–2 / LH by OAAD	3.1	15	>227
LH by RIA–3 / LH by OAAD	5.5	11–24	118–>208

DISCUSSION

Although neither analytical exclusion chromatography nor disc electrophoresis at an alkaline pH afforded any evidence of heterogeneity of the three human LH preparations (HLH-I, -II and -III) described above, the homogeneity of these preparations must remain in question for two reasons. First, all three components showed significant TSH activity in the McKenzie mouse assay, and, while this may be due to intrinsic TSH activity of each form of the LH molecule, an equally plausible explanation is contamination of the LH preparations with small amounts of TSH which have escaped detection (or have not been separated) by the analytical systems employed as criteria of purity. Second, preliminary work has shown that at least one of the LH components, HLH-II, exhibits some degree of heterogeneity in the acidic disc electrophoresis system of Reisfeld *et al.* (1962). Unfortunately, the severe limitations imposed by the extremely small amounts of material available at this time do not allow the presentation of more definitive information concerning the homogeneity of these LH preparations.

In order to compare the biological activity of HLH-I, -II and -III with that of other human LH preparations, it is necessary to express the LH potencies in common terms. Potencies (or relative potencies) of protein hormones are usually based on dry weight determination of the hormone; however, because of the minute quantities of the purified LH preparations, our potency estimates are based on protein analysis by the Lowry procedure using bovine serum albumin as standard. Nevertheless, rough comparisons in potency can be made, for Kathan *et al.* (1967) have found that on an equal weight basis the Lowry colour yield for their best preparations of human LH is 60 per cent of that for bovine serum albumin. If one applies this figure to the LH potency estimates based on protein determination for HLH-I, -II and -III, the potency estimates based on dry weight are: HLH-I, 4.56 u./mg; HLH-II, 6.42 u./mg; and HLH-III, 5.10 u./mg, values which are in the same range as those found by other workers for their most active preparations. The potencies of some of these human LH preparations, as determined by the ovarian ascorbic acid depletion assay, are: 3.2 u./mg, by Reichert and Parlow (1964); 3 u./mg, by Parlow *et al.* (1965); 4.8 u./mg, by Hartree (1966); 3.35 u./mg, by Ryan (1968); 4 u./mg, by Shome *et al.* (1968); and 5.5 u./mg, by Bates *et al.* (1968).

While the degree of electrophoretic heterogeneity is not known for the LH preparations of Parlow *et al.* (1965), of Hartree (1966) or of Bates *et al.* (1968), this information is available for the other preparations mentioned above. The LH described by Shome *et al.* (1968) was resolved into four discrete protein bands following disc electrophoresis at pH9.5; the biological activity of the individual bands was not investigated. Ryan (1968), upon examination of his most potent LH by starch gel electrophoresis at pH 8.0, found two discrete bands and some smudging of stainable protein. Extraction and assay of protein from each zone in the gel slab indicated that where there was stainable protein there was also LH activity. Recently, Reichert *et al.* (1968) have shown that a single stained band is found following disc electrophoresis at pH 9.5 of human LH (potency = 3.2 u./mg) prepared by the method of Reichert and Parlow (1964). It is not clear why this preparation exhibits electrophoretic homogeneity in comparison to the pronounced heterogeneity seen in several other preparations (Ryan, 1968; Shome *et al.* 1968) isolated by similar methods and possessing similar biological potencies.

The existence of polymorphism in glycoproteins is not unique to human LH, having been observed in TSH from a number of species including man (Condliffe *et al.* 1960; Carsten and Pierce, 1960; Wynston *et al.* 1960; Shome *et al.* 1968) and in HCG. Using ion exchange and exclusion chromatography, either alone or in combination, as the final purification step, preparations of HCG with potencies ranging from 12,000 to 18,000 i.u. per mg have been obtained (Reisfield and Hertz, 1960; Got, 1965; Wilde and Bagshawe, 1965; van Hell *et al.* 1966; Bell *et al* 1969; Bahl, 1969) which on careful examination usually proved to be heterogeneous. There is evidence that the heterogeneity seen in these highly purified preparations of HCG as well as the differences in potency of the same preparations may be due to differences in the carbohydrate units, probably in sialic acid content, attached to a common polypeptide core (van Hell *et al.* 1966; Bell *et al.* 1969). Similarly, Reinhold *et al.* (1968) have shown that porcine pancreatic ribonuclease consists of a mixture of at least eight electrophoretically distinguishable glycoprotein components which appear to differ from each other only in respect to the polysaccharide portion of the enzyme. Although it appeared that the two active components isolated from sheep pituitary glands by Ward *et al.* (1959) represented polymorphic forms of ovine LH, it was later shown by the same group that these two components could be accounted for by a reversible association of a single form of LH with an inactive glycoprotein (Ward *et al.* 1961). Similarly, the multiplicity of active ovine LH components differing in chromatographic and electrophoretic behaviour reported by Li's laboratory (Jutisz and Squire, 1958; Squire and Li, 1959) were later attributed, at least in part, to protein-protein interactions (Squire *et al.* 1962). The same authors also found three distinct LH fractions from CM-cellulose chromatography of human pituitary material. With two of these fractions, it was possible to separate LH activity from a large quantity of inactive protein by zone electrophoresis at pH 3.6, from which the authors infer that the active fractions resolved by CM-cellulose chromatography are interaction products of LH, presumably with different inactive proteins (Squire *et al.* 1962). It seems unlikely that a similar explanation is applicable to the three components which we have designated HLH-I, -II and -III, for in our hands disc electrophoresis was carried out under 'stacking' conditions highly favourable to the disruption of complexes (Disc Electrophoresis Newsletter, 1964), and thus it appears that what we have observed is indeed polymorphism. At present it is not known whether this polymorphism is artifactual in origin or represents a true diversity of luteinizing hormones as they are synthesized and secreted by the pituitary gland. That these multiple components may not represent artifacts formed by the action of proteolytic enzymes present in pituitary extracts is suggested by the experiment described above which indicates that the number and quantitative relationship of the LH components is unchanged by early treatment of human pituitary extracts with DFP, a procedure having a pronounced inhibitory effect on the number and quantity of human growth hormone components (Peckham, 1967; Chen and Wilhelmi, 1964). The development of an extremely sensitive and unequivocally specific radioimmunoassay for human LH might prove of value for investigating the heterogeneity of LH which is secreted by the pituitary gland, using an approach similar to that employed by Berson and Yalow (1966) for human growth hormone.

Our finding that there is in extracts of human pituitary glands an abundance of material with little or no LH activity by bioassay but considerable activity in the HLH radioimmunoassay poses an interesting question concerning its identity. Even though the extracts were treated immediately with DFP, the possibility that degradation of native LH by proteolytic enzymes took place before extraction cannot be ruled out, for the time *post mortem* before the glands were frozen was not controlled. Indeed, the fact that the material in question was eluted after biologically active LH suggests that it may consist of degradation products of native LH. Regardless of the origin or identity of this immunologically active but biologically inactive 'HLH', these experiments clearly demonstrate the danger inherent in using radioimmunoassay as the sole means of determining the LH content of pituitary extracts or fractions derived from them, especially at early stages in the purification process.

ACKNOWLEDGMENTS

These investigations were supported by NIH Grant AM 05654 to E. Knobil. The advice, help and encouragement of Prof. Knobil are gratefully acknowledged.

I am most grateful to Dr A.F. Parlow for all of the LH and FSH bioassays, and to Dr J.M. McKenzie for the TSH bioassays. I should also like to thank Drs S.E. Monroe, J.D. Neill and J.K. Datta of this laboratory for the immunological studies with radioiodinated HLH-I, -II and -III as well as most of the HLH radioimmunoassays, and Dr Parlow for several confirmatory HLH radioimmunoassays.

REFERENCES

Bahl, O.P. (1969). *J. biol. Chem.* **244,** 567.

Bates, R.W., Garrison, M.M., Cooper, J.A. & Condliffe, P.G. (1968). *Endocrinology,* **83,** 721.

Bell, J.J., Canfield, R.E. & Sciarra, J.J. (1969). *Endocrinology,* **84,** 298.

Berson, S.A. & Yalow, R.S. (1966). *J. biol. Chem.* **241,** 5745.

Carsten, M.E. & Pierce, J.G. (1960). *J. biol. Chem.* **235,** 78.

Chen, H.C. & Wilhelmi, A.E. (1964). *Abstracts of the Endocrine Society Meeting, San Francisco,* p. 80.

Condliffe, P.G., Bates, R.W., Garrison, M.M. & Howard, T.B. (1960). *Biochim. biophys. Acta,* **37,** 150.

Disc Electrophoresis Newsletter October 1964, **6,** 4. Bethesda, Maryland: Canal Industrial Corporation.

Got, R. (1965). *Eur. Rev. Endocrin.* Supp. **1,** 191.

Greenwood, F.C., Hunter, W.M. & Glover, J.S. (1963). *Biochem. J.* **89,** 114.

Hartree, A.S. (1966). *Biochem. J.* **100,** 754.

Hell, H. van, Goverde, B.C., Schuurs, A.H.W.M., de Jager, E., Matthijsen, R. & Homan, J.D.H. (1966). *Nature, Lond.* **212,** 261.

Jutisz, M. & Squire, P.G. (1958). *Bull. Soc. Chim. biol.* **40,** 1875.

Kathan, R.H., Reichert, L.E. Jr. & Ryan, R.J. (1967). *Endocrinology,* **81,** 45.

Lowry, O.H., Rosebrough, N.J., Farr, A.L. & Randall, R.J. (1951). *J. biol. Chem.* **193,** 265.

McKenzie, J.M. (1958). *Endocrinology,* **63,** 372.

McKenzie, J.M. & Williamson, A. (1966). *J. clin. Endocr. Metab.* **26,** 518.

Midgley, A.R. Jr. (1966). *Endocrinology,* **79,** 10.

Neill, J.D., Peckham, W.D. & Knobil, E. (1967). *Nature, Lond.* **213,** 1014.

Parlow, A.F. (1961). In *Human Pituitary Gonadotropins,* p. 201. Ed. Albert, A. Springfield, Illinois: Thomas.

Parlow, A.F., Condliffe, P.G., Reichert, L.E. Jr. & Wilhelmi, A.E. (1965). *Endocrinology,* **76,** 27.

Peckham, W.D. (1967). *J. biol. Chem.* **242,** 190.

Porath, J. (1962). In *International Symposium on Pharmaceutical Chemistry,* p. 233. London: Butterworth.

Reichert, L.E. Jr., Kathan, R.H. & Ryan, R.J. (1968). *Endocrinology,* **82,** 109.

Reichert, L.E. Jr. & Parlow, A.F. (1964). *Endocrinology,* **75,** 815.

Reinhold, V.N., Dunne, F.T., Wriston, J.C., Schwarz, M., Sarda, L. & Hirs, C.H.W. (1968). *J. biol. Chem.* **243,** 6482.

Reisfeld, R.A. and Hertz, R., (1960). *Biochim. biophys. Acta,* **43,** 540.

Reisfeld, R.A., Lewis, U.J. & Williams, D.E. (1962). *Nature, Lond.* **195,** 281.

Ryan, R.J. (1968). *J. clin. Endocr. Metab.* **28,** 886.

Shome, B., Parlow, A.F., Ramirez, V.D., Elrick, H. & Pierce, J.G. (1968). *Archs Biochem. Biophys.* **103,** 444.

Squire, P.G. & Li, C.H. (1959). *J. biol. Chem.* **234,** 520.

Squire, P.G., Li, C.H. & Andersen, R.N. (1962). *Biochemistry, Wash.* **1,** 412.

Steelman, S.L. & Pohley, F.M. (1953). *Endocrinology,* **53,** 604.

Steelman, S.L., Segaloff, A. & Andersen, R.N., (1959). *Proc. Soc. exp. Biol. Med.* **101,** 452.

Steelman, S.L., Segaloff, A. & Mays, M.G. (1958). *Archs Biochem. Biophys.* **78,** 262.

Ward, D.N., Adams-Mayne, M. & Wade, J. (1961). *Acta endocr., Copenh.* **36,** 73.

Ward, D.N., McGregor, R.F. & Griffin, A.C. (1959). *Biochim. biophys. Acta,* **32,** 305.

Wilde, C.E. & Bagshawe, K.D. (1965). *Ciba Fdn Study Grps,* **22,** 46.

Wynston, L.K., Free, C.A. & Pierce, J.G. (1960). *J. biol. Chem.* **235,** 85.

Purification of Human Pituitary FSH and LH Controlled by Disc Electrophoresis and Carbohydrate Analysis

D. Grasslin, Y. Yaoi and G. Bettendorf

SUMMARY

A crude gonadotrophic material extracted from acetone-dried human pituitaries was subjected to CM and QAE-Sephadex chromatography. With the cation exchanger no effective separation of FSH and LH occurred, but in a single run with the strongly basic anion exchanger QAE-Sephadex A 50, a highly purified material was obtained with LH activity of 2110 i.u./mg and less than 2.5 i.u./mg of FSH. Two different LH-species could be shown by disc electrophoresis. The yield of the potent LH was about 10 per cent of the crude material or about 100 μg per pituitary. In the methods described no purification of FSH was possible, but after chromatography on QAE-Sephadex the total FSH activity could be recovered almost completely. All potency estimations were based on dry weight. There was no loss of LH-activity either by dialysis or by lyophylisation.

MATERIALS AND METHODS

We have been endeavouring to find a simple purification method for FSH and LH from human pituitaries which would result in a good yield of highly purified material. The crude gonadotrophin was obtained as described earlier (Bettendorf *et al.* 1962; Bettendorf *et al.* 1968) by a modification of the method of Koenig and King (1950). Usually 100 pituitaries were taken for one extraction procedure. The mean yield of the crude gonadotrophin (E_3) was between 70 and 150 mg. FSH activity was measured by the augmentation test (Steelman and Pohley, 1953), LH by the OAAD test according to Karg (1957) and Parlow (1961). All potency estimations were based on dry weight. The biological activities of E_3 varied between 100 and 200 i.u. FSH/mg and 100 to 1000 i.u. LH/mg.

The crude material (E_3) was subjected to ion exchange chromatography on CM-Sephadex and on QAE-Sephadex. A column of 1.5 x 22 cm was used for all procedures. In all experiments between 40 and 85 mg of E_3 were applied to the column. The columns were eluted at a flow rate of about 10 ml/hr. A constant flow was maintained with a gravity feed set-up. The absorbance of the eluate was recorded at 280 mμ through a Uvicord. All procedures were carried out at 4°C. The purification steps were controlled by disc electrophoresis and sialic acid estimations.

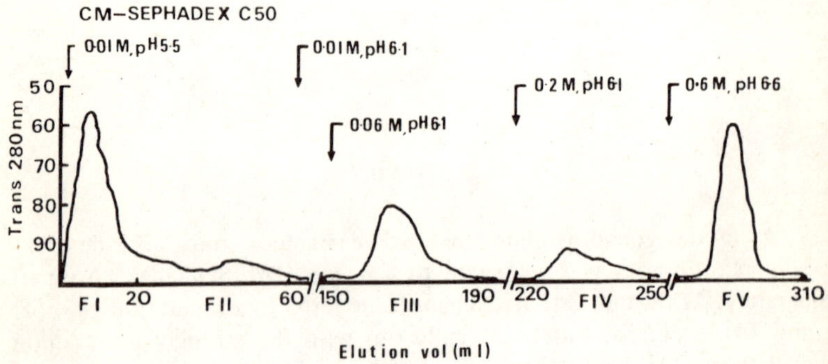

Figure 1

	Fraction					
	F I	F II	F III	F IV	F V	Starting material
LH i.u./mg	–	397	19	225	855	340
FSH i.u./mg	–	35	41	150	115	250
Protein yield (mg)	11	14	6.5	1.5	7.0	41.0

Elution pattern from chromatography of crude gonadotrophin (E_3) on CM-Sephadex C 50. Buffer solution: ammonium acetate-acetic acid.

Ion exchange chromatography on CM-Sephadex (Fig. 1). The anion exchanger was equilibrated with 0.01 M ammonium acetate pH 5.5, as the starting buffer. Approximately 50 per cent of the applied material was unadsorbed. Fraction I was inactive with respect to FSH and LH. F-II contained most of the albumin, low FSH activity and nearly the same LH activity as the starting material. F-III, eluted with 0.06 M buffer, pH 6.1, contained only small amounts of biological activity, but there was a 2.5-fold increase in LH activity in the last fraction (F V), when compared with E_3. The ratio of FSH: LH in this fraction V was 0.13, compared with the starting material, in which it was 0.73. We found more than 80 per cent loss of total FSH, while 85 per cent of LH was recovered. All fractions were dialysed against distilled water and lyophilised.

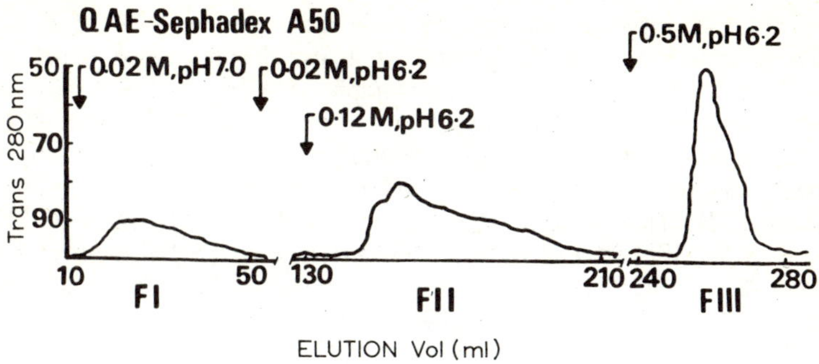

Figure 2

Fraction

	F I	F II	F III	Starting material
LH i.u./mg	1060	700	47	530
FSH i.u./mg	<4	370	25	83
Protein yield (mg)	5	29	19	85

Elution pattern from chromatography of crude gonadotrophin (E_3) on QAE-Sephadex A 50 (starting buffer pH 7.0: imidazol-HC1)

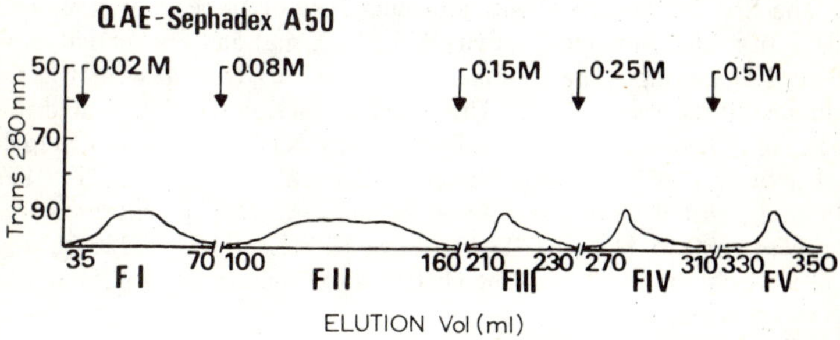

Figure 3

Fraction

	F I	F II	F III	F IV	F V	Starting material
LH i.u./mg	2100	1080	330	125	–	460
FSH i.u./mg	<2.5*	160	225	36	–	107
Protein yield (mg)	7.0	6.8	22.5	5.0	<1.0	65.0

* Negative with 150 μg.

Elution pattern from chromatography of crude gonadotrophin (E_3) on QAE-Sephadex A 50 (starting buffer pH 6.4: imidazol-HC1)

Ion exchange chromatography on QAE-Sephadex (Figs. 2 & 3). The column size and the quantity of material applied was the same as used for CM-Sephadex chromatography. QAE-Sephadex A 50 was equilibrated with 0.02M imidazol-buffer, adjusted to pH 7.0 with HC1. Elution with the starting buffer resulted in the emergence of a broad peak containing 6 per cent of the protein applied to the column (Figure 2). This fraction (QAE_2F_1) contained highly purified LH with a specific activity of 1060 i.u./mg and an FSH activity lower than 4 i.u./mg. This was a two-fold increase in LH compared with the starting material and at least a twenty-fold decrease with respect to FSH.

After increasing the molarity with NaC1 to 0.1 M pH 6.2, a second fraction (F II) was eluted with LH activity of 700 i.u./mg and FSH activity of 370 i.u./mg, which represents a 4.5-fold increase of FSH activity.

The next fraction eluted from the column with 0.5 M NaC1 (F III) had low LH (47 i.u./mg) and FSH (25 i.u./mg) potencies and contained most of the albumin. More than 60 per cent of the protein applied to the column was recovered. There was a loss of about 40 per cent of the total LH activity, while FSH was completely recovered.

In a second experiment we repeated the ion exchange chromatography on QAE-Sephadex, but altered the starting imidazol buffer from pH 7.0 to 6.4 at the same molarity of 0.02 (Figure 3).

The first fraction (F I) with a protein yield of 10 per cent contained LH with a high biological activity (2110 i.u./mg) and FSH activity less than 2.5 i.u./mg. There was no increase in ovarian weight after administration of 150 μg. The second fraction (F II), eluted by increasing the concentration to 0.08 M with NaC1, also contained high LH activity (1080 i.u./mg), but contained 160 i.u. FSH/mg. Further fractions, eluted from the column by increasing NaC1 concentrations stepwise to 0.15 M and 0.25 M, showed FSH activity of 225 i.u./mg and LH activity of 330 i.u./mg (F III), and 36 i.u./mg and 125 i.u./mg (F IV) respectively.

All fractions were dialysed, and freeze-dried, and calculations of potencies were based on dry weight. The total recovery of FSH and LH activity was almost complete, 65 per cent of the applied material was recovered.

Disc electrophoresis. The chromatographic purification was controlled by analytical disc electrophoresis in 7.5 per cent acrylamide gel at pH 8.9 and 3mA per tube (Davis, 1962). Staining was with amidoblack 10B and 100 to 200 μg of protein was applied to each gel.

The heterogeneity of the starting material (E_3) was demonstrated bv disc-electrophoresis; at least 10 fractions were obtained (Figure 4_2).

The CM-Sephadex fractions were also subjected to disc-electrophoresis (Fig. 4). In CM_2F_5 with a high LH activity three distinct slow moving bands and a faster moving zone developed. The first are probably LH and the second zone, FSH. In CM_2F_2 with LH activity similar to the starting material and low FSH activity a single slower moving band occurred, which is probably due to LH. The faster moving heavily stained band is albumin. In CM_2F_1 only one single band appeared in the albumin area, which is a PAS-positive glycoprotein without biological activity.

The fractions of the QAE separation also were subjected to polyacrylamide electrophoresis (Figure 5).

In QAE_3F_1 the highly purified LH showed two sharp bands and a weak band, migrating little from the origin, with complete removal of contaminating proteins. This corresponds to the LH-rich fraction of CM-Sephadex (CM_2F_5). Fraction II (QAE_3F_2), containing FSH and LH activity, showed three lightly stained bands in the 'LH-region' in addition to two distinct bands, moving behind a small quantity of albumin. Fractions QAE_3F_3 and QAE_3F_4 both containing the two hormones, revealed albumin as the major component and several slower bands.

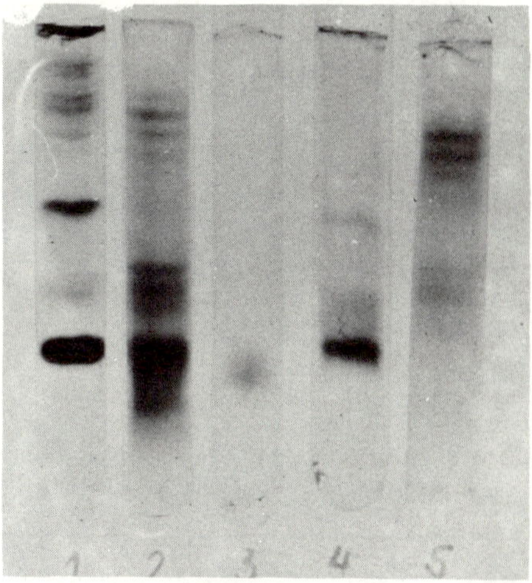

Figure 4

Disc electrophoretic patterns after Fractionation on CM-Sephadex.
1 = normal serum, 2 = E_3H_{69}

3 = CM_2F_1, 4 = CM_2F_2, 5 = CM_2F_5

Determination of N-acetyl neuraminic acid. The sialic acid content of the crude gonadotrophin and of the fractions after separation on CM-Sephadex and on QAE-Sephadex was determined according to the procedure of Warren (1959) (Table I). The highly purified LH (QAE_3F_1) was found to contain only 0.5 per cent sialic acid, while QAE_3F_2 with LH and FSH had 2.3 per cent. In the biologically inactive glycoprotein CM_2F_1 neuraminic acid was estimated to be present in an amount of 10.1 per cent.

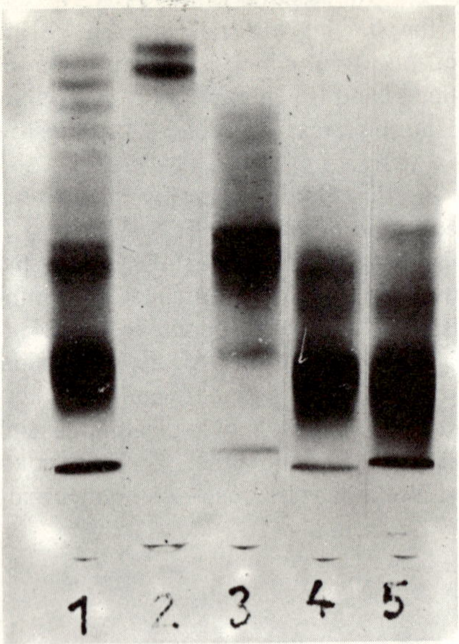

Figure 5

Disc electrophoretic patterns after Fractionation on QAE–Sephadex.

1 = E_3M_{69}, 2 = QAE_3F_1, 3 = QAE_3F_2

4 = QAE_3F_3, 5 = QAE_3F_4

TABLE I

Neuraminic Acid Content of Crude Gonadotrophin, CM–Sephadex and QAE–Sephadex Fractions*

	E_3H_{69}	CM_2F_1	CM_2F_2	CM_2F_3	CM_3F_5
N-acetyl neuraminic acid	4.1%	10.1%	2.3%	1.9%	2.9%
	E_3M_{69}	QAE_3F_1	QAE_3F_2	QAE_3F_3	QAE_3F_4
N-acetyl neuraminic acid	4.5%	0.5%	2.3%	3.4%	3.7%

* according the procedure of Warren (1959)

DISCUSSION

In CM-Sephadex chromatography the biological activity was spread over all the adsorbed fractions. There was no increase in specific FSH activity and a great loss of total FSH activity. In spite of the 2.5-fold increase in the specific LH activity of the last fraction it still contained FSH of half the original specific activity.

In QAE-Sephadex chromatography there was again a wide spreading of biological activity. In contrast to the cation exchanger the strongly basic anion exchanger resulted in a differential adsorption of FSH and LH. The main LH activity appeared in the first fraction, in which no FSH could be detected. In the following fractions a gradual diminution of LH activity was observed. There was no remarkable increase in FSH specific activity and also no separation from LH. The main FSH fraction appeared with 0.15 M buffer.

In examining the different chromatographic fractions by disc electrophoresis a localization of LH activity was possible. From the disc patterns we have to conclude that two LH species exist: the most potent material appears as two distinct slow moving bands. The second LH species is a faster moving protein, probably with three sharp bands, but still contaminated with FSH and inert proteins. These observations are in agreement with earlier results obtained using electrofocusing (Bettendorf *et al.* 1968). A corresponding characterisation of FSH was not possible.

We thought that it would be useful to follow up the purification steps by sialic acid analysis, but we did not find a correlation between potency and sialic acid content.

ACKNOWLEDGMENT

This work was supported by a grant from Deutsch Fouschungsgemeinshaft, SFB 37.

REFERENCES

Bettendorf, G., Apostolakis, M. & Voigt, K.D. (1962). *Acta endocr. Copenh.* **41,** 1.

Bettendorf, G., Breckwoldt, M., Czygan, P.-J., Fock, A. & Kumasaka, T. (1968). *In Gonadotropins 1968,* p. 13. Ed. Rosemberg, E. Los Altos, California: Geron-X.

Davis, B.F. (1962). *Pre-print 'Disc Electrophoresis',* Rochester, NY: Distillation Prod. Div. Eastman Kodak Co.

Karg. H. (1957). *Klin. Wschr.* **35,** 643.

Koenig, V.L. & King, E. (1950). *Arch. Biochem.* **26,** 219.

Parlow, A.F. (1961). In *Human Pituitary Gonadotropins,* p. 300. Ed. Albert, A. Springfield, Illinois: Thomas.

Steelman, S.L. & Pohley, F.M. (1953). *Endocrinology,* **53,** 604.

Warren, L. (1959). *J. biol. Chem.* **234,** 1971.

Comparison of the Characteristics of Human Pituitary and Urinary FSH and LH During Purification

Vernon C. Stevens, David G. Anderson and John E. Powell

Similar procedures have been used for the purification of pituitary and urinary gonadotrophins (Reichert and Parlow, 1964; Roos, 1968; Stevens, 1967). Few studies have been reported in which a comparison of the properties of urinary and pituitary FSH and LH was made under identical conditions. This report contains some observations of the properties of pituitary and urinary gonadotrophins during purific-ation using conditions the same for each.

MATERIALS AND METHODS

Gonadotrophin preparations. Purification of pituitary FSH and LH was accomplished from 2.0 g of a crude extract containing 87.0 i.u. FSH and 102.0 i.u. LH per mg by bioassay kindly supplied by the National Pituitary Agency, Baltimore, Md. This material was precipitated from an ethanol-ammonium acetate (10 per cent NH_4Ac in 40 per cent ethanol, pH 5.1) extract of approximately 100 g of acetone-dried pituitaries with two volumes of ethanol. The precipitate was washed with ethanol and ether and dried.

Two separate lots of urinary gonadotrophin of 1 g each were combined for purification (obtained from the Instituto Farmacologico, Serono, Rome). The combined fractions had a biological potency of 57.0 i.u. FSH and 90 i.u. LH per mg. This material had been prepared from menopausal urine by procedures described by Donini *et al.* (1964).

Since none of the fractions during purification were lyophilised, estimates of solids were made by protein determinations using the method of Lowry *et al.* (1951), using bovine serum albumin as standard. Each 2.0 g preparation (pituitary and urinary) was dissolved in dilute phosphate-borate buffer, pH 8.0 and a protein determination performed. The estimated protein in the pituitary fraction was 1278 mg and 1671 in the urinary fraction. Both pituitary and urinary fractions were divided into two equal portions and frozen at -20°C until fractioned.

DEAE - cellulose chromatography. Fractionation of both pituitary and urinary gonadotrophins on DEAE - cellulose was performed essentially as described by Reichert and Parlow (1964). One of the two equal portions of urinary or pituitary material was fractionated at a time. Thus, two DEAE fractionations were performed for urinary and two for pituitary gonadotrophins.

For each fractionation, a column 32.5 x 2.5 cm was packed with DEAE–cellulose equilibrated in a 0.032M phosphate – 0.02M borate buffer, pH 8.0, diluted 1 : 5 with distilled water. A flow rate of 1.2 ml/min was used and approximately 12 ml fractions were collected. Adsorbed materials were removed from the column stepwise with 0.03M, 0.10M and 0.50M NaCl in the diluted phosphate-borate buffer. Elution of fractions was observed by means of an ultraviolet monitor with absorption at 280 mμ. The contents of each tube were then measured at 280 mμ spectrophotometrically. Eluted fractions were pooled according to protein peaks, dialysed, and aliquots taken for protein determination and assay. The remaining pool was concentrated by per-evaporation in dialysis tubing, and then stored at -20°C until further fractionated.

Gel filtration chromatography. A column of Sephadex G-100, 90.0 x 2.5 cm, was packed and used repeatedly for the various filtrations. Each fraction from DEAE–C was passed over the column once in each of two buffer systems. The column, equilibrated with 0.1 M K_2HPO_4–0.3 M NaCl, pH 7.0, was used during the first chromatography of each fraction. Pooled samples from the first filtration were passed over the same column after re-equilibration of the Sephadex with 0.05 M NH_4HCO_3, pH 8.0. The flow rate was adjusted to 0.80 ml per minute and fractions of 9.5 ml were collected.

Following each filtration, the absorbancy at 280 mμ was measured and aliquots taken for protein determination and assay. The remaining contents of each tube were frozen (-20°C) until the assays were completed. Fractions were then pooled based on immunological activity, dialysed, aliquots taken for assay, and the remainder concentrated by per-evaporation. Pools were frozen between filtration steps.

Preparative polyacrylamide gel electrophoresis. This procedure was used for the final step of purification for pituitary and urinary FSH. The electrophoresis apparatus used had a gel surface area of 10 x 0.5 cm. A lower gel 4.0 cm in height, contained 7.0 g of acrylamide, 0.125 g of bis-acrylamide, 57.5 μl of N, N, N^1, N^1-tetramethylethylenediamine and 70 mg of ammonium persulfate per 100 ml of 0.40 M Tris-HCl buffer, pH 8.9. The 1.0 cm upper gel contained 3.0 g of acrylamide, 0.063 g of bisacrylamide and 0.50 mg of riboflavin per 100 ml of 0.10 M Tris-HCl buffer, pH 6.7. The lower and upper electrode buffers contained 2.5mM Tris-20 mM glycine, pH 8.3. Cross flow elution was accomplished with a 0.1 M Tris-HCl buffer, pH 8.3. Electrophoresis was performed at 300 volts and 20 mA at 2°C. Five ml fractions were collected at a flow rate of 0.042 ml per minute. Liquid samples were made 6 per cent with sucrose and applied to top of the polymerized upper gel. Electrophoresis was continued for 26 hours. Aliquots were taken from each tube and assayed by radioimmunoassay. Tubes were pooled according to the FSH activity in each, dialysed and the pool assayed again.

Assays. All assays were performed on liquid samples and the dose based on protein determination by the Lowry method. Pooled fractions were bioassayed for FSH by the ovarian augmentation assay (Steelman and Pohley, 1953) and LH by the ovarian ascorbic acid depletion assay (Parlow, 1961) using a 3 x 3 design on each. Radioimmunoassays were performed on each tube from gel filtration and polyacrylamide electrophoresis as well as on all pooled fractions. Three doses of unknowns were assayed in all immunoassays. In assays of tube fractions, buffer blanks were included to detect non-specific responses. Assay procedures have been described elsewhere (Stevens, 1969). Reference preparations of pituitary origin were used for assaying pituitary fractions and potencies were converted to i.u. 2nd IRP for convenience in reporting. The second IRP standard was used for assays of urinary FSH and LH.

RESULTS

DEAE–cellulose chromatography. The protein elution profiles of 1.0 g of pituitary and 1.0 g of urinary gonadotrophin are shown in Figure 1. While protein elution was nearly equal in all urinary gonadotrophin fractions, less protein was eluted with 0.03 M and 0.10 M NaCl than the unadsorbed and 0.50 M NaCl elutes with the pituitary preparation.

Table 1 shows recoveries of protein and relative potency estimates for FSH and LH of the pooled fractions from two urinary and two pituitary gonadotrophin columns. In the pituitary fractionation, FSH was purified four times, while the LH increased about 3.5 times as measured biologically. Urinary FSH and LH were purified approximately three times in this procedure as determined by bioassay. In both fractionations, immunological estimates were in general agreement with the biological. The most potent pituitary LH preparation was found in the unadsorbed fraction while the most potent urinary LH was in the 0.03 M fraction, however, relatively more LH was found in the pituitary FSH fraction (0.10 M NaCl) than the urinary FSH fraction.

Gel filtration chromatography. The most potent LH fractions from DEAE–C chromatography were first subjected to filtration on a

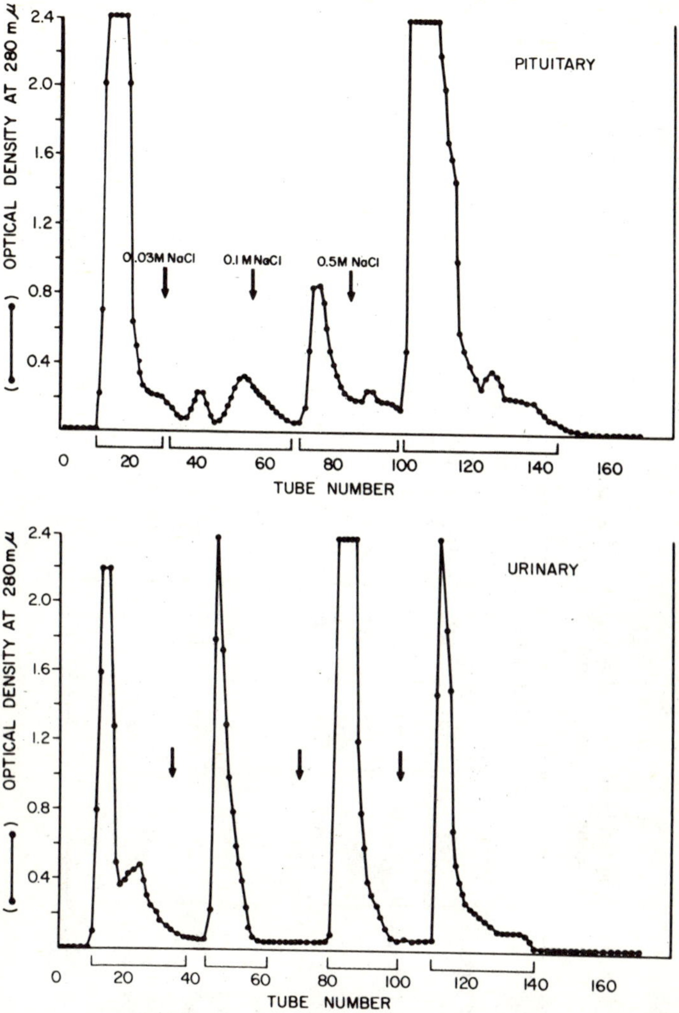

Figure 1. Protein elution profile of 1.0 g of pituitary gonadotrophin (top) and 1.0 g of urinary gonadotrophin (bottom) after DEAE - cellulose chromatography.

Sephadex G-100 column equilibrated with the phosphate-sodium chloride buffer. The elution profile for the pituitary and urinary LH fractions are illustrated in Figure 2. A single major peak was seen with the pituitary fraction while two major peaks were observed with the urinary fraction. Assays of each tube indicated that urinary LH was contained only in the first protein peak. FSH activity was located in the same areas as LH activity for both pituitary and urinary materials.

TABLE 1

Biological and Immunological Activities of Pituitary and Urinary Gonadotrophin Fractions after DEAE–Cellulose Chromatography

FRACTION	PROTEIN (mg)	BIOLOGICAL ACTIVITY (i.u./mg)			IMMUNOLOGICAL ACTIVITY (i.u./mg)		
		FSH	LH	FSH/LH	FSH	LH	FSH/LH
Starting Pituitary	1278.0	87	102	0.85	120	100	1.02
Unadsorb	165.1	135	351	0.38	130	333	0.39
0.03 M NaCl	90.2	51	232	0.22	85	211	0.40
0.10 M NaCl	160.4	420	189	2.22	540	157	3.43
0.50 M NaCl	265.8	30	72	0.31	13	75	0.17
% Recovery	53.2						
Starting Urinary	1671.0	57	90	0.63	62	87	0.71
Unadsorb	285.1	10	108	0.09	10	137	0.07
0.03 M NaCl	265.4	100	272	0.37	115	250	0.46
0.10 M NaCl	335.0	185	75	2.46	180	45	4.00
0.50 M NaCl	165.2	17	49	0.35	25	42	0.59
% Recovery	62.8						

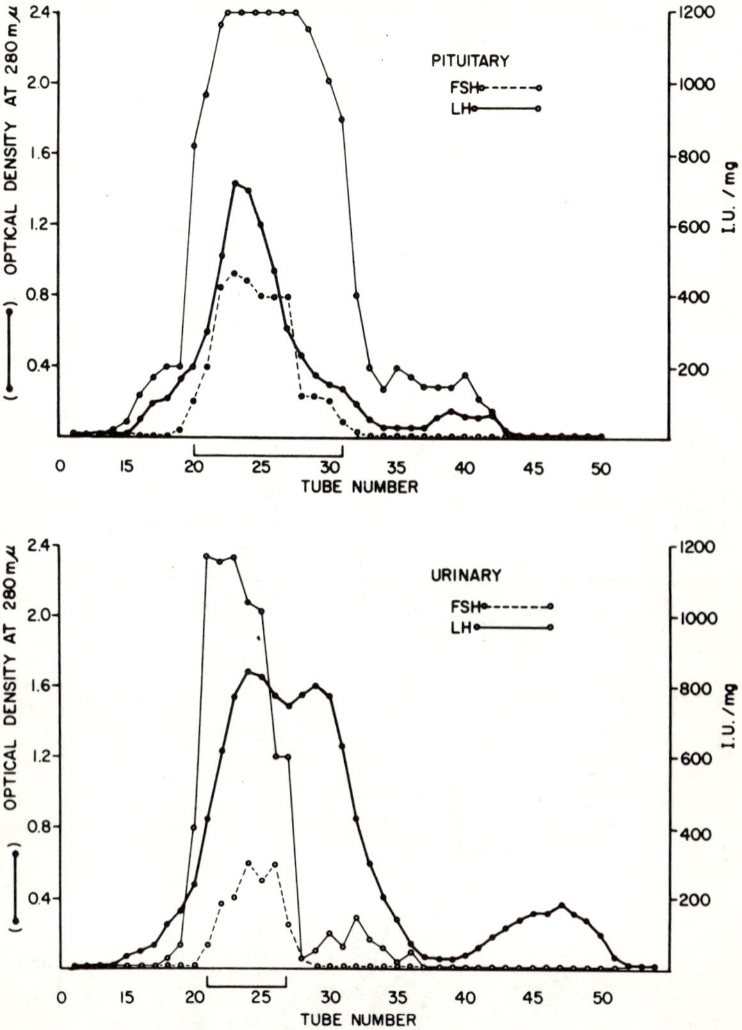

Figure 2. The distribution of FSH and LH after gel filtration of principal LH fractions on Sephadex G-100 equilibrated with 0.1 M K_2HPO_4 - 0.3 M NaC1, pH 7.0.

Figure 3 illustrates the elution profile of the most potent pituitary and urinary FSH fractions from DEAE–C on the first gel filtration column. The major protein peak emerged in the same volume for pituitary and urinary FSH fractions. Several smaller peaks were seen following the major peak with the pituitary material which were not present in the urinary fraction. No significant resolution of FSH and

LH was obtained by gel filtration with this buffer system, although LH emerged in two different areas of the profile.

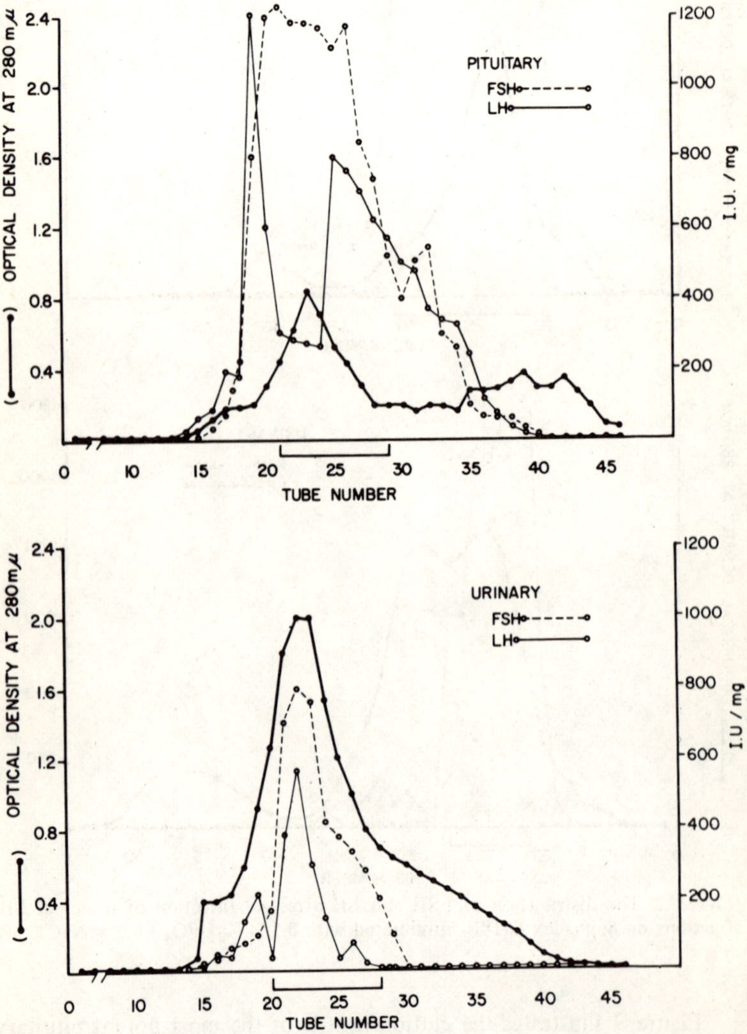

Figure 3. The distribution of FSH and LH after gel filtration of principal FSH fractions on Sephadex G-100 equilibrated with 0.1 M K_2HPO_4 - 0.3 M NaC1, pH 7.0.

TABLE II

Biological and Immunological Activities of Pituitary and Urinary Gonadotrophin Fractions after the First Gel Filtration on Sephadex G–100

FRACTION	PROTEIN (mg)	BIOLOGICAL ACTIVITY (i.u./mg)			IMMUNOLOGICAL ACTIVITY (i.u./mg)		
		FSH	LH	FSH/LH	FSH	LH	FSH/LH
Pituitary LH	70.1	137	750	0.18	127	720	0.18
Urinary LH	115.0	140	532	0.26	100	426	0.23
Pituitary FSH	47.0	1036	400	2.59	1025	417	2.45
Urinary FSH	138.0	350	123	2.84	325	100	3.25

The tubes on the first gel filtration were pooled as indicated in Figures 2 and 3. The biological and immunological activity of these fractions is shown in Table II. Pituitary and urinary LH increased in potency about twice by the bioassay and immunoassay following this purification step. By bioassay the pituitary FSH potency increased 2.5 times while the urinary FSH increased less than twice. Immunoassays showed similar increases in potency for both fractions.

Each of the LH and FSH preparations from the first gel filtration were then passed over the same column equilibrated with the bicarbonate buffer system. The protein elution profile for the second filtration of pituitary and urinary LH fractions are shown in Figure 4. A single major protein peak was observed with each, however, the pituitary material emerged from the column slightly ahead of the urinary preparation. FSH activity was located with pituitary LH activity while the FSH in the urinary fraction emerged slightly ahead of the peak of LH activity.

The elution profile of the principal pituitary and urinary FSH fractions after the second Sephadex is shown in Figure 5. Again, one major protein peak was observed. FSH and LH activity was located throughout the peak in the pituitary material. In the urinary fraction, FSH was located principally in the first portion of the peak and LH activity in the latter. FSH activity was located in about the same area of the peak for both pituitary and urinary materials although the protein peak of pituitary material emerged earlier from the column.

Following the second gel filtration, pituitary LH increased in relative potency 2.1 and 2.2 times by bioassay and immunoassay respectively (Table III). Urinary LH increased in potency 1.9 times by immunoassay but only 1.4 times by bioassay. Better resolution of the contaminating FSH was obtained with the urinary LH in this step than with the pituitary LH.

The pituitary FSH fraction increased in potency about one and one-half times by bioassay and immunoassay. The LH contamination in this preparation, as reflected by the FSH-LH ratio, was not significantly reduced. However, the urinary FSH doubled in potency by bioassay and by immunoassay with a significant reduction in LH contamination.

Preparative polyacrylamide gel electrophoresis. Each of the FSH fractions obtained from the second gel filtration on Sephadex shown in Figure 5 were further purified by polyacrylamide gel electrophoresis. Electrophoresis conditions were kept the same for both fractionations although the amount of starting material was greater for the urinary fraction.

The elution profile of materials absorbing light at 280 mμ revealed similar patterns for the two experiments. The large, well defined peak that emerged first during electrophoresis was nearly identical in both runs. This material of unknown composition moves at the ion front.

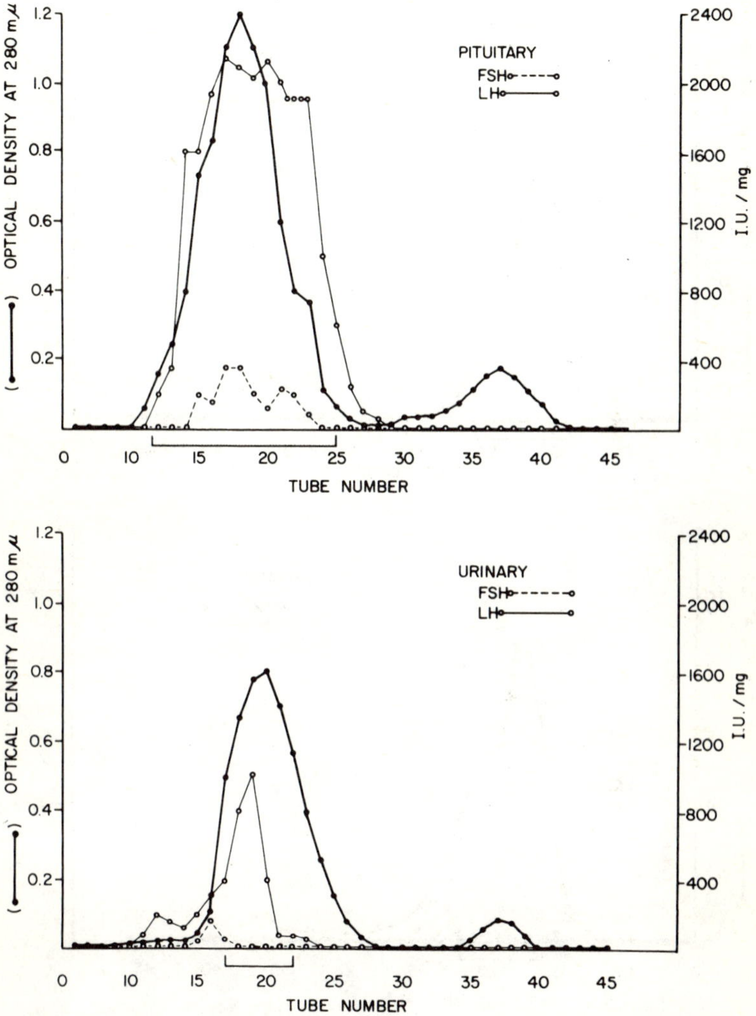

Figure 4. Profiles of FSH and LH after a second gel filtration of principal LH fractions on Sephadex G-100. The column was equilibrated with 0.05M NH_4HCO_3.

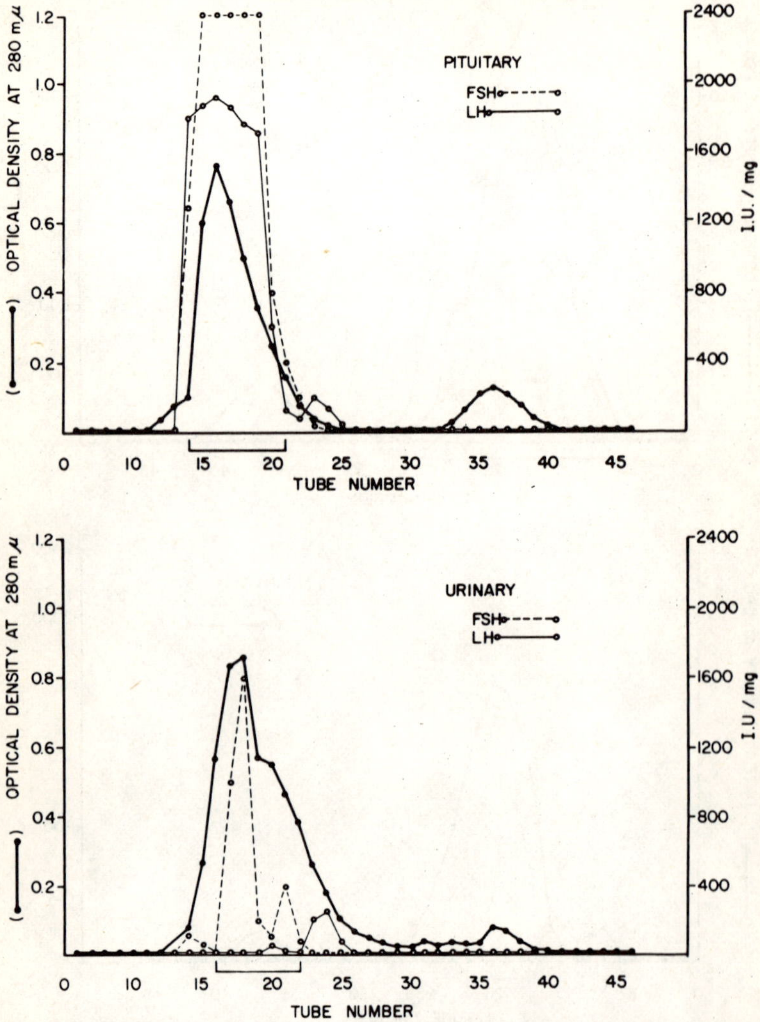

Figure 5. Profiles of FSH and LH after a second gel filtration of principal FSH fractions on Sephadex G-100. The column was equilibrated with 0.05 M NH_4HCO_3.

TABLE III

Biological and Immunological Activities of Pituitary and Urinary Gonadotrophin Fractions after the Second Gel Filtration on Sephadex G–100

FRACTION	PROTEIN (mg)	BIOLOGICAL ACTIVITY (i.u./mg)			IMMUNOLOGICAL ACTIVITY (i.u./mg)		
		FSH	LH	FSH/LH	FSH	LH	FSH/LH
Pituitary LH	31.2	150	1550	0.10	161	1577	0.10
Urinary LH	47.1	15	740	0.02	13	810	0.02
Pituitary FSH	26.0	1637	700	2.33	1800	900	2.00
Urinary FSH	68.0	708	42	16.89	800	57	14.03

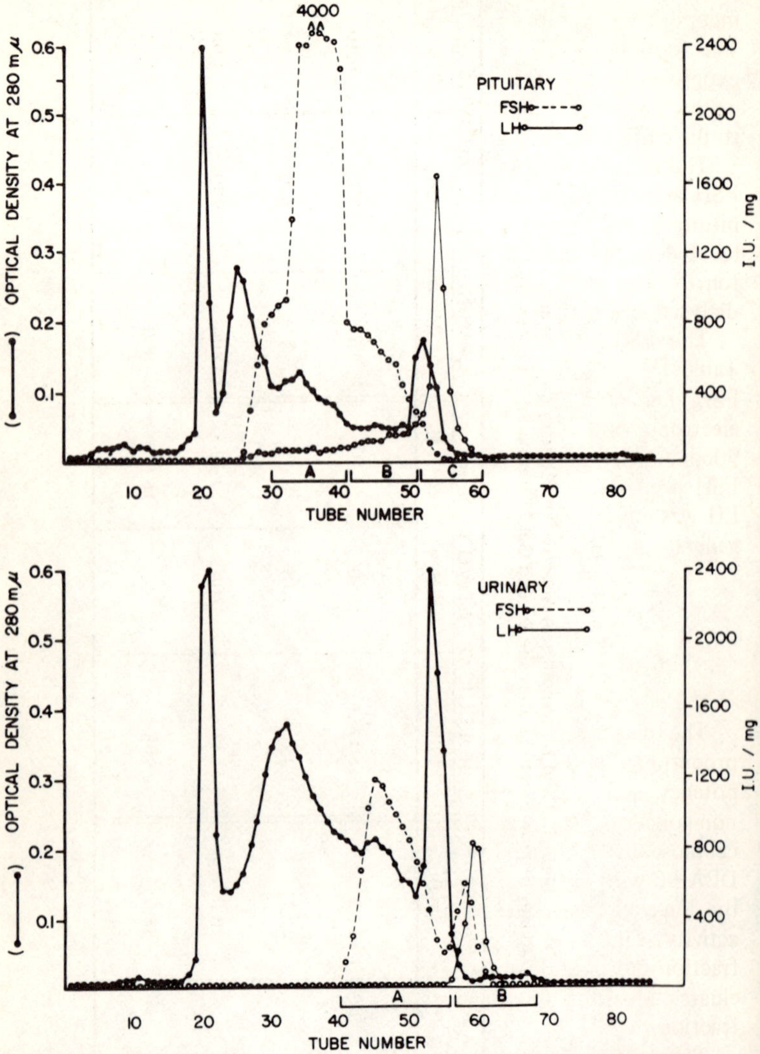

Figure 6. The elution profile of FSH and LH following polyacrylamide gel electrophoresis of the principal FSH fractions from the second gel filtration.

Following this peak, a broader second peak was observed with both FSH preparations. The second peak emerged from the gel sooner during electrophoresis of the pituitary FSH. A third protein peak, that was incompletely resolved from the second, was observed in both patterns. The fourth and final protein peak appeared at the same area in both experiments. The protein eluted in this fraction was accompanied by a yellow colour. A distinct yellow band could be observed in the gel prior to the elution of the fourth protein peak.

The distribution of FSH and LH activities is shown in Figure 6. The FSH activity was located in the third peak and fourth peak for both pituitary and urinary fractions. Although there was little LH activity located in the third protein peak, high LH activity was found in the fourth. The areas indicated by solid bars in Figure 6, were pooled, dialysed and assayed.

The FSH and LH activities of the pooled fractions are shown in Table IV. A pituitary FSH preparation with a potency of 3705 i.u. FSH and 205 i.u. LH per mg was obtained by bioassay after electrophoresis. The principal urinary FSH fraction had an activity of 906 i.u. FSH and 10 i.u. LH per mg. Although a more potent urinary FSH preparation was found in a later fraction, its contamination with LH was greater. Bioassays and immunoassays of these fractions were in general agreement.

DISCUSSION

The data obtained in this study revealed some differences in the properties of pituitary and urinary gonadotrophins. Although the relative potency of the starting materials was nearly the same, the non-gonadotrophin materials were different as revealed by DEAE – cellulose chromatography. Considerably more protein was adsorbed to DEAE-C with the urinary gonadotrophins. FSH activity was eluted from the ion-exchange column similarly for both preparations but not LH activity. The most potent pituitary LH was eluted in the unadsorbed fraction but considerable material was still present in the 0.10 M NaCl eluate. Most of the urinary LH was eluted in the 0.03 M NaCl fraction.

The behaviour of FSH and LH during gel filtration on Sephadex G-100 equilibrated with the phosphate-sodium chloride buffer was quite similar for both pituitary and urinary hormones. This observation was previously reported by Roos (1968). During chromatography on the second Sephadex column (bicarbonate buffer system), urinary FSH was more retarded than the pituitary gonadotrophin. Also, urinary FSH and LH were better resolved during this fractionation.

TABLE IV

Biological and Immunological Activities of Pituitary and Urinary Gonadotrophin Fractions after Preparative Polyacrylamide Electrophoresis

FRACTION	PROTEIN (mg)	BIOLOGICAL ACTIVITY (i.u./mg)			IMMUNOLOGICAL ACTIVITY (i.u./mg)		
		FSH	LH	FSH/LH	FSH	LH	FSH/LH
Starting Pituitary Gonadotrophin	24.1	1637	700	2.33	1800	900	2.00
Pool A	8.2	3705	205	18.15	3850	300	12.83
Pool B	3.2	1861	1050	1.77	1760	900	1.95
Pool C	4.6	300	1850	0.16	225	1765	0.12
Starting Urinary Gonadotrophin	60.2	708	42	16.89	800	57	14.03
Pool A	22.0	906	10	90.00	890	12	74.10
Pool B	12.0	1050	212	4.95	925	255	3.62

Pituitary gonadotrophins generally increased in relative potency more than urinary gonadotrophins during ion-exchange chromatography and during the first gel filtration. Urinary FSH increased in potency more than pituitary FSH after the second gel filtration although urinary LH increased less than its pituitary counterpart in this step. The principal difference in pituitary and urinary FSH and LH during these procedures appears to be that urinary materials have more substances with no gonadotrophic activity retained with gonadotrophins during purification and that urinary FSH and LH are better resolved in a buffer of low molarity.

The protein elution patterns of FSH preparations during polyacrylamide electrophoresis indicated that pituitary FSH migrated faster than urinary FSH. Roos (1968) observed similar mobilities for urinary and pituitary FSH. Perhaps the larger sample of urinary FSH in this study accounts for the difference in mobility.

The presence of the yellow colour migrating through the gel during electrophoresis and appearing just before the fourth protein peak has not been observed before by us or reported by others performing similar experiments (Roos, 1968; Stevens, 1967; Donini, *et al.* 1968; Saxena and Rathnam, 1967). It is postulated that under the conditions used, the pH of the buffer was altered during electrophoresis and that slower migrating proteins were eluted after several hours. These same conditions could have been responsible for making riboflavin electrophoretically mobile and it migrated as a yellow band. The real reason for these differences from earlier studies is not known. It is also significant that the protein peak associated with this colour was a mixture since both FSH and LH were found.

There are some obvious discrepancies between the assays of individual tubes during purification and the pooled fractions from the same tubes. Tube contents were assayed only once and no attempt was made to remove buffers from samples. The pooled fractions were dialysed and assayed several times. Assays of pooled fraction should therefore be considered more accurate.

Pooled fractions throughout purification were analyzed for heterogeneity by disc electrophoresis. Although a reduction in the number of components present in FSH and LH preparations was observed, none of the final preparations were homogeneous. This observation is significant considering the high biological and immunological activity of these preparations.

REFERENCES

Donini, P., Montezemolo, R. & Puzzuoli, D. (1964). *Acta endocr. Copenh.* **45**, 321.

Donini, P., Puzzuoli, D., D'Alessio, I. & Donini, S. (1968). In *Pharmacology of Hormonal Polypeptides and Proteins*, p. 229. Ed. Martini, D. New York: Plenum.

Lowry, O.H., Rosebrough, N.J., Farr, A.L. & Randall, R.J. (1951). *J. biol. Chem.* **193**, 265.
Parlow, A.F. (1961). In *Human Pituitary Gonadotropins*, p. 300. Ed. Albert, A. Springfield, Illinois: Thomas.
Reichert, L.E. & Parlow, A.F. (1964). *Endocrinology*, **74**, 236.
Roos, P. (1968). *Acta endocr. Copenh.* **59**, Suppl. 131.
Saxena, B.B. & Rathnam, P. (1967). *J. biol. Chem.* **242**, 3769.
Steelman, S.L. & Pohley, F.M. (1953). *Endocrinology*, **53**, 604.
Stevens, V.C. (1967). In *Recent Research on Gonadotrophic Hormones*, p. 142. Ed. Bell, E.T. & Loraine, J.A., Edinburgh: Livingstone.
Stevens, V.C. (1969). *J. clin. Endocr. Metab.* **29**, 904.

Purification and Partial Chemical-physical Characterization of FSH from Menopausal Urine

P. Donini, D. Puzzuoli, I. D'Alessio, G. Bergesi and S. Donini.

SUMMARY

A highly purified FSH preparation was obtained by batchwise purification of human menopausal gonadotrophin (HMG) on DEAE - C, followed by chromatography on a DEAE - C column, gel filtration on Sephadex G-100 and by preparative polyacrylamide gel electrophoresis. This preparation, which assayed 1,255 i.u. FSH/mg and 3.2 i.u. LH/mg, was investigated for its physical and chemical properties. The following findings were obtained :

1. The preparation was homogeneous in ultracentrifugation and disc electrophoresis studies.
2. The molecular weight calculated on the $S_{20,w}$ value and on the diffusion constant ($D_{20,w}$) was 31,600.
3. The results of aminoacid analysis agreed fairly well with those previously reported by other investigators.
4. The molecular sieve chromatography on Sephadex G-100 revealed significant differences between FSH, LH and HCG.
5. The carbohydrate components of the urinary FSH were as follows: the hexoses, hexosamine and sialic acid were evaluated to be 10.4, 9.6 and 8.6 per cent respectively, as calculated on the dry weight basis.
6. The ultraviolet absorption spectrum of neutral and alkaline solutions was studied. The $A_{1cm}^{1\%}$ at 277 mμ and pH 7.0 was calculated to be 6.85.
7. The immunoelectrophoretic studies showed that the FSH preparation contained two protein components with the same electrophoretic mobility, but no albumin contamination was revealed.

INTRODUCTION

Several studies on the purification and physico-chemical characterization of human pituitary follicle-stimulating hormone have been reported in the last few years by Roos and Gemzell (1964a), Roos (1968), Amir *et al.* (1966), Papkoff *et al.* (1967), Reichert *et al.* (1968), Saxena and Rathnam (1967), Peckham and Parlow (1969), but relatively few studies appeared on the purification and characterization of human urinary FSH: Roos and Gemzell (1964b), Roos (1968), Stevens (1967), Donini *et al.* (1968a).

In this paper we wish to describe some studies on the chemical and physical properties of a highly purified human urinary FSH preparation.

MATERIALS AND METHODS

Preparation of FSH fraction for further purification. One batch of HMG (Pergonal, E134) was used as the starting material for the purification. This material, prepared as described by Donini *et al.* (1964) was purified according to the method of Roos and Gemzell (1964b). The DEAE-Cellulose (DEAE-C) chromatography and the gel filtration on Sephadex were carried out at 4°C. Effluents from columns were monitored for protein content by reading at 280 mμ in a Beckman DU II spectrophotometer. Dialysis was performed against distilled water at 4°C. The final desalting was carried out on Sephadex G-25 columns and the salt-free solutions were lyophilized. The preparative polyacrylamide gel electrophoresis was carried out according to the procedure previously reported by Donini *et al.* (1968a). The instrument used was from Buchler Instruments, Inc., Fort Lee, N.J., U.S.A. The analytical disc electrophoresis was performed as described by Davies (1964). The Sephadex was supplied by Pharmacia, Uppsala, Sweden.

Preparation of highly purified luteinizing hormone (LH). This material was obtained from HMG according to the method of Donini *et al* (1968b). This preparation assayed 1166 i.u./mg, as determined by radioimmunoassay and 982 i.u./mg by O.A.A.D. bioassay. The FSH activity was undetectable, as biologically determined at a dose level of 1.26 mg.

Preparation of highly purified chorionic gonadotrophin (HCG). This material was prepared by Donini and Donini (1969) from the commercial HCG (3,400 i.u./mg) by DEAE-C column chromatography and gel filtration on Recychrom (LKB, Instruments, Stockholm, Sweden) packed with Sephadex G-100. The immunochemical grade HCG (12,000 i.u./mg) was shown to be homogeneous by disc electrophoresis and immunoelectrophoresis.

Biological assay. FSH activity was determined by the HCG augmentation assay of Steelman and Pohley (1953). The luteinizing activity was measured by the ovarian ascorbic acid depletion method of Parlow (1961) and by radioimmunoassay according to the method of Donini and Donini (1969). The reference preparation used was the 2nd IRP-HMG. FSH and LH activities were expressed as international units of the 2nd IRP-HMG. The experimental design was 3+3 assay. Statistical calculation was performed according to the method of Borth *et al.* (1957). Other methods used in this study are described herein.

RESULTS

Purification. A flow diagram (Figure 1) illustrates the purification of urinary FSH.

5.46 g of HMG (Pergonal, 82.7 i.u.–FSH and 58.3 i.u.–LH per mg)

Batchwise separation of FSH and LH on DEAE–C

FSH fraction (2.88 g, 133 i.u.–FSH/mg) | LH fraction (1.74 g, 86 i.u.–LH/mg)

DEAE–C column chromatography

Eluted with borate-phosphate buffer pH 8.0 + 0.1 M NaCl (828 mg, 351 i.u.–FSH/mg) | Other fractions containing LH or impurities

Gel filtration on Sephadex G–100 column

Symmetrical part of the 2nd protein peak (323.7mg, 692 i.u.–FSH/mg) | Other fractions containing LH or impurities

Preparative polyacrylamide gel electrophoresis

Part of 2nd protein peak (see Fig. 4)
Highly purified FSH
104.3 mg – 1255.6 i.u.–FSH/mg
3.2 i.u.–LH/mg
| Other fractions containing LH or impurities

Figure 1. Purification of follicle-stimulating hormone from human menopausal gonadotrophin. The biological activities are expressed in terms of i.u. (2nd IRP-HMG).

The first step of further purification was the chromatography on a DEAE-C column, as described by Reichert and Parlow (1964) and slightly modified by Donini *et al.* (1966). 2.88 g of partially purified FSH (E146-2) was applied to a 2 x 100 cm column packed with DEAE-C, Whatman DE-50, previously equilibrated with 0.00213 M sodium borate and 0.00351 M sodium phosphate buffer, pH 8.0. The chromatography was carried out by stepwise changing of solvents at a flow rate of 35 ml/hr. The elution pattern and FSH activities are shown in Figure 2.

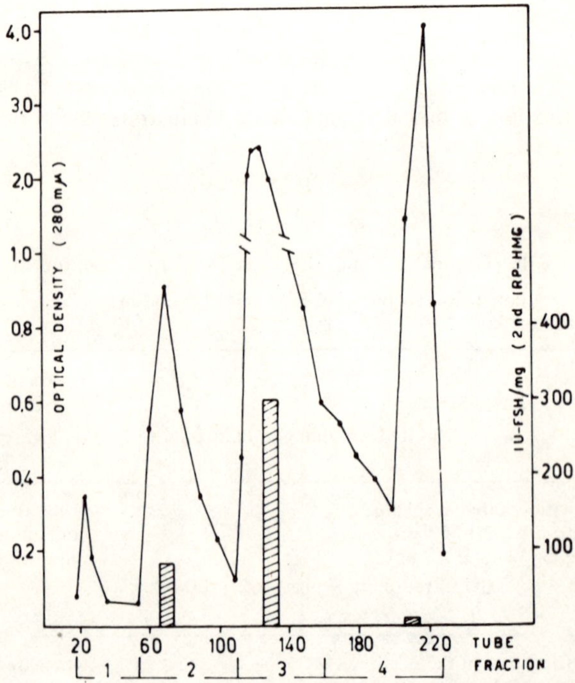

Figure 2. First step. Chromatography of partially purified urinary FSH (E146-2) on DEAE-C column 2 x 100 cm. 2.88 g of protein applied to the column. Fraction 1 unadsorbed proteins on the column equilibrated with 0.00426 M sodium borate and 0.00702 M sodium phosphate buffer, pH 8. Fractions 2, 3 and 4 proteins eluted with doubled molarity of the same buffer and containing 0.05 M, 0.1 M and 0.5 M sodium chloride, respectively. Samples 10 ml/tube. Recovery total protein 87.4 per cent, FSH 86.7 per cent.

Fraction 1 was unadsorbed proteins. Fractions 2, 3 and 4 were eluted with 0.00426 M sodium borate and 0.00702 M sodium phosphate buffer, pH 8.0 containing 0.05, 0.1 and 0.5 M sodium chloride respectively. We wish to point out that most of the FSH activity was found in fraction 3, obtained by pooling the tubes corresponding to the third protein peak. It is of interest to note that the FSH activity of fraction 3 was 2.3 times as high as the precursor. The most potent fraction obtained (E174-3) was purified in the second step by chromatography on Sephadex G-100 carried out according to the method of Roos and Gemzell (1964b).

Partially purified FSH (827 mg) was applied to the 3.0 x 150 cm column, packed with Sephadex G-100 previously equilibrated with 0.1 M potassium phosphate buffer, pH 7.0, containing 0.3 M sodium chloride. In this chromatography, as shown in Figure 3, we obtained three protein peaks, but the effluent was pooled in five fractions. It is of interest to point out that the FSH activity was mainly found in the sharp and symmetric part of the second peak, which contained 77.1 per cent of the FSH activity applied to the column. It should be noted that the FSH potency of this fraction was approximately doubled as compared to the precursor used in this step.

In the last step, further purification was achieved by the preparative polyacrylamide gel electrophoresis, carried out as described by Donini *et al.* (1968a). The electrophoresis was performed at 1.5°C, 300 V, 45 mamp.

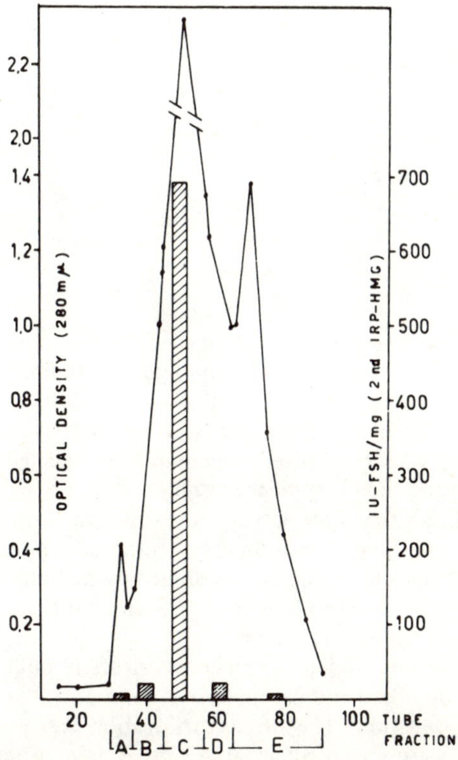

Figure 3. Second step. Gel filtration of partially purified urinary FSH (E174-3) on Sephadex G-100 column 3 x 150 cm. 827 mg of protein applied to the column. Samples: 10 ml/tube. Recovery total protein 95.2 per cent, FSH 79 per cent.

Of the partially purified FSH obtained by the second step (E174bis-C) 322 mg was divided into four aliquots which were purified in four runs. A typical disc electrophoresis experiment is shown in Figure 4.

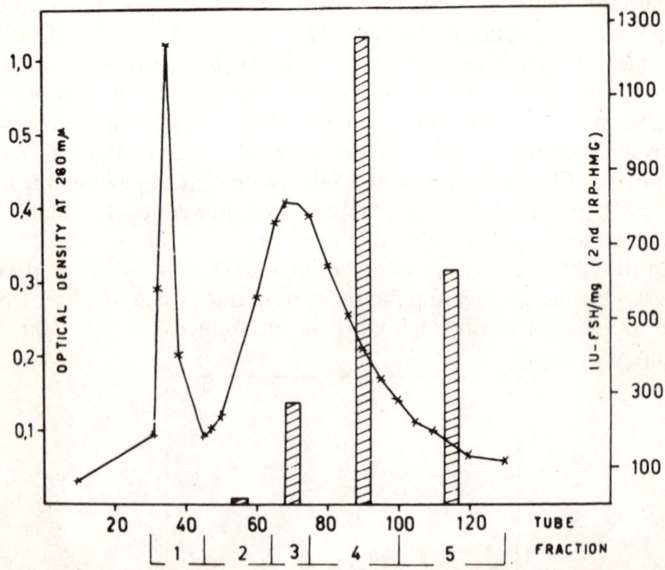

Figure 4. Third step. Preparative polyacrylamide electrophoresis of partially purified urinary FSH (E174bis-C). 86 mg of protein applied to the gel column. Samples: 4 ml/tube. Recovery total protein 89.3 per cent, FSH 84.3 per cent.-

The first peak was unpolymerized acrylamide, which absorbed at 280 mμ, but contained a negligible amount of protein. The second peak was divided into four fractions by pooling the tubes 56/64, 65/75, 76/100 and 101/130. The effluents, after overnight dialysis against distilled water, were lyophilized and completely desalted on a Sephadex G-25 column and then lyophilized again.

Table I shows that FSH activity was very low in the first part of the effluent and that it was mainly recovered in fraction 4, which assayed 1,255 i.u.- FSH/mg, as determined by the biological assay. The LH contamination was very low, that is, 3.26 i.u.-LH/mg were detected, when tested by radioimmunoassay.

Characterization studies. The characterization studies were carried out on the lyophilized and dry material.

Ultracentrifugation. The ultracentrifugation was performed in the Spinco Mod. 3 ultracentrifuge at 59,780 r.p.m., using 0.6 per cent solution of FSH in 0.1 M phosphate buffer, pH 6.5.

TABLE I

Relative potencies and recoveries of the fractions obtained by preparative polyacrylamide gel electrophoresis of partially purified FSH (E174bis–C). The biological activities are expressed as i.u. of 2nd IRP–HMG.

Fraction	Protein mg+	Bio – FSH i.u./mg	Bio – FSH F. L. (P = 0.95)	Radioimmuno – LH i.u./mg	Radioimmuno – LH F. L. (P = 0.95)	FSH/LH
Precursor (E174bis–C)	322	692	555.3 – 890.1	——	——	——
1	8.0	4.6	——	——	——	——
2	63.8	15.6	10.8 – 20.0	0.35	0.25 – 0.45	44.5
3	65.9	276.8	197.5 – 340.1	0.53	0.40 – 0.70	52.2
4	104.3	1255.6	898.5 – 1713.6	3.26	2.3 – 5.1	385.1
5	45.7	630.4	506.9 – 772.7	6.5	5.5 – 7.7	96.9
	287.7					

+Dry weight after desalting and lyophilization.

Figure 5 shows that the substance gave a single symmetric peak. The $S_{20,w}$ value was calculated to be 1.94 S. The diffusion constant $(D_{20,w})$ was found to be 5.98×10^{-7}. The molecular weight, calculated on the basis of these results was 31,600.

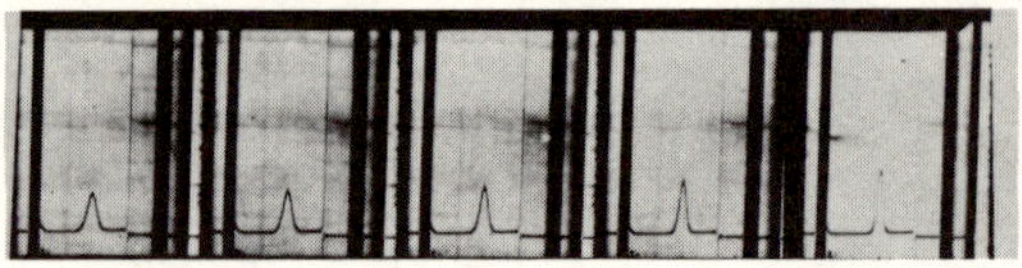

Figure 5. Sedimentation in the ultracentrifuge at 59780 rpm of highly purified FSH at a concentration of 0.6 per cent in 0.1 M phosphate buffer, pH 6.5. The direction of sedimentation is towards the left.

Electrophoretic studies. The FSH preparation subjected to the analytical polyacrylamide gel electrophoresis at pH 8.3 showed a single band, as illustrated in Figure 6.

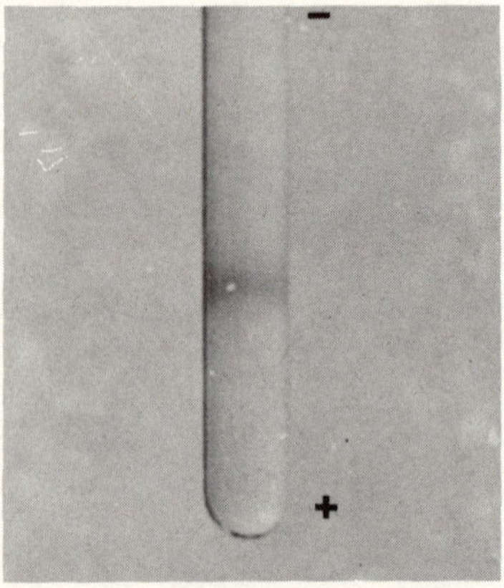

Figure 6. Disc electrophoresis of urinary FSH on polyacrylamide gel 7.5 per cent, pH 8.3, 5 x 40 mm column at 3 mA/tube for 1hr. Protein load was 100 ug.

Spectrophotometric studies. The ultraviolet absorption spectra of FSH dissolved in 0.01 M phosphate buffer, pH 7.0 and in 0.1 N NaOH were determined in the range 240 to 310 mμ. The ultraviolet spectra of neutral and alkaline solutions are shown in Figure 7.

The $A_{1cm}^{1\%}$ at 277 mμ and at pH 7.0 was calculated to be 6.85, a value slightly lower than that previously reported by Donini *et al.* (1968a).

Molecular Sieve Chromatography on Sephadex Column. Analytical gel filtration studies were carried out on a Sephadex G-100 column 2 x 110 cm at 3°C, according to the method of Reichert *et al.* (1968).

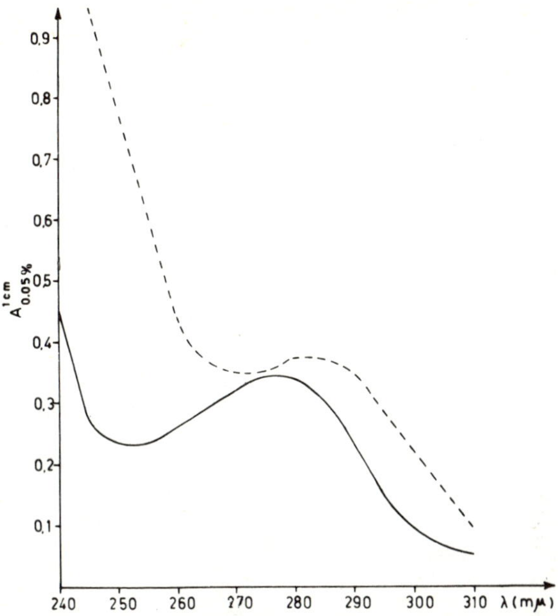

Figure 7. Ultraviolet absorption spectra of urinary FSH in 0.01 M potassium phosphate buffer pH 7.0 (——) and in 0.1 N sodium hydroxide (- - - -).

The details for equilibration of the gel and packaging of the column were as described by Andrews (1965). The void volume (Vo) of the column, as determined by filtration of Dextran Blue (Pharmacia), was 117.0 ml. The gel filtration was carried out at a flow rate of 8.0 ml/hr. and samples of 2.3 ml/tube were collected by the refrigerated Beckman fraction collector.

In a total volume of 1 ml of buffer, 2 mg of each protein were applied to the column and the effluent was scanned for protein content by spectrophotometric readings at 230 mμ. The ratios between the void volume (Vo) and the elution volume (Ve) for various purified proteins and for urinary FSH, LH and HCG preparations are reported in Table II.

Figure 8 shows the relationship between the Ve/Vo ratio and the logarithm of molecular weight of the proteins studied. From the relationship it was possible to calculate the FSH molecular weight, which was estimated to be 61660.

Amino acid analysis. The amino acid analysis was performed in a Spackman-Moore and Stein, 120 C Analyser Spinco, Beckman, under the following conditons: (1) Native hormone hydrolyzed with hydrochloric acid at 105°C for 22 hr. (2) Same as in (1) for 48 hr. (3) Hydrolysis carried out for 22 hr. with the addition of phenol to protect

TABLE II

Gel filtration of highly purified urinary FSH, LH and HCG preparations on Sephadex G–100 column according to the method of Andrews (1965). 2 mg of each material in 1 ml of total volume were applied to the column 2 x 110 cm, flow rate 8.0 ml/hr. Samples: 2.3 ml/tube. Spectrophotometric reading at 230 mμ.

Run no.	Filtered Material	Ve/Vo
1	Dextran Blue (Pharmacia) Vo = 117.0	
2	Bovine Serum Albumin (Behringwerke A.G., Germany, M.W. 67,000)	1.36
	Cytochrome (Seravac Labs., Maidenhead, Berks., England, M.W. 12,400)	2.17
3	Bovine Serum Albumin	1.36
	Chicken Egg Albumin (Koch Light Labs., Colnbrook, Bucks, Eng., M.W. 45,000)	1.61
	Chymotrypsinogen A (Seravac Labs., M.W. 25,000)	1.84
	Cytochrome	2.17
4	Urinary FSH	1.42
5	Urinary LH	1.50
6	Urinary HCG	1.26

tyrosine. (4) Same as in (3) for 48 hr. (5) Hydrolysis performed for 22 hr. after oxidation with performic acid and lyophilization. Table III shows the amino acid residues of urinary FSH preparations, as found by Roos (1968), Butt (1967) and us. The tryptophan content of our FSH preparation, estimated by spectrophotometric readings, and according to the method of Goodwin and Morton (1946), was 1.48 per cent.

Carbohydrate analysis. The sugars present in human urinary FSH were identified by thin-layer chromatography. The hormone was hydrolyzed in 1N H_2SO_4 at 100°C for 6 hours, as described by Papkoff *et al.* (1967). A layer of Alusil (Silicagel and aluminium oxide 1:1) was used which was developed with butanol, acetic-acid, and H_2O (60:30:10). Sugars were detected by aniline phosphate-diphenylamine in ethanol. The chromatogram showed the presence of galactose, mannose and fucose. The hexoses were determined by the orcinol-sulphuric acid procedure of Winzler (1955) using mannose galactose as a standard. Hexosamines were quantitated by the method of Rondle and Morgan (1955), using glucosamine as standard. Sialic acid was estimated by the thiobarbituric acid method of Warren (1959) after

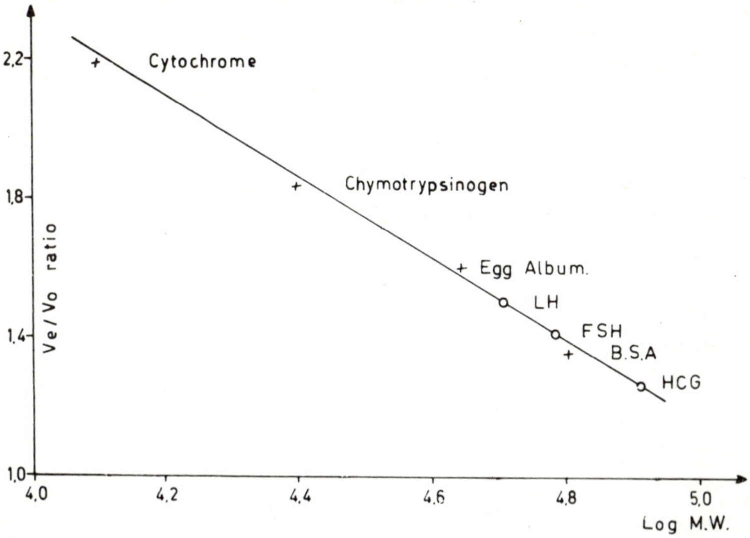

Figure 8. Molecular sieve chromatography on Sephadex G-100 column 2 x 110 cm. Void volume Vo (Dextran Blue) = 117.0 ml. 0.05 M Tris-HCl + 0.1 M KCl buffer, pH 7.5. The Ve/Vo ratios of urinary FSH, LH, HCG and of various proteins plotted against log of molecular weights.

hydrolysis for 1 hr. at 80°C with 0.1 N sulphuric acid. The standard was the synthetically pure N-acetylneuraminic acid supplied by Koch-Light Laboratories, Colnbrook, Bucks, England. In Table IV, the carbohydrate content of our human urinary FSH is reported and compared to that of other urinary and pituitary FSH preparations.

Stability studies. The stability of the highly purified urinary FSH was studied. The biological potency of fraction 4, obtained by polyacrylamide gel electrophoresis in the last step of purification, was determined before desalting and lyophilizing the solution. The dose levels for FSH bioassay were calculated on the protein concentration, as determined by spectrophotometric reading at 277 mμ, assuming that the $A^{1\%}_{1cm}$ of highly purified FSH was 6.85.

After overnight dialysis against distilled water, fraction 4 was lyophilized and completely desalted on a Sephadex G-25 column and then lyophilized again. The FSH biological activity of the desalted and lyophilized fraction 4 was determined by calculating the dose levels on the dry weight basis. No significant difference was found between the potency estimates before or after lyophilization.

Furthermore, the stability of our FSH preparation was demonstrated by assaying the biological potency as soon as the FSH was lyophilized and after one year of storing at 4°C in a dry state, as shown by Table V.

TABLE III

Amino acid composition of three different urinary FSH preparations

AMINO ACID	a (1)	a (2)	a (3)
Lysine	7	10	8
Histidine	4	7	7
Arginine	8	7	8
Aspartic acid	27	14	17
Threonine	12	11	13
Serine	16	7	13
Glutamic acid	21	24	23
Proline	19	7	17
Glycine	23	11	11
Alanine	12	11	13
Half cystine	32	20	18
Valine	11	7	9
Methionine	1	2	3
Isoleucine	5	6	4
Leucine	11	12	12
Tyrosine	4	7	6
Phenylalanine	7	7	13
Tryptophan	2	+	2

a (1) Residues to the nearest integer calculated for M. W. of 23,600 of the protein moiety of urinary FSH prepared by Roos (1968).

a (2) Residues to the nearest integer calculated for M. W. of 30,000 of urinary FSH reported by Butt (1967).

a (3) Residues to the nearest integer calculated for M. W. of 21,500 of the protein moiety of urinary FSH described in this paper.

\+ Not estimated.

TABLE IV

Carbohydrate components of human pituitary and urinary FSH

	Pituitary FSH			Urinary FSH	
	Papkoff *et al.* (1967)[a]	Reichert *et al.* (1968)[a]	Roos (1968)[b]	Donini (this paper)[a]	Roos (1968)[b]
	%	%	%	%	%
Hexose	3.9	11.6	19.6	10.4	14.5
Hexosamine	2.4	9.1.	15.2 +	9.6	8.2+
Fucose	0.4	——	——	——	——
Sialic acid	1.4	5.2	7.0	8.6	3.1

a – Calculated on the dry weight basis

b – Calculated on the weight of protein, as determined spectrophotometrically at pH 7.0.

+ – Calculated on the residues per mole.

TABLE V

Stability of highly purified urinary FSH stored for 1 year at 4°C.

Conditions	R. P. (Steelman & Pohley) in terms of 2nd IRP–HMG, 95% fiducial limits
Lyophilized after dialysis, desalted on Sephadex G–25 column and lyophilized again. Bioassay at the time of preparation.	1255 i.u./mg (745–1468)
After 12 months at 4°C.	1213 i.u./mg (913–1575)

The potency was estimated on the basis of dry weight.

Immunoelectrophoretic studies. The immunoelectrophoretic studies were carried out in agar plate, according to the method of Scheidegger (1955).

The rabbit anti-HMG and anti-HCG sera were obtained by immunizing the animals with partially purified HMG and HCG preparations respectively. Rabbit anti-human albumin serum was supplied by Behringwerke A. G., Germany.

Unabsorbed and absorbed anti-HMG sera were used. The absorption of antiserum was performed with HCG (3,000 i.u./mg) and then with kaolin-acetone extract of urine from a hypophysectomized patient, according to the method of Tamada *et al.* (1967). Any fractions obtained by the preparative polyacrylamide gel electrophoresis did not show any precipitin line, when assayed with anti-human albumin serum.

Figure 9 shows the immunoelectrophoretic pattern of the purest FSH fraction. The unabsorbed anti-HMG gave two lines against this FSH preparation, but only a single arc was visible when the absorbed anti-serum was used. The same FSH fraction, tested with unabsorbed anti-HCG, gave one line which was probably due to antibodies directed against FSH or FSH-like, as reported by Robyn *et al.* (1968 and 1969).

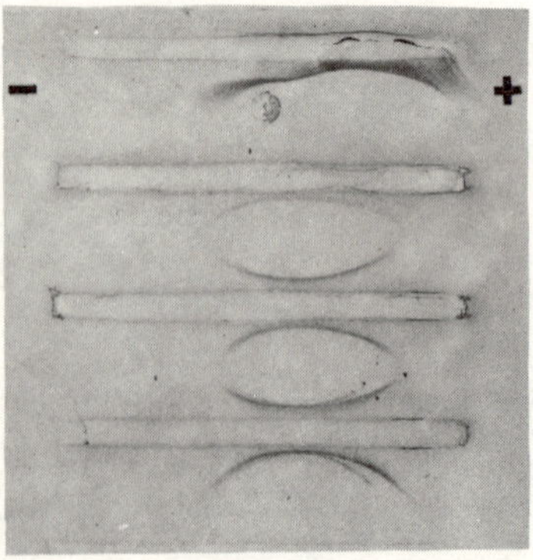

Figure 9. Agar immunoelectrophoresis. Wells A filled with human normal serum, B, C and D with 100 μg of highly purified urinary FSH. Troughs 1 and 5 filled with anti-human albumin serum, 2 with unabsorbed anti-HCG serum, 3 and 4 with absorbed and unabsorbed anti-HMG serum respectively.

DISCUSSION

As previously reported by us (Donini *et al.*, 1968a), only the proteins containing the highest FSH activity and corresponding to the symmetric peaks obtained by the first and second steps of purification, were subjected to the preparative disc electrophoresis. By this procedure we were able to prepare a human urinary FSH, which assayed 1255 i.u.-FSH/mg and 3.2 i.u.-LH/mg, as determined by bioassay and radioimmunoassay respectively. We wish to point out that this FSH preparation was 15 times as potent as the starting material and that the FSH/LH ratio rose from 1.42 to 385. The ultracentrifugal studies showed that the FSH was monodispersed, and the $S_{20,w}$ value was calculated to be 1.94 S, which agrees well with the value obtained by Roos (1968). The molecular weight calculated on the $S_{20,w}$ and on the diffusion constant ($D_{20,w}$) was 31,600, whereas Roos (3) has found a value of 29,400 calculated on a $S_{20,w}$ value of 1.96 S.

The amino acid analysis of urinary FSH reported by Butt (1967) and by Roos (1968) agreed fairly well with our results, except the aspartic acid, glycine and half-cystine residues were significantly higher in Roos's preparation. It is noteworthy that the two preparations, of Roos (1968) and Donini (this paper), whose FSH activity was 780 i.u. and 1,255 i.u./mg respectively, have an amino acid composition similar to that of the FSH preparation reported by Butt (1967), whose FSH activity was only 12 i.u./mg. Although these three preparations had a similar grade of homogeneity and a similar amino acid composition, a strong difference between the biological activity does exist. This could be explained by a different composition of the carbohydrate moiety, i.e. a lower content of sialic acid in the preparation reported by Butt (1967). Unfortunately, this hypothesis cannot be proven because the carbohydrate analysis was not performed. The homogeneity of our FSH preparation was demonstrated also by the single band obtained in the analytical disc electrophoresis. In an earlier report, a highly purified FSH was described by Stevens (1967). However, this preparation was not tested for homogeneity. Another urinary FSH with a similar grade of purity was described by Roos and Gemzell (1964b) and by Roos (1968). This preparation, which has not been isolated in a dry state, was claimed to be homogeneous both in the ultracentrifuge and free-zone electrophoresis. Roos (1968) reported that in the bioassay 'all doses of unknowns were based on absorbance measurements at 280 mμ, assuming $A^{1\%}_{1cm}$ = 10. Also in the carbohydrate analysis Roos states that, 'the data is expressed as percent by weight of protein (as determined spectrophotometrically at pH 7.0)'. We would like to point out that our dry FSH preparation dissolved in 0.01 M potassium phosphate buffer, pH 7.0 i.e. in the same experimental conditions as described by Roos (1968), gave a value of $A^{1\%}_{1cm}$ = 6.85. In view of this data it seems possible to us, that some values reported by Roos have been overestimated.

Concerning the results reported in this paper and regarding the molecular sieve chromatography on the Sephadex G-100 column, it is worth remembering the limitations inherent in the correlations between the molecular weight and the Ve/Vo ratio of proteins in general and of glycoproteins in particular, as emphasized by Reichert *et al.* (1968).

In this connection it is of interest to note the difference in the mobility of the highly purified FSH, LH and HCG preparations, when filtered through a Sephadex G-100 column in well standardized experimental conditions. The molecular weights of FSH and LH, as calculated from the Ve/Vo ratio were 61,660 and 51,290 respectively. Regarding the carbohydrate components of urinary FSH, we should point out that there is good agreement between our results and those reported by Reichert *et al.* (1968) for the pituitary FSH as far as the hexose and hexosamine contents are concerned. On the other hand, some discrepancies exist between our results and those previously reported by others (Table IV) on the sialic acid content. Our value of 8.6 per cent of sialic acid, which represents the average of three determinations, is much higher than those reported by other investigators, as far as the urinary FSH preparation is concerned.

Regarding the stability of the highly purified FSH preparations from human pituitary glands and from menopausal urine, it seems that some differences exist between the two hormones obtained by different sources. It was demonstrated by Amir *et al.* (1966), and recently pointed out again by Peckham and Parlow, (1969), that the freeze-drying can destroy, to a large extent, the biological activity of human pituitary FSH. On the other hand, we have demonstrated that when the protein concentration of FSH solution was determined assuming $A^{1\%}_{1cm} = 6.85$ and the dose level for bioassay was calculated on the spectrophotometric reading, no difference was found between the FSH potencies if the level doses were calculated on the basis of dry weight. This result seems to demonstrate that lyophilization does not affect the bioactivity of urinary FSH. It is also interesting to note that the urinary FSH has good stability when stored in a dry state at 4°C, as compared to the stability of pituitary FSH in the same storage condition.

As far as the immunoelectrophoretic findings of our urinary FSH are concerned, we would like to point out that a precipitin line was obtained when the FSH was tested with unabsorbed anti-HCG serum. The immunological reaction was probably due to the antibodies directed against the FSH-like antibodies contained in the anti-HCG serum. This finding could be explained by another immunoelectrophoretic plate, where the anti-HCG serum was previously absorbed with FSH and no precipitin arc was visible when the urinary FSH was tested.

Although the immunoelectrophoretic studies showed that our FSH preparation seems to contain more than a single protein we wish to point out that it can be considered one of the purest and most biologically active urinary follicle-stimulating preparation so far described.

ACKNOWLEDGEMENTS

We are indebted to and would like to thank Dr B. Lunenfeld and Mrs A. Eshkol, Institute of Endocrinology, Tel-Hashomer Government Hospital, Tel-Aviv, Israel; and Dr A. Patchornik and Dr Y. Burstein, Weizman Institute, Rehovot, Israel, for performing the ultracentrifugal studies and amino acid analysis.

We are also indebted to Dr R. Bangham, National Institute for Medical Research, London, for supplying the 2nd IRP-HMG and the 1st International Standard of HCG.

REFERENCES

Amir, S.M., Barker, S.A., Butt, W.R. & Crooke, A.C. (1966). *Nature, Lond.* **209,** 1092.

Andrews, P. (1965). *Biochem, J.* **96,** 595.

Borth, R., Diczfalusy, E. & Heinrichs, M.D. (1957). *Arch. Gynaek.* **188,** 377.

Butt, W.R. (1967). In *The Chemistry of Gonadotrophins* p. 59 Ed. Kugelmass, I.N. Springfield, Illinois: Thomas.

Davies, B.J. (1964). *Ann. N.Y. Acad. Sci.* **121,** 404.

Donini, S., D'Alessio, I., and Donini, P. (1968b). In *Gonadotropins 1968* p. 37 Ed. Rosemberg, E., Los Altos, California: Geron-X.

Donini, S. and Donini, P. (1969). *Acta Endocr. Copenh.* Suppl. **142,** 257.

Donini, P., Montezemolo, R. & Puzzuoli,D. (1964). *Acta Endocr. Copenh.* 45, 321.

Donini, P., Puzzuoli, D., D'Alessio, I., Lunenfeld, B., Eshkol, A. & Parlow. A.F. (1966). *Acta Endocr.* **52,** 169.

Donini, P., Puzzuoli, D., D'Alessio, I. & Donini, S. (1968a). In *Pharmacology of Hormonal Polypeptides and Proteins.* p. 229.Ed. Martini, L. New York: Plenum.

Goodwin, T.W. & Morton, R.A. (1946). *Biochem. J.* **40,** 628.

Papkoff, H., Mahlmann, L.J. & Li, C.H. (1967). *Biochemistry N.Y.* **6,** 3978.

Parlow, A.F. (1961). *In Human Pituitary Gonadotropins.* p. 300. Ed. Albert, A. Springfield, Illinois, Thomas.

Peckham, W.D. and Parlow, A.F. (1969). *Endocrinology,* **84,** 953.

Reichert, L.E., Kathan, R.H. & Ryan, R.I. (1968). *Endocrinology,* **82,** 109.

Reichert, L.E. & Parlow, A.F. (1964). *J. Clin. Endocr. Metab.* **24,** 1040.

Robyn, C., Diczfalusy, E. & Finney, D.J. (1968). *Acta Endocr. Copenh.* **58,** 593.

Robyn, C., Petrusz, P. & Diczfalusy, E. (1969). *Acta Endocr. Copenh.* **60,** 137.

Rondle, C.J.M. & Morgan, W.T.G. (1955). *Biochem. J.* **61,** 586.

Roos, P. (1968). *Acta Endocr. Copenh.* **59,** Suppl. 131, 5-93.

Roos, P. & Gemzell, C.A. (1964a). *Biochim. Biophys. Acta,* **82,** 218.

Roos, P. & Gemzell, C.A. (1964b). *Biochim. Biophys. Acta,* **93,** 217.

Saxena, B.B. & Rathnam, P. (1967). *J. biol. Chem.* **242,** 3769.

Scheidegger, J.J. (1955). *Int. Arch. Allergy. appl. Physiol.* **1,** 103.

Steelman, S.L. & Pohley, F.M. (1953). *Endocrinology,* **53,** 604.

Stevens, V.C. (1967). In *Recent Research on Gonadotrophic Hormones* p. 139 Ed. Bell, E.T. & Loraine, J.A. Edinburgh; Livingstone.

Tamada, T., Soper M. & Taymor, M.L. (1967). *J. clin. Endocr. Metab.* **27,** 379.

Warren, L. (1959). *J. biol. Chem.* **234,** 1971.

Winzler, R.J. (1955). In *Methods of Biochemical Analysis.* Vol. II, p. 279 Ed. Glick, D. New York; Interscience publishers.

Studies on the State of Human Urinary FSH in a Crude Extract and on its Purification

Madelaine Théoleyre and Marian Jutisz

It now seems well established that FSH, after its release into the blood stream and passage through the renal filter, undergoes some transformations and that the hormone recovered from urine has not exactly the same chemical form as the FSH present in the pituitary gland.

Taymor, *et al.* (1967) and Mori (1967) published, independently, some data concerning the immunological relationship between human urinary and pituitary FSH. Using somewhat different techniques, the two groups reached a very similar conclusion, viz. that pituitary and urinary FSH are immunologically dissimilar.

On the other hand, it is well known by all biochemists who intend to purify human urinary gonadotrophins that this purification gives rise to many more problems and difficulties than the purification of corresponding pituitary gonadotrophins. Indeed, in an apparently very similar state of purification, pituitary gonadotrophins are much more potent hormones than the urinary hormones. To our knowledge nobody has prepared a potent urinary FSH preparation such as that of some of the pituitary FSH preparations. This led some authors to consider that urinary FSH is complexed with other proteins or substances of a different nature present in urine (Donini, *et al.*, 1966; Jiang and Albert, 1968; Théoleyre and Jutisz, 1969).

The purpose of the present work was to get rid of the complexes or aggregates of FSH with other proteins using different types of substances known to be denaturing or dissociating agents of proteins, before undertaking the purification of this hormone. Among those substances three were tried: metaphosphoric acid, sulphosalicylic acid and n-butanol.

These reagents have been used previously for the purification of gonadotrophins and other proteins. Metaphosphoric acid has been employed by Ellis (1958) in the course of purification of sheep pituitary hormones, and for the purification of ovine FSH (Papkoff, *et al.* 1964 and 1967) and LH (Papkoff, *et al.*, 1965). Precipitation with a sulphosalicylate buffer at pH 3.6 has been proposed by Squire and Li (1959) as a step in the purification of ovine LH. The use of n-butanol was introduced by Morton (1955) for dissociating lipoprotein complexes and for releasing undenatured protein. SE-Sephadex (sulphoethyl Sephadex) was used by Papkoff *et al.* (1964; 1965;1967)for the purification of ovine FSH and LH.

In this paper we submit some preliminary results of a systematic exploration of the possibilities offered, for the purification of urinary gonadotrophins, by the treatment of a crude postmenopausal urinary extract with dissociating or denaturing agents.

MATERIALS

All chemical products and materials for chromatography were reagent grade, obtained from usual commercial sources. Sulphosalicylic acid (R.P. grade), urea (R.P. grade), pyridine and n-butanol were supplied by Prolabo, Paris. Pyridine was boiled for three hours with solid NaOH (50g/litre) and redistilled. Synthetic n-butanol was freshly redistilled. Metaphosphoric acid (Pro analysis) was obtained from Merck AG., Darmstadt (Germany) and guanidine hydrochloride (A-grade) from Calbiochem, Lucerne (Switzerland). SE-Sephadex, C–50 (particle size: 40 – 120 u) and Sephadex G–200 (particle size: 40 – 120 u) were purchased from Pharmacia, Uppsala (Sweden) and DEAE-cellulose (0.87 mEq/g) from Brown Company, Berlin, N.H. (U.S.A.).

In most experiments, a crude preparation of HMG called the 'alcohol fraction' (AF) was used as starting material for the first step of purification. This fraction was obtained from urine of post-menopausal women, by precipitation with zinc acetate at pH 6.0, extraction of activity with ammonium acetate 0.2 M at pH 9.5 and precipitation with ethanol, as already described (Jutisz *et al.*, 1962). Its mean biological activity was 0.78 2nd IRP–HMG–FSH Units. In some experiments, a more purified fraction, obtained by filtration of the AF-fraction on a Sephadex G–100 column (Jutisz *et al.*, 1964) was used (AFSx), with a mean biological activity of 2.5 2nd IRP–HMG–FSH Units.

Biological assays were carried out by the augmentation test of Steelman and Pohley (1953) for FSH and by the ovarian ascorbic acid depletion test of Parlow (1961) for LH. The results were expressed in terms of the reference preparation 2nd IRP with 95 per cent confidence limits.

EXPERIMENTAL PROCEDURE

Treatment of crude FSH preparations

Metaphosphoric acid. 0.1 M HPO_3 was added at 4°C to a 2 per cent water solution of AFSx fraction (100 mg) until pH 3.5 was obtained. The resulting precipitate was removed, after 15 min. by centrifugation and put aside for assay (PI). The supernatant was adjusted to pH 4 (0.1 N NaOH), treated with 3 volumes of cold 95 per cent ethanol (– 10°C) and after one night at – 30°C, centrifuged. The precipitate was redissolved in water, dialysed against a solution of ammonium bicarbonate 0.05 M (Am Bi) and lyophilyzed (PS).

TABLE I

Treatment of AFSx fraction with metaphosphoric acid

Fraction	Weight (mg)	FSH activity/mg*	% of activity
AFSx (starting mat.)	100	2.5 (2.0–3.1)	100
PI (precipitate)	55	1.5 (0.9–2.2)	32
PS (supernatant)	22	4.4 (2.2–9.0)	45

*In terms of 2nd IRP reference preparation with 95% confidence limits.

Table I shows the results obtained in a typical experiment. This procedure did not make it possible to concentrate the FSH in one of the two fractions obtained and only yielded a low recovery of the activity; it was considered to be unsuitable for our purpose.

Sulphosalicylic acid. (Théoleyre and Jutisz, 1969) 0.2 M sulphosalicylic acid was added at 4° to a 2 per cent water solution of AF fraction until pH 3.6 was obtained. The resulting precipitate was removed by centrifugation (SP) and the supernatant was adjusted to pH 4 (0.1 N NaOH) and treated with 3.5 volumes of cold 90 per cent ethanol (– 10°C). The precipitate was collected, redissolved in water, dialysed against Am Bi solution and lyophilized (SS).

Table II shows the results obtained in three experiments with AF-fraction as starting material; 75 to 90 per cent of starting activity was concentrated in the supernatant SS-fraction with an approximately three-fold purification. Products thus obtained were further examined by chromatography on DEAE-cellulose. Only one experiment was performed with AFSx-fraction (Exp. Sa) and the result was considered to be uninteresting.

TABLE II

Treatment of AF– or AFSx-fractions with sulphosalicylic acid

Fraction	Weight (g)	FSH activity/mg*	% of activity
AF (starting mat.)	10.00	0.78(0.6–1.2)	100
Experiment S4			
SP4 (precipitate)	3.13	none	0
SS4 (supernatant)	3.05	2.2(1.9–3.8)	87
Experiment S5			
SP5 (precipitate)	3.02	0.6(0.3–1.1)	23
SS5 (supernatant)	2.76	2.1(1.8–2.5)	75
Experiment S9			
SP9 (precipitate)	2.84	none	0
SS9 (supernatant)	2.61	2.5(1.9–3.3)	84
Experiment Sa			
AFSx (starting mat.)	0.100	2.5(2.0–3.1)	100
SPa (precipitate)	0.046	0.9(0.4–2.1)	16
SSa (supernatant)	0.025	4.5(2.7–7.5)	45

* In terms of 2nd IRP reference preparation with 95% confidence limits.

TABLE III

Treatment of AF-or AFSx-fractions with n-butanol

Fraction	Weight (mg)	FSH activity/mg*	% of activity
Experiment BU 1			
AFSx (starting mat.)	150	2.5 (2.0–3.1)	100
Bu 1 (final prod.)	118	2.9 (2.0–4.3)	93
Experiment Bu 2			
AFSx (starting mat.)	1500	2.5 (2.0–3.1)	100
Bu 2 (final prod.)	1208	2.0 (1.5–2.8)	66
Experiment Bu 3			
AF (starting mat.)	7000	0.78 (0.6–1.2)	100
Bu 3 (final prod.)	3586	1.3 (1.2–2.2)	85
Experiment Bu 4			
AF (starting mat.)	7000	0.78 (0.6–1.2)	100
Bu 4 (final prod.)	3753	1.4 (1.0–1.9)	96

* In terms of 2nd IRP reference preparation with 95% confidence limits.

n-Butanol. This treatment was carried out either with AF-fractions or AFSx-fractions. A 2 per cent solution of one of these fractions in 0.05 M ammonium bicarbonate pH 8, in a separating funnel, was gently stirred for 75 min. at 4°C, with 1.2 volumes of n-butanol. The aqueous phase was separated after equilibration at 4°C, dialysed against Am Bi solution and lyophylized (Bu).

Table III shows the results obtained in four experiments, Two different starting materials were treated with n-butanol: a less purified AF-fraction and a fraction AFSx approximately three times more potent, derived from the preceding AF-fraction. In the first case, a two-fold purification was obtained (Exp. Bu 3 and 4), with a good recovery of the activity. In the second case (Exp. Bu 1 and 2), no purification was observed, but the recovery of the activity was quite good. The products thus obtained were further studied by chromatography on DEAE-cellulose.

Purification of materials obtained after treatment with sulphosalicylic acid and n-butanol.

Chromatography on DEAE-cellulose column. In a standard procedure (Théoleyre and Jutisz, 1969) a DEAE-cellulose column (4.3 x 36 cm) was equilibrated with a 0.008 N HCl-Tris pH 7.3 buffer: 3 g of the material dissolved in the same Tris buffer (1.2% solution) were applied to the column. The column was developed (flow rate 90 ml/hr.), first with the starting buffer, and then with NaCl solutions at gradually increased concentrations (0.04, 0.15 and 0.50 M) in the same buffer. Fractions (15 ml) were dialysed and lyophilized.

Figure 1 shows patterns of chromatography on DEAE-cellulose columns of three different products: a crude HMG fraction (AFSx-fraction), a fraction obtained after treatment of the AF-fraction with sulphosalicylic acid (SS-fraction) and a fraction obtained after treatment of the AFSx-fraction with n-butanol (Bu2 fraction). The outlines of the three patterns were similar but the sizes of the 0.10 or 0.15 M NaCl peaks, containing FSH activity, were different, depending on the starting material.

Table IV shows the yields and activities of FSH fractions recovered from the 0.10 – 0.15 M NaCl peaks after chromatography of different crude FSH preparations. It seems evident from these experiments that the treatment of crude AF- or AFSx-fractions with sulphosalicylic acid or n-butanol improved their behaviour throughout the chromatography as well as their final activity. Starting directly from an AFSx-fraction, only a 2.5 fold purification was obtained (DASx 12). A four-fold purification was observed in the case of sulphosalicylic fractions SS 4 and SS 5 (DS 4 and DS 5) and about 25-fold in the case of the first n-butanol fraction (DBu 2). Chromatography of the second butyl alcohol fraction, having as starting material an AF product (Bu 3), (Table III), gave only a two-fold purification (DBu 3).

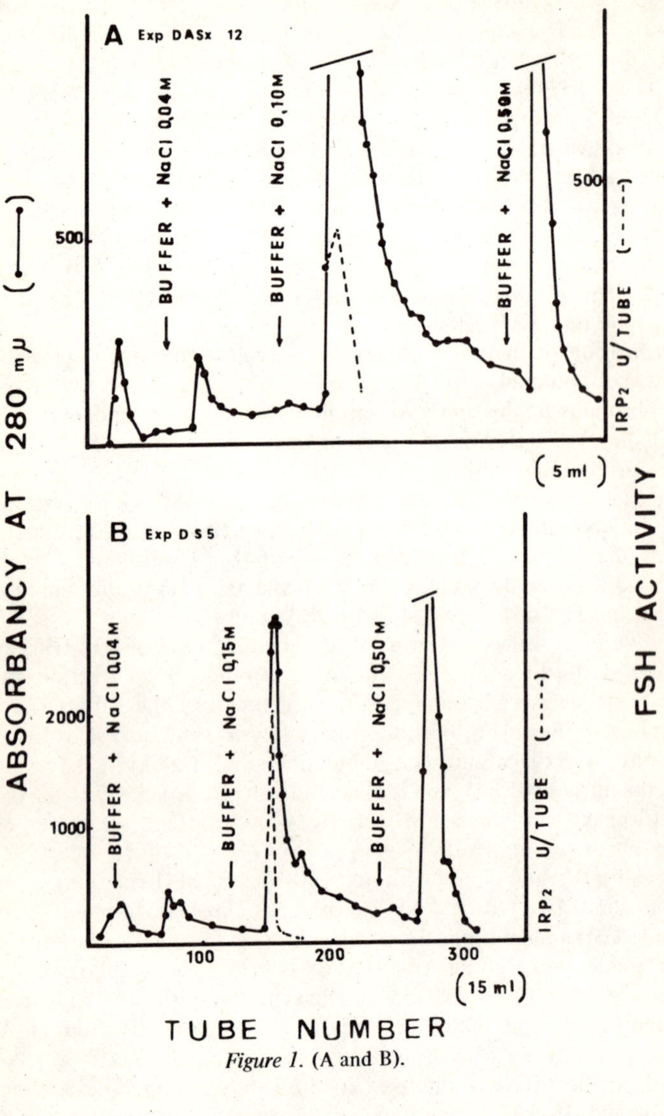

Figure 1. (A and B).

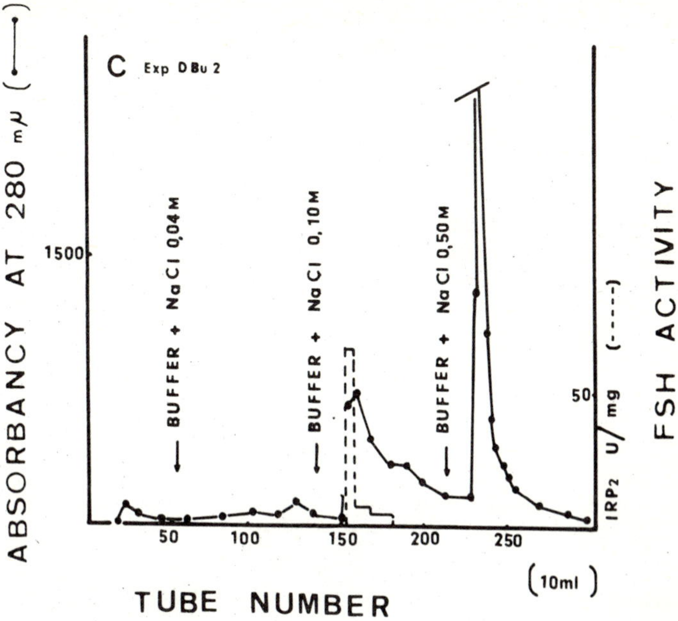

Figure 1. (C).

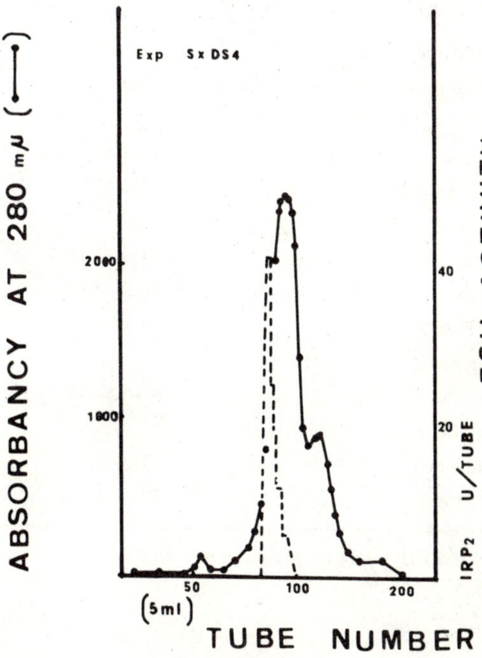

Figure 2.

To sum up, fraction Bu 3 excepted, all the fractions serving as starting material for chromatography had very similar FSH potency. In spite of this, final results were different, depending on the treatment preceding chromatography. This fact is well illustrated in the case of experiment DBu 2 where the final product had an activity of 67.3 2nd IRP U/mg. In experiment DASx 12, a similar starting material, but not treated with n-butanol, produced a product with a FSH activity of only 5.4 2nd IRP U/mg. Though the treatment with n-butanol had not purified fraction Bu 2 (Table III), a 25-fold purification of this fraction was obtained by chromatography on a DEAE column.

TABLE IV

Purification of different crude FSH-fractions by chromatography on DEAE-cellulose

Fraction	Weight (mg)	FSH activity/mg*	% of activity
Experiment DASx 12			
AFSx (starting mat.)	304	2.1(1.2–3.6)	100
DASx 12 (final prod.)	100	5.4(3.2–9.4)	85
Experiment DS 4			
SS4 (starting mat.)	3050	2.2(1.9–3.8)	100
DS4 (final prod.)	445	9.8(7.5–12.7)	64
Experiment DS 5			
SS5 (starting mat.)	2760	2.1(1.8–2.5)	100
DS5 (final prod.)	338	8.5(6.9–10.5)	49
Experiment DBu 2			
Bu2 (starting mat.)	1118	2.0(1.5–2.8)	100
DBu2 (final prod.)	25	67.3(46.8–95.9)	73
Experiment DBu 3			
Bu3 (starting mat.)	3521	1.3(1.2–2.2)	100
DBu3 (final prod.)	974	2.7(2.3–3.1)	58

*In terms of 2nd IRP reference preparation with 95% confidence limits.

Filtration on Sephadex G-200 column of materials obtained by treatment with sulphosalicylic acid followed by chromatography on DEAE-column. Further purification of sulphosalicylic acid – DEAE – fractions were attempted by gel filtration.

Figure 2 shows the patterns obtained when 300 mg of DS 5 fraction were filtered on a 3.0 x 100 cm Sephadex G-200 column, equilibrated with 0.05 M pyridinium acetate buffer pH 5.0. The column was developed with the same buffer, at a rate of 25 ml/hr. and 5 ml fractions were collected. The FSH activity was identified in the ascending part of the main peak.

TABLE V

Activities of FSH fractions obtained after treatment with sulphosalicylic acid followed by chromatography on a DEAE column and filtration on Sephadex G-200 column†.

Fraction	Weight (mg)	FSH activity/mg*	LH activity/mg*
Experiment SxDS 4			
DS 4 (starting mat.)	445	9.8(7.5–12.7)	-
SxDS 4 (final prod.)	11	41.3(18.4–92.5)	1.26(0.5–3.1)
Experiment SxDS 5			
DS 5 (starting mat.)	338	8.5(6.9–10.5)	-
SxDS 5 (final prod.)	26	28.8(21.2–39.2)	1.36(0.5-3.7)

†The biological activity of the starting fraction AF, expressed in 2nd IRP Units/mg: FSH: 0.78(0.6–1.2), LH: 4.31(1.8–7.4).

*In terms of 2nd IRP reference preparation with 95% confidence limits.

Table V shows FSH and LH activities of final products obtained after filtration on Sephadex G-200 columns. The FSH activity of these fractions (SxDS) lies between 30 and 40 2nd IRP Units and their LH activity, between 1.26 and 1.36 2nd IRP Units; the yield was dependent on the number of tubes collected in the active fraction.

Direct chromatography of Fractions obtained with sulphosalicylic acid on SE-Sephadex column. As chromatography of sulphosalicylic acid fractions on DEAE-cellulose gave a rather low purification (4x), another ion exchanger, SE-Sephadex, was tried. This chromatography was carried out under conditions similar to those employed by Papkoff *et al.* (1967). In a typical experiment, an SE-Sephadex column (4.2 x 42 cm) was equilibrated with 0.03 M ammonium acetate buffer pH 4.0. A sample of 3.0 g of the material dissolved in the same buffer (1.2 per cent solution) was applied to the column. The column was developed with a flow rate of 80 ml/hr, first with the starting buffer and than successively with 0.01 M Na_2HPO_4 pH 6.0 and with 0.01 M Na_2HPO_4 pH 8.6.

Figure 3 shows the patterns obtained when a sulphosalicylic acid fraction SS 9 was applied to the SE-Sephadex column. The FSH activity was eluted with the buffer pH 8.6.

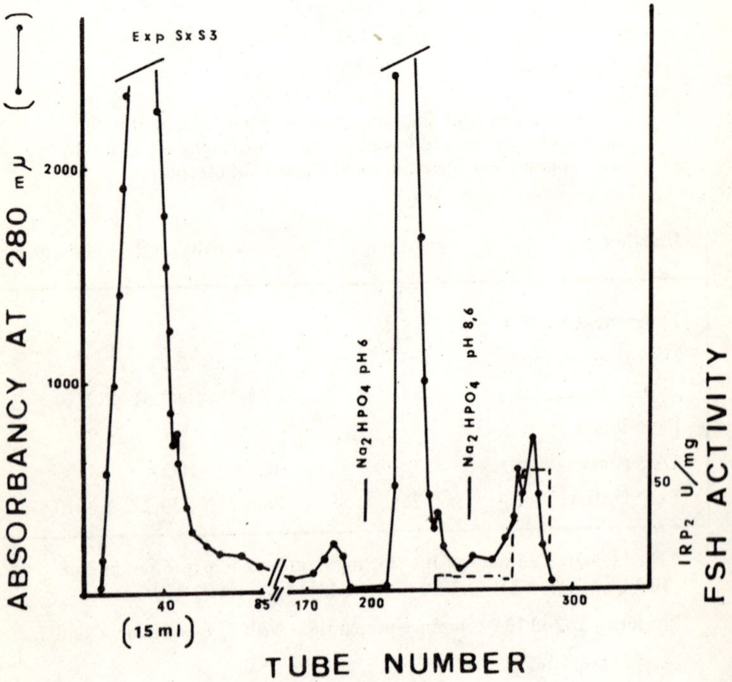

Figure 3.

Table VI shows two results of chromatography on an SE-Sephadex column. This chromatography appears to be much more interesting for purification of sulphosalicylic acid fractions than chromatography on the DEAE-cellulose column. In a single step, a more than 20-fold purification of FSH was observed. Using one step less than in the case of DBu 2 product (Table IV), a material of very similar potency was obtained.

TABLE VI

Purification of sulphosalicylic acid fractions by chromatography on SE-Sephadex column.

Fraction	Weight (mg)	FSH activity/mg*	%of activity
Experiment Se S2			
SS 7 (starting mat.)	3000	2.4(1.9–3.0)	100
Se S2 (final prod.)	111	49.8(37.9–65.8)	77
Experiment Se S3			
SS 9 (starting mat.)	2580	2.5(1.9–3.3)	100
Se S3 (final prod.)	87	55.4(37.7–81.4)	75

*In terms of 2nd IRP reference preparation with 95% confidence limits.

DISCUSSION

Several attempts to purify human urinary FSH have been made in recent years (Roos and Gemzell, 1964; Reichert and Parlow, 1964; Blatt *et al,* 1966; Théoleyre, 1966; Roos, 1967; Donini *et al.*, 1968; Jiang and Albert (1968). By incorporating sodium dodecyl sulphate in the hormone have been obtained, depending on the potency of the starting materials. In our opinion, purification of human urinary FSH gives rise to two main problems: (1) at the early stage of the purification process, dissociation of the hormone bound to other proteins, forming a complex or aggregates and (2) complete separation from LH contaminant.

To our knowledge, the problem of dissociating urinary FSH from a complex with other proteins has been considered only in a work of Jiang and Albert (1968). By incorporating sodium dodecyl sulphate in the buffer solution to carry out gel filtration of a partly purified FSH, these authors were able to separate a protein mixture which was otherwise inseparable in the absence of this agent.

Inactivation of LH in urinary FSH preparations by urea has been attempted by Schmidt-Elmendorf *et al.* (1962) and more recently by Butt *et al.* (1965). As pointed out by these authors, this treatment inactivated most of the LH but also destroyed some of the FSH. Donini *et al.* (1966) proposed a selective inactivation of LH contaminant in urinary FSH based on the binding of LH with an anti-HCG serum.

The results presented in the present work show that the treatment of crude urinary FSH preparations with n-butanol or sulphosalicylic acid allowed elimination of some contaminant proteins and lipoproteins, strongly bound with FSH, and thus facilitated further purification of this hormone.

When the treatment with n-butanol was applied to a starting material with an FSH activity of approx. 2.5 2nd IRP Units, subsequent chromatography on a DEAE column of the treated product, produced a 25-fold purification and quite a good yield of a preparation with 67 2nd IRP Units.

Two different procedures were used following treatment of crude FSH preparations with sulphosalicylic acid.

In the first case, the treated sample was further purified by chromatography on a DEAE column, followed by filtration on a Sephadex G–200 column. Though FSH activity of the final products (30 to 40 2nd IRP Units) was somewhat lower than that of some recent preparations of urinary FSH (Roos *et al.* 1964; Jiang and Albert, 1968; Donini *et al.*, 1968), their LH activity (1.3 to 1.4 2nd IRP 4 Units) was very low as compared to most of these preparations. Only Donini *et al.* (1966) claimed that after the binding of LH with anti-HCG serum, their urinary FSH preparations were freed from LH (LH was undetectable at the highest dose tested). Thus, treatment with sulphosalicylic acid appears to be an important step in the purification of FSH, if it is required to remove LH.

In the second case, sulphosalicylic acid fractions were further purified by the chromatography on SE-Sephadex. The results obtained showed that this procedure of chromatography was a very interesting step in the purification of FSH, enabling at least a 20-fold purification of the material. It would be interesting to assay LH in the final product thus obtained, and to examine the possibility of its further purification.

REFERENCES

Blatt, W.F., Park, M.Y., Taymor, M. L. & Todd, R. (1966). *J. clin. Endocr.* **26**, 189.

Butt, W.R., Crooke, A.C. & Wolf, A. (1965). *Gonadotropins: Physico-chemical and immunological properties.* p.85 Ciba Fdn Study Groups No.22.

Donini, P., Puzzuoli, D., D'Alessio, I. & Donini, S. (1968). *Adv. exp. Med. Biol.*, **2**, 229.

Donini, P., Puzzuoli, D., D'Alessio, I., Lunenfeld, B., Eshkol, A. & Parlow, A.F. (1966). *Acta Endocr. Copenh.*, **52**, 186.

Ellis, S. (1958). *J. biol. Chem.*, **233**, 63.

Jiang, N.S. & Albert, A. (1968). *J. clin. Endocr.* **28**, 1295.

Jutisz, M., Théoleyre, M., Colonge, A. & Courrier, R. (1962). *Bull. Soc. Chim. biol.* **44**, 83.

Jutisz, M., Théoleyre, M., Colonge, A. & Courrier, R. (1964). *C.r. hebd. Séanc. Acad. Sci., Paris,* **259**, 1195.

Mori, K.F. (1967). *Endocrinology,* **81**, 1241.

Morton, R.K. (1955), In *Methods in Enzymology, Vol. 1* p.25 Ed. Colowick, S.P. and Kaplan, N.O. New York: Academic Press.

Papkoff, H., Candiotti, A. & Li, C.H. (1964). *Fedn Proc. Fedn Am. Socs exp. Biol.* **23**, Abst. 1853.

Papkoff, H., Gospodarowicz, D., Candiotti, A. & Li, C.H. (1965). *Archs Biochem. Biophys.*, **111**, 431.

Papkoff, H., Gospodarowicz, D. & Li, C.H. (1967). *Arch. Biochem. Biophys.*, **120** 434.

Parlow, A.F. (1961), In *Human Pituitary Gonadotropins*, p.330. Ed. Albert A. Springfield, 111., Thomas.

Reichert, L.E. Jr. & Parlow, A.F. (1964). *J. clin. Endocr.* **24.** 1040.

Roos, P. (1967). Thèse de l'Université d'Uppsala. Almqvist & Wiksells.

Roos, P. & Gemzell, C.A. (1964) *Biochim. Biophys. Acta*, **93**, 217.

Schmidt-Elmendorff, H., Loraine, J.A. & Bell, E.T. (1962). *J. Endocr.* **24**, 153.

Squire, P.G. & Li, C.H. (1959) *J. biol. Chem.* **234**, 520.

Steelman, S.L. & Pohley, F.M. (1953). Endocrinology, **53**, 604.

Taymor, M.L., Tamada, T., Soper, M. & Blatt, W.F. (1967). *J. clin. Endocr.* **27**, 709.

Théoleyre, M. (1966). Thèse de l'Université de Paris.

Théoleyre, M. et Jutisz, M. (1969). *C.r. hebd. Séanc. Acad. Sci.*, **268**, Série D, 1994.

Further Studies on the Purification and Characterisation of Human Chorionic Gonadotrophin

H. van Hell and A.H.W.M. Schuurs

Much work has already been performed on the purification and characterization of HCG isolated from urine either of healthy pregnant women or of patients with trophoblastic tumours. The reader may be referred to review articles by Lunenfeld & Eshkol (1967) and by Loraine (1967). Homogeneity data published by various workers appear to be controversial. Highly potent preparations were obtained which were shown to be heterogeneous while other preparations with much lower potencies were found to be homogeneous. This discrepancy can be partly explained by the different criteria of homogeneity applied by the various authors. In our laboratories HCG was isolated from urine of healthy pregnant women and purified by means of various chromatographic procedures the final one being gel filtration on Sephadex G100. The products were characterized by various analytical techniques and found to contain a number of components, all of which possessed HCG properties (van Hell *et al.* 1968; Goverde *et al.* 1968; Schuurs *et al.* 1968).

The present communication deals with the application of isoelectric focusing as the final step in the purification of HCG. With this technique the previously found heterogeneity of HCG could be confirmed.

EXPERIMENTAL

HCG with a potency of about 9000 i.u./mg (bioassay) was obtained by methods described earlier (van Hell *et al.* 1968).

Isoelectric focusing was performed as described in the LKB Instruction Manual 1-8100-E01. The carrier ampholyte pH ranged from 3 to 5.

Starch gel electrophoretic analysis and bioassay of the preparations obtained were performed as described earlier (van Hell *et al.* 1968). For the reagents used for the immunoassays reference is made to a previous paper (Schuurs *et al.* 1968). The assays were subdivided in three or four blocks, in each of which all unknown preparations were compared with the standard, the latter in triplicate or quadruplicate. Within each block dilution series of standards and unkowns were randomised with respect to time and place. The dilution series were prepared separately, so that each of them resulted in one independent observation. The latter was defined as the logarithm of the concentration corresponding with a chosen response level (critical concentration and critical response level). Each relative potency was calculated as the antilogarithm of the mean log critical concentrations of unknown and standard. An HCG laboratory standard (Pr. 412), assayed immunochemically against the 2nd International Standard for HCG was used in all assays. In the haemagglutination inhibition test, dilution series of 1.41 were used. The determination of the critical response level from the agglutination patterns has been described elsewhere (Schuurs, 1970). In the complement fixation assay dilution series of 1.25 were used; the critical response level was 50 per cent haemolysis.

N-acetylneuraminic acid (NANA) was estimated as described by Goverde *et al.* (1968).

RESULTS

The results of a typical isoelectric focusing experiment are shown in Figure 1. The fractions obtained were pooled as indicated in the figure and analysed.

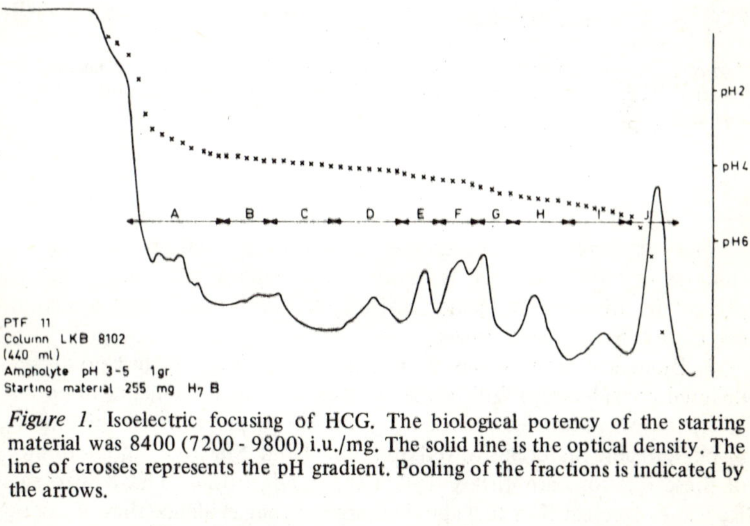

Figure 1. Isoelectric focusing of HCG. The biological potency of the starting material was 8400 (7200 - 9800) i.u./mg. The solid line is the optical density. The line of crosses represents the pH gradient. Pooling of the fractions is indicated by the arrows.

Starch gel electrophoretic analysis (Table I) showed that the fractions A, B and J contained a large number of bands. The fractions D to I inclusive proved to contain one major band and 30 per cent or less of other material.

In Table II the potencies found by means of the seminal vesicle test and in two immunoassay methods are recorded. The highest biological potencies are found in the fractions B, C, D and E. The immunochemical potencies as found with the haemagglutination inhibition test do not differ significantly from one another. In the complement fixation test the fractions C to I inclusive have a significantly higher potency than A, B or J.

The highest NANA contents were found in the fractions C, D, E, F and G. Immunodiffusion (Fig. 2) and starch gel immunoelectrophoresis showed that the major components of the various fractions are immunochemically identical. In immunodiffusion experiments with antigen concentrations from 0.5 to 4 mg/ml against an antiserum mixture containing precipitating antibodies to HCG and its impurities, the fractions D and E behaved as homogeneous while A and J produced two strong lines and the other fractions produced a weak line in addition to that of HCG. However, at antigen concentrations of 8 and 16 mg/ml an additional precipitin line was seen with fractions D and E.

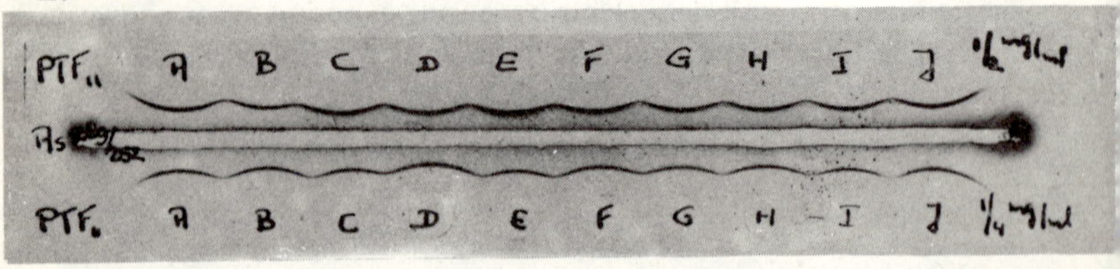

Figure 2. Immunodiffusion of all fractions against a mixture of two antisera to purified HCG (potencies 11,000 (9,600 - 13,800) and 10,800 (10,000 - 12,800) i.u./mg).

DISCUSSION

The experiments with isoelectric focusing confirm our earlier findings that HCG activity as found in the urine of pregnant women is present in different molecular entities. The evidence for this assertion can be summarised as follows:

1. Biological activity has been found in all fractions but potencies differed considerably. With some fractions obtained by means of either gel filtration or isoelectric focusing, up to 90 per cent of the amido black stainable material showed up in one band. Sincethe isoionicpoints of these fractions are different, and the bands proved to have different R_a values (see caption to Table I) there is strong evidence that physicochemically different components carry biological activity.

TABLE I

Semiquantitative evaluation of the starch gel electrophoretic analysis of a separation obtained by isoelectric focusing.

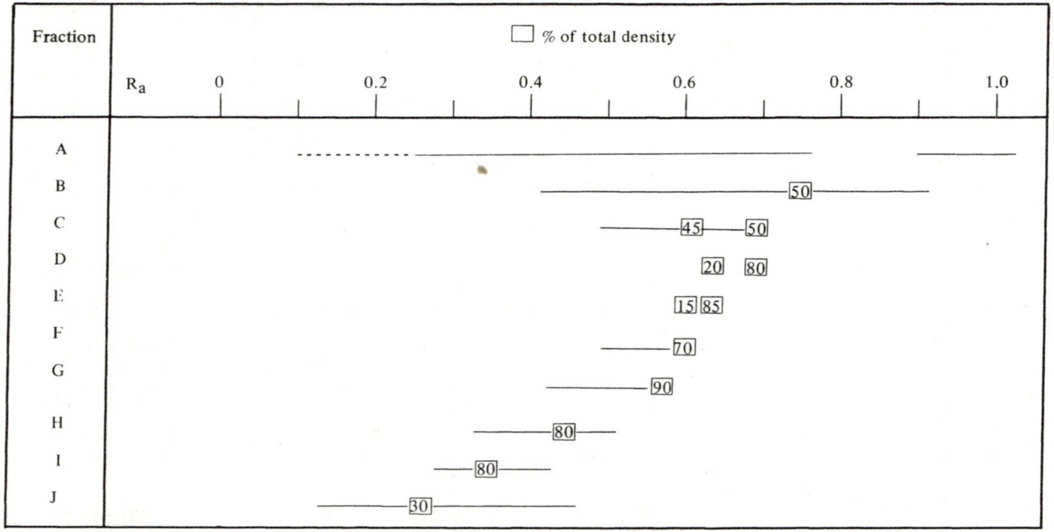

Table 1. The fractions isolated are recorded in the first column.
The bands present in the electropherograms are characterized by their R_a values, the R_a value being the quotient of the distance covered by a particular band and the distance travelled by bovine serum albumin (van Hell *et al.* 1968). The amount of amido black stainable material in each band is expressed as a percentage of the total amount of stainable material in the whole gel pattern. This percentage is given in the Table, the position of the figure indicating its R_a value. The data of the main bands are given only. The position of other stainable material is represented by a line or a dotted line where the stain is weak.

TABLE II
Biological, immunochemical and physicochemical properties of HCG fractions obtained by isoelectric focusing.

Fraction	Isoionic points pH	NANA %	Potencies in i.u. HCG/mg with 95% confidence limits			Recovery	
			By bioassay (1)	By immunoassay (2) Haemaggl. inhib. test (3)	By immunoassay (2) Complement fixation test (4)	Biol. act. %	Weight %
A	1.10–3.73	6.1	5800(4600–7400)	3700(2600–5100)	3660(3450–3890)	2.0	2.9
B	3.73–3.88	6.8	11400(9600–13500)	4100(3000–5700)	5350(5040–5680)	2.7	2.0
C	3.88–4.02	7.2	13300(11100–15900)	5300(3800–7300)	6600(6200–7000)	5.3	3.3
D	4.04–4.22	7.9	12400(9900–15400)	4800(3400–6600)	7700(7300–8200)	15.7	10.7
E	4.26–4.35	7.6	12900(11100–15000)	6400(4600–8900)	8300(7800–8800)	18.8	12.2
F	4.38–4.52	7.7	8400(7000–10000)	5500(3900–7600)	7200(6800–7700)	17.1	17.1
G	4.59–4.75	7.3	6300(5300–7600)	5200(3700–7200)	7900(7400–8300)	10.0	13.3
H	4.80–5.02	6.4	6200(5000–7700)	4800(3400–6600)	6900(6500–7300)	8.4	11.4
I	5.04–5.30	5.6	4100(3400–4900)	4100(3000–5700)	6500(6100–6900)	2.6	5.4
J	5.39–>7	4.1	1800(1500–2300)	3900(2800–5400)	4380(4130–4660)	0.8	3.5
					total recovery	83	82

(1) seminal vesicle test.
(2) Results expressed in i.u. assuming that one vial of the laboratory standard contains 10,000 i.u. by immunoassay.
(3) $\lambda = 0.104$
(4) $\lambda = 0.023$

2. The strong immunochemical relatedness or, probably, identity between the fractions, and between the bands coming from individual fractions was substantiated both on the preparations obtained by gel filtration and on those obtained from isoelectric focusing.

3. Another common feature of the various fractions is their immunochemical activity, as determined by both the haemagglutination inhibition test and in the complement fixation test.

Contrary to the biological activity, the immunochemical potencies do not show much difference, although significantly higher potencies were found in the middle fractions (C to I inclusive) with the more precise complement fixation test. This may reflect a higher degree of immunochemical homogeneity of these fractions, as suggested by the findings in the immunodiffusion experiments. However, it could also be interpreted as being caused by a population of molecules with a higher affinity for the antibodies. The higher immunochemical potencies found with the complement fixation test as compared with the haemagglutination inhibition test cannot yet be explained. These points deserve further investigation.

The most active preparations obtained by isoelectric focusing appeared to have lower NANA contents than the most active preparations from the G-100 gel filtration (van Hell *et al.* 1968). This discrepancy cannot easily be explained. Starch gel electrophoretic analysis did not reveal much difference between the preparations obtained with these methods. The recoveries appeared to be very satisfactory (Table II) so that a loss of activity during electrofocusing does not seem likely. The presence of low molecular weight material can be virtually excluded because all preparations were finally desalted by means of gel filtration with Sephadex G-50. A tentative explanation could be a somewhat different quality of the starting material.

Several literature data confirm that HCG is a multicomponent system. Hamashige *et al.* (1967) have reported the finding of at least three distinct components with HCG activity which they obtained after chromatography with DEAE - Sephadex and DEAE - cellulose.

Bell *et al.* (1969) have shown the heterogeneity in charge and sialic acid content of a product homogeneous in size and amino acid composition.

Despite electrophoretic heterogeneity HCG appears to be homogeneous in the ultracentrifuge (van Hell *et al.* 1968). Therefore, the homogeneity found by Schlumberger (1968) by ultracentrifugal analysis is not contradictory to electrophoretic heterogeneity.

REFERENCES

Bell, J.J., Canfield, R.E. & Sciarra, J.J. (1969). *Endocrinology*, **84**, 298.

Goverde, B.C., Veenkamp, F.J.N. & Homan, J.D.H. (1968). *Acta Endocr. Copenh.* **59**, 105.

Hamashige, S., Astor, M.A., Arquilla, E.R. & van Thiel, D.H. (1967). *J. clin. Endocr. Metab.* **27,** 1690.

Hell, H. van, Goverde, B.C., Schuurs, A.H.W.M., Jager, E. de, Matthijsen, R. & Homan, J.D.H. (1966). *Nature, Lond.* **212,** 261.

Hell, H. van, Matthijsen, R. & Homan, J.D.H. (1968). *Acta Endocr. Copenh.* **59,** 89.

Loraine, J.A. (1967). *In Hormones in Blood,* 2nd ed., Vol. 1, p. 313. Ed. Gray, C.H. & Bacharach, A.L., New York: Academic Press.

Lunenfeld, B. & Eshkol, A. (9967). *Vitams Horm.* **25,** 137.

Schlumberger, H.D. (1968). *Z. Naturf.* **23B,** 1412.

Schuurs, A.H.W.M. (1970). *Acta Endocr. Copenh.,* **63,** Suppl. 142, 95.

Schuurs, A.H.W.M., Jager, E. de & Homan, J.D.H. (1968). *Acta Endocr. Copenh.* **59,** 120.

Chemical Studies on a Pituitary FSH Preparation

M.F. Chaplin, C.J. Gray and J.F. Kennedy

SUMMARY

A reproducible preparation of follicle stimulating hormone (FSH), and its N-acetylneuraminic acid-free derivative, have been treated with various glycosidases, galactose oxidase, proteolytic enzymes, diazo-sulphanilic acid, chloramine T, alkali, alkaline borohydride, acid and periodate, in order to obtain information about the structure of the molecule. The nature and sequence of some of the glycosidic linkages have been postulated, and FSH appears to have a complex branched chain structure. A method of fragmentation which is described should prove useful in further structural investigations.

INTRODUCTION

There are several reports on the qualitative analysis of the human gonadotrophins but extensive structural studies have only been reported for human chorionic gonadotrophin (HCG), (Bahl, 1969). Results are presented here of preliminary structural studies on a preparation of human pituitary follicle stimulating hormone (FSH).

EXPERIMENTAL AND RESULTS

Preparation of FSH. The preparation of FSH (CP 150) was obtained by gel filtration of the fraction called CP 1 (Butt, 1967) on Bio-Gel P-150 in 0.01 M phosphate buffer, pH 7.0. (Barker *et al.* 1969a). The CP 150 is homogeneous on Bio-Gel P-150 and sediments as a single component in the same buffer with $S_{20,w}$ = 2.8s. Sedimentation equilibrium studies by Dr S. P. Spragg, University of Birmingham, gave a mean value of 33,900 (S.E.M. 1,120) for the molecular weight. Some aggregation appeared to have occurred, the aggregate having a calculated molecular weight of 51,500 (S.E.M. 3,700). Higher molecular weight material was also present in a preparation which had been stored for a long period.

The preparation, CP 150, was reproducible in both specific activity and composition; the average specific activity of four preparations was 1725 i.u./mg. It was more stable than preparations isolated using DEAE - Sephadex columns in the last stages of purification. (Amir *et al.* 1966).

Component Analysis. Several preparations of CP 150 gave essentially similar amino-acid compositions. The average numbers of residues per 33,900 molecular weight are shown in Table I.

TABLE I

Component Analysis of the CP 150 Preparation of FSH

No. of residues[a] per 33,900 mol. wt.

Aspartic Acid	18.5	Tyrosine	7.7
Threonine	17.1	Phenylalanine	8.1
Serine	11.7	Lysine	12.2
Glutamic Acid	24.4	Histidine	5.7
Proline	13.0	Arginine	7.2
Glycine	16.1	Tryptophan	6.8
Alanine	11.9	D–Galactose	9.0
Valine	14.0	D–Mannose	19.7
Cystine	13.2	L–Fucose	2.1
Methionine	2.4	2–Acetamido–2–deoxyglucose	18.8
Isoleucine	7.2	2–Acetamido–2–deoxygalactose	1.0
Leucine	12.3	N–Acetylneuraminic acid	5.7

[a] Obtained from the average compositions of four preparations of CP 150

Except for tryptophan, the amino-acid analyses were carried out after hydrolysis *in vacuo* in 5.9 M hydrochloric acid at 108°C for 24 hr. Tryptophan was estimated after hydrolysis of the CP 150 in barium hydroxide solution. Carbohydrates were determined as described by Barker *et al.* (1969a).

Action of neuraminidase. When CP 150 (batch FM9) was incubated with neuraminidase, the N-acetyl neuraminic acid released (Warren, 1959) corresponded to 102 per cent of the estimate obtained after acid hydrolysis (0.1 M hydrochloric acid at 80°C for 1 hr., Table I).

The product was fractionated on CM - cellulose to remove neuraminidase. The non-adsorbed material was detected by its u.v. absorption at 280 nm and was subsequently separated from N-acetyl neuraminic acid by fractionation on Bio-Gel P-150, (Fig. 1).

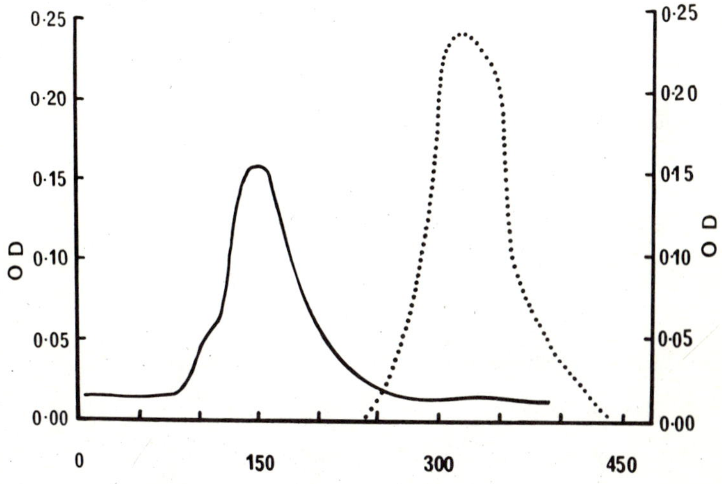

Figure 1. Fractionation of neuraminidase-treated CP 150 on Bio-Gel P-150. (————) O.D. 220 nm; (......) Warren assay, O.D. 549 nm.

The N-acetyl neuraminic acid released from CP 150 was hydrolysed in 1M sulphuric acid at 100°C for 1 hr. It contained less than 0.5 per cent glycollic acid estimated by the method of Klenk and Ullenbruck (1966).

The N-acetyl neuraminic acid-free derivative (CP 150 NF) contained no neuraminidase since incubation with the FSH preparation, CM1, (Hartree, 1966), which is a good substrate for neuraminidase, released no detectable neuraminic acid.

Action of other glycosidases. Ram enzymes were prepared from an acetic acid extract of testes (Fisons Pharmaceuticals Ltd.) by precipitation with ammonium sulphate. *Coccobacillus* enzymes were obtained by ultrasonic disruption of a gram negative soil *Coccobacillus* (Chase, 1938) and were purified by ammonium sulphate precipitation (Dr A. Harboe). Yeast galactosidase (β-D-galactosidase, B.D.H. Ltd.) was purified by fractionation on Bio-Gel P-150. Jack-bean mannosidase

(α-D-mannosidase) was extracted from Jack-bean meal (Sigma Chemical Co. Ltd.) by a combination and adaptation of the methods of Li (1967) and Snaith and Levvy (1968), and was finally purified by fractionation on Bio-Gel P-150. The enzyme preparations were each tested against a series of o- and p-nitrophenylglycosides and their relative activities are given in Table II.

TABLE II

Relative Activities of Enzymes Used

	Ram	*Cocco-bacillus*	Yeast	Jack Bean
α–D–galactosidase	4	50	0	0
β–D–galactosidase	3	50	250	0
β–2–acetamido–2–deoxy–D–glucoside acetamido–deoxyglucohydrolase	300	450	0	0
β–2–acetamido–2–deoxy–D–galactoside acetamido–deoxygalactohydrolase	40	220	0	0
α–D–mannosidase	1	0	0	33,000
α–L–fucosidase	16	10	0	0
β–L–fucosidase	0	0	0	0

CP 150 NF was incubated under sterile conditions with each of the enzymes in turn at 37°C for 48 hr. and the incubates were separately fractionated on Bio-Gel P-150. Material which was totally included was fractionated on a pre-calibrated ion-exclusion column of AG 50WX2 Li^+ form) as described by Barker *et al.* (1969b). Fractions corresponding to the elution positions of monosaccharide and di- and tri-saccharides were separately bulked. After hydrolysis of the di- and tri-saccharides in 2M HC1 at 100°C for 2-4 hrs. the hydrolyzates and the monosaccharides were analysed for carbohydrate by gas-liquid chromatography as the trimethylsilyl derivatives (Kennedy and Butt, 1969).

The *Coccobacillus* enzymes released 10 per cent of the total L-fucose of CP 150 NF as monosaccharide. The other enzymes did not apparently release any carbohydrate although the ram enzymes were shown to be effective in releasing some D-galactose and 2-acetamido-2-deoxy-D-glucose from a less pure preparation of FSH (CP1 NF). The ram enzymes also failed to release carbohydrate from untreated CP 150.

Action of galactose oxidase. CP 150 and CP 150 NF were assayed for D-galactose before and after hydrolysis with 2.0M hydrochloric acid for 4 hr. at 100°C by the enzymic method using D-galactose oxidase (Barker *et al.* 1969a). The hydrolyzed preparations were both estimated to contain 9 galactose residues per molecular weight of 33,900. The estimate for unhydrolyzed CP 150 NF corresponded to 106 per cent of this value, but unhydrolyzed CP 150 gave no reaction for D-galactose.

Partial acid hydrolysis. Hydrolyses were carried out in 0.5-2 M sulphuric acid or 1-4M hydrochloric acid at 80 to 100°C for 0.5 - 9 hr. or in 0.1 M acetic acid at 100°C for 5 to 18 hr. The neutralized hydrolyzates were fractionated on the ion exclusion column (Barker *et al.* 1969b) for neutral sugars and on AG 50 WX8 (H^+ form) for basic sugars, using the automated fluorimetric scan for formaldehyde released on periodate oxidation (Cho Tun *et al.* 1969). The milder conditions gave only neutral monosaccharides while the stronger hydrolysis conditions released basic monosaccharides in addition.

Alkaline hydrolysis. CP 150 was hydrolyzed *in vacuo* in 0.2M sodium hydroxide at 26.5°C for 96 hr. and in 0.5M sodium hydroxide at 37°C for 24 hr. The hydrolyzates were analysed by a modification of the Morgan-Elson assay (Morgan and Elson 1934; Reissig *et al.* 1955) for pre-formed chromogen (Kuhn and Krüger, 1956). Standards carried through the procedure gave their usual responses, but no chromophore was observed from the alkaline hydrolysed preparation.

Alkaline borohydride treatment. CP 150 was hydrolysed in 0.2M sodium hydroxide containing 0.1 per cent sodium borohydride at 26.5°C for various times. The neutralized hydrolyzates were fractionated on an ion exclusion column (Barker *et al.* 1969b) and scanning of the eluates for formaldehyde released on periodate oxidation (Cho Tun *et al.* 1969) showed two major peaks corresponding to oligomeric and polymeric materials (Fig. 2). A peak corresponding to a trace of monosaccharide appeared after 96 hr. hydrolysis (Fig. 2). Similar results were obtained using 0.1M sodium hydroxide under the same conditions.

A 96 hr.-hydrolysate was desalted, after neutralization, on Bio-Gel P-10 and an amino-acid analysis carried out on the high molecular weight material (Table III).

Reaction with chloramine-T (N-chlorotoluenesulphonamide). CP 150 (batch CL10) and CP 150 NF (batch FM9) were treated with chloramine-T (220 μg/mg protein) and immediately fractionated on columns of Bio-Gel P-150 at 4°C. A high molecular weight peak (Figs. 3 & 4) containing virtually all of the protein applied was obtained from CP 150. Amino-acid compositions of this material, and of the starting material, obtained from 24 hr. acid hydrolyses (5.9 M hydrochloric acid at 108°C) are shown in Table IV.

Two peaks were obtained from CP 150 NF corresponding to totally

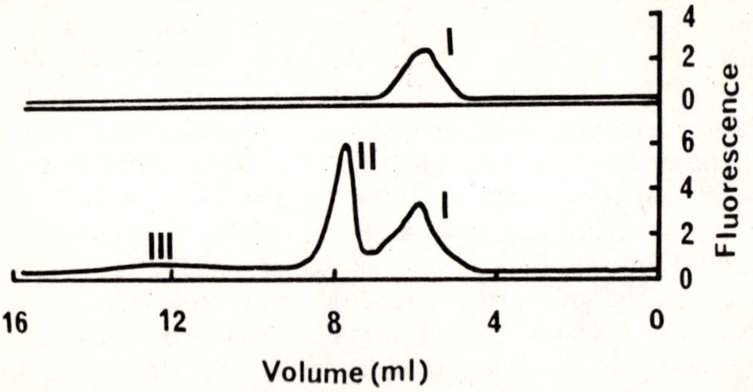

Figure 2. Ion exclusion fractionation of CP 150 (top) and alkaline borohydride treated CP 150 (bottom). I, polymer; II, oligomer; III, monomer.

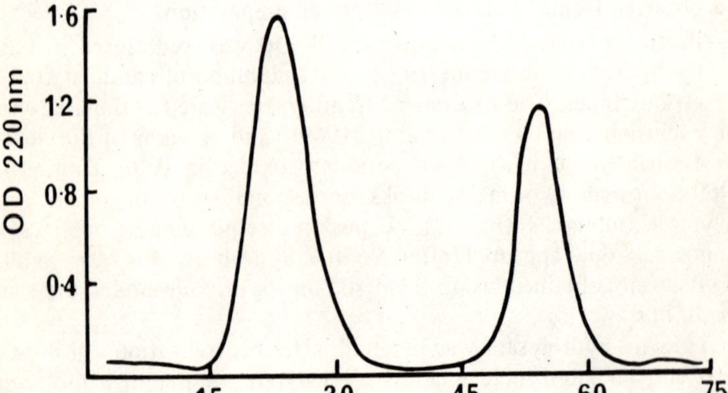

Figure 3. Chromatography on Bio-Gel P-150 of CP 150 (1.54 mg) after treatment with chloramine-T. Area of first peak = 5.2 O.D. ml. The second peak is attributed to reaction products from chloramine-T.

TABLE III

Amino-Acid Composition of CP 150 Preparation of FSH Before and After Treatment with Alkaline Borohydride

Amino-acid	Amino-acid ratios	
	Before	After
Cysteic acid	0	2.0
Aspartic acid	15.9	18.3
Threonine	14.2	13.6
Serine	10.1	17.2
Glutamic acid	21.7	22.6
Proline	9.8	10.3
Glycine	11.9	17.6
Alanine	9.7	13.6
'A'	–	16.0
Valine	11.4	11.0
Cystine	9.4	0.0
Methionine	1.8	0.0
Isoleucine	6.6	7.4
Leucine	10.4	10.4
Tyrosine	6.9	6.4
Phenylalanine	7.2	7.0
Lysine	10.9	13.1
Histidine	4.7	5.1
Arginine	6.3	5.6

'A' appears at the position of glucosamine or α-amino-butyric acid. In the starting material there is a peak corresponding to 2-amino-2-deoxy-D-glucose with a value, based on these ratios, of 7.8.

TABLE IV

Effect of chloramine–T on the CP 150 (batch CL10) preparation of FSH

Amino-acid	Amino-acid ratios (Based on Lysine = 10) Before	After
Cysteic acid	0.0	4.0
Aspartic acid	14.6	15.7
Threonine	13.6	14.0
Serine	10.6	9.5
Glutamic acid	20.2	20.1
Proline	11.0	10.4
Glycine	13.7	14.7
Alanine	9.8	9.5
Valine	10.3	10.5
Cystine	10.9	1.7
Methionine	2.0	1.0
Isoleucine	5.5	5.3
Leucine	10.5	9.2
Tyrosine	5.9	5.5
Phenylalanine	6.4	6.4
Lysine	10.0	10.0
Histidine	4.9	5.1
Arginine	6.0	6.1
Tryptophan	5.5[a]	0.0

[a]Not determined on this preparation, but on another of otherwise similar composition and activity (batch CL9).

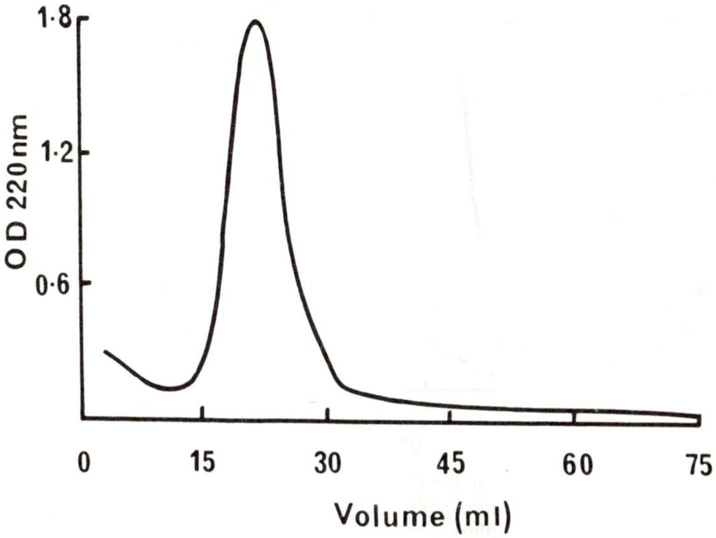

Figure 4. Chromatography on Bio-Gel P-150 of CP 150 (1.54 mg). Area of peak = 5.5 O.D. ml.

excluded and partially included material (Fig. 5). The latter material was similarly fractionated on Bio-Gel P-30 to give two peaks (Fig. 6) corresponding to excluded and partially included material. This partially included material was refractionated on Bio-Gel P-2 giving two peaks (Fig. 7) of which the major was in the totally excluded position. The product was therefore separated into three fractions, P-150 excluded, P-30 excluded and P-2 excluded, and amino-acid analyses were carried out, as described above, on these fractions, and on the original starting material (batch FM9), (Table V). The relative protein contents of the three fractions were 1.0 : 5.5 : 10.5 respectively. The ultra-violet absorption spectrum of the product from the chloramine-T oxidation of CP 150 NF is shown in Figure 8.

Photo-oxidation. Photo-oxidation was carried out by illuminating solutions of CP 150 (batch CL9) in the presence of methylene blue with a 500W photoflood lamp at a distance of 4 in. for various times. Amino acid analyses were carried out after 24 hr. acid hydrolyses except for tryptophan which was determined separately after 48 hr. hydrolysis with barium hydroxide solution, (Table VI).

Diazosulphanilic acid. CP 150 (batch CL10) was treated at pH 7.0 with diazosulphanilic acid (270 μg/mg protein) and low molecular weight material removed on Bio-Gel P-10. The visible spectrum of the high molecular weight product is shown in Figure 9.

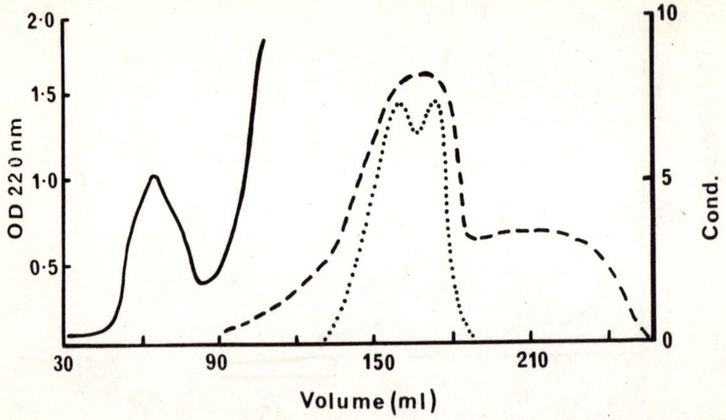

Figure 5. Fractionation on Bio-Gel P-150 of Chloramine T treated CP 150 NF. (————) O.D. 220 nm, scale x 10^{-1}; (– – – –) O.D. 220 nm, scale x 1; (........) conductivity (mmho). Lower molecular weight materials (72 – 256 ml) refractionated - Figure 6.

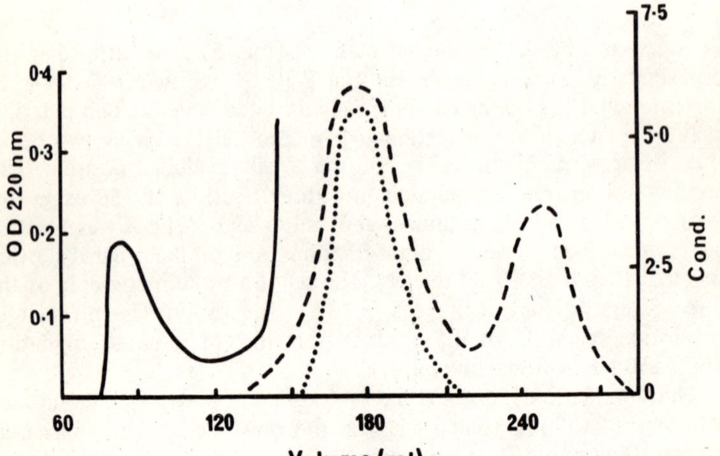

Figure 6. Fractionation on Bio-Gel P-30 of lower molecular weight materials from fractionation on Bio-Gel P-150 of chloramine-T treated CP 150 NF. (————) O.D. 220 nm, scale x 1; (– – – –) O.D. 220 nm, scale x 5; (........) conductivity (mmho). Lower molecular weight materials (126 - 224 ml) refractionated - Figure 7.

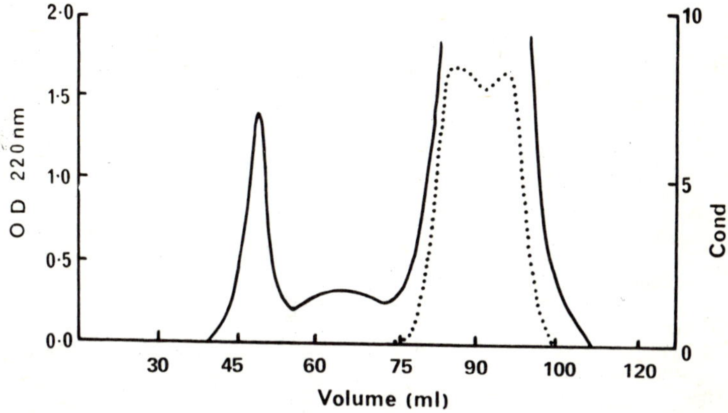

Figure 7. Fractionation on Bio-Gel P-2 of lower molecular weight materials from fractionation on Bio-Gel P-30 of chloramine-T treated CP 150 NF. (————) O.D. 220 nm; (........) conductivity (mmho).

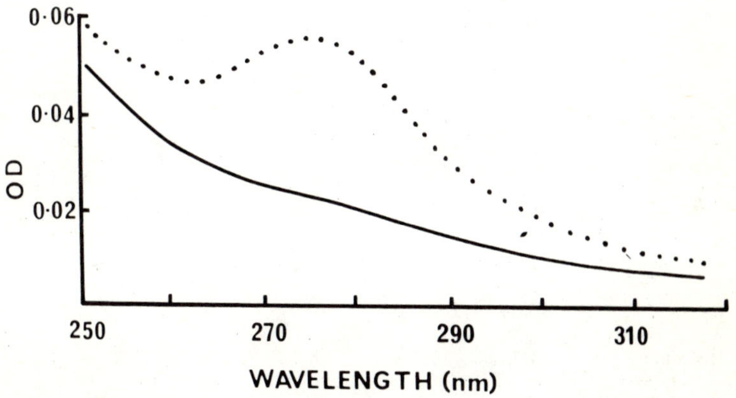

Figure 8. Change in u.v. spectrum of CP 150 NF on treatment with chloramine-T. (........) before; (————) after.

TABLE V

Compositions of products obtained from the chloramine–T oxidation of CP 150 NF

Amino-acid	P–150 excluded	Fraction P–30 excluded	P–2 excluded	Total composition	CP 150 NF (batch FM9)
Cysteic acid	4.7	n.d.	3.0	3.0[a]	0.0
Aspartic acid	17.6	13.1	15.3	14.4	n.d
Threonine	21.3	14.0	11.5	12.3	13.6
Serine	19.7	17.4	10.8	13.3	10.6
Glutamic acid	25.9	18.9	18.4	18.8	17.5
Proline	15.6	20.3	7.9	11.7	10.7
Glycine	30.1	13.5	17.5	16.9	16.7
Alanine	19.0	11.3	10.8	11.2	10.9
Valine	11.6	12.9	11.4	11.8	10.9
Cystine	0.0	1.4	0.1	0.5	12.2
Methionine	0.0	1.2	0.2	0.5	2.0
Isoleucine	5.0	5.3	5.6	5.5	5.4
Leucine	10.3	8.0	8.7	8.6	9.3
Tyrosine	0.5	8.3	3.8	5.6	5.7
Phenylalanine	1.6	8.1	6.5	6.7	6.5
Lysine	10.0	10.0	10.0	10.0	10.0
Histidine	4.7	4.8	4.7	4.7	5.8
Arginine	2.3	6.4	n.d.	6.2[a]	5.2

n.d. not determined. [a] estimated.

TABLE VI

Effect of Photo-oxidation on Amino-Acid Composition of CP–150

	Amino acid ratios		
Time of exposure	0 min	2 min	20 min
Aspartic acid	15.2	15.2	15.2
Threonine	13.4	13.2	13.6
Serine	9.7	9.5	9.4
Glutamic acid	20.8	21.0	21.2
Proline	9.4	8.9	10.0
Glycine	11.6	12.2	13.6
Alanine	9.2	9.6	9.5
Valine	10.8	10.8	10.6
Cystine	9.0	9.1	6.9
Methionine	1.8	1.4	1.4
Isoleucine	6.3	7.1	6.2
Leucine	10.4	10.9	10.2
Tyrosine	6.6	5.9	3.3
Phenylalanine	6.9	6.7	6.4
Lysine	10.4	10.0	10.4
Histidine	4.5	3.3	1.9
Arginine	6.0	6.7	6.3
Tryptophan	5.5	0.0	0.0

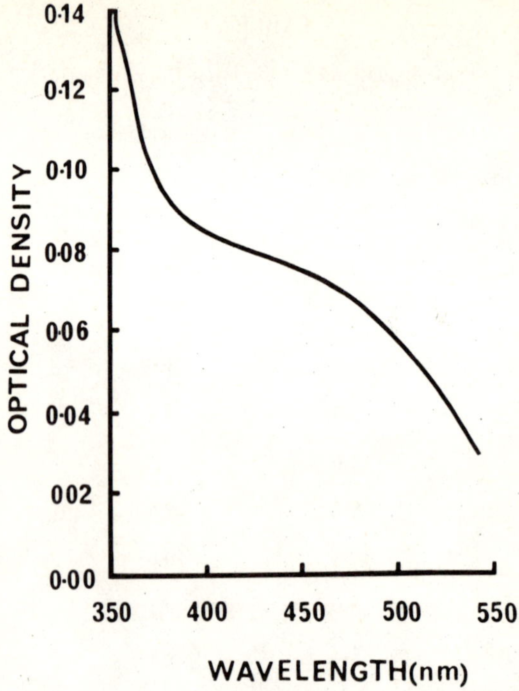

Figure 9. Visible spectrum, in 0.1N NaOH, of CP 150 after treatment with diazosulphanilic acid.

Proteolytic enzymes. Chymotrypsin had no effect on CP 150 (batch CG110/1) while trypsin appeared to cause a degree of degradation of the protein chain as judged by gel filtration on Bio-Gel P-150.

Periodate oxidation. The results of earlier periodate oxidation studies (Kennedy and Butt, 1969) on CP1, using 29 mM unbuffered sodium metaperiodate at 4°C for 28 hr., are summarised in Table VII.

DISCUSSION

The preparation of FSH used in these experiments (CP 150) contains approximately 33 per cent carbohydrate (Barker *et al.*, 1969a) which is in good agreement with the analyses of Roos (1968) and Reichert *et al.* (1968). The tryptophan content of 6.8 residues per molecular weight of 33,900 was estimated on an Auto-analyzer after hydrolysis with barium hydroxide (Noltmann *et al.*, 1962). This result is much higher than that calculated using the equation of Beaven and Holiday (1952) which is based on the u.v. absorption at 280 and 294 nm. When this method was

TABLE VII

Periodate Oxidation of CP1

Component	n moles/mg CP1	
	Before	After
L–fucose	39	0.0
D–galactose	125	52
D–mannose	394	79
N–acetylneuraminic acid	155	37*
2–amino–2–deoxy–D–galactose	29	6
2–amino–2–deoxy–D–glucose	824	110
ethylene glycol (≡ glycolaldehyde)	–	87
propane – 1,2–diol	–	44
glycerol	–	93
tetritol	–	0.0
cysteic acid	0.0	142
aspartic acid	312	242
threonine	222	182
serine	223	176
glutamic acid	409	395
proline	225	168
alanine	218	183
valine	154	161
methionine	46	27
isoleucine	96	73
leucine	229	156
tyrosine	118	54
phenylalanine	131	108
lysine	220	111
histidine	67	67
arginine	159	98
glycine	199	185
½ cyst	224	148

*Based on Warren positive results

applied to CP 150 the result was 2.6 which is nearer the value of 3.6 obtained by Roos (1968) using a different equation. The estimated molecular weight of 33,900 is in good agreement with the values obtained by gel filtration and radiation inactivation studies, (Odell *et al.*, 1964). The higher calculated values for the molecular weight of some of the material at the bottom of the ultracentrifuge cell, and in the preparation stored over a long period, indicates that some association takes place, and that a population of molecules containing monomer, dimer and possibly higher aggregates is present.

The estimate of N-acetylneuraminic acid released from CP 150 by incubation with neuraminidase is the same as that following acid hydrolysis. This suggests that either all the neuraminic acid units occupy terminal non-reducing positions or that at least one N-acetylneuraminic acid residue occupies such a position and the remainder form chains linked to this residue. The N-acetylneuraminic acid cannot be linked at C-9, otherwise the unhydrolyzed preparation, CP 150, would respond in the Warren (1959) assay, since the corresponding heptulosaminic acid would be released by oxidation of the C-7 - C-8 bond. β-Formylpyruvic acid is obtained on periodate oxidation only after hydrolysis of CP 150 or periodate oxidised CP 150 (Fig. 10).

Linkage at C-2 would involve the anomeric carbon atom in a ketal and this would appear, therefore, to be the more likely glycosidic linkage. If the neuraminic acid is terminal and C-2 linked, the periodate oxidation would cleave the C7-C8 bond, giving N-acetylheptulosaminic acid as the terminal unit. On subsequent hydrolysis, the N-acetylheptulosaminic acid would give a positive reaction in the Warren assay. Kuhn and Gauche (1965) have shown, however, that when acidic periodate is used in the similar thiobarbituric assay of Aminoff (1961), the molar extinction co-efficient of N-acetylheptulosaminic acid is considerably less than that of N-acetylneuraminic acid. The results of the periodate oxidation studies therefore are consistent with the C-2 linked, terminal structure since the product gave 24 per cent of the original Warren value (Table VII).

Since CP 150 is not acted upon by the ram enzymes, it may be that the terminal N-acetylneuraminic acid units block the degradation of the carbohydrate chains, or that the ionic nature of the molecule afforded by the N-acetylneuraminic acid residues prevents the action of the glycosidases which would otherwise act upon other terminal non-reducing carbohydrate residues. However, since CP 150 NF is also resistant to these enzymes, the preparation is either a very poor substrate for these enzymes - perhaps it is too highly branched - or β-mannose is the key carbohydrate to the chains. Since the *Coccobacillus* fucosidase removes a small amount of the L- fucose, it appears that at least one of the L- fucose units is terminal non-reducing

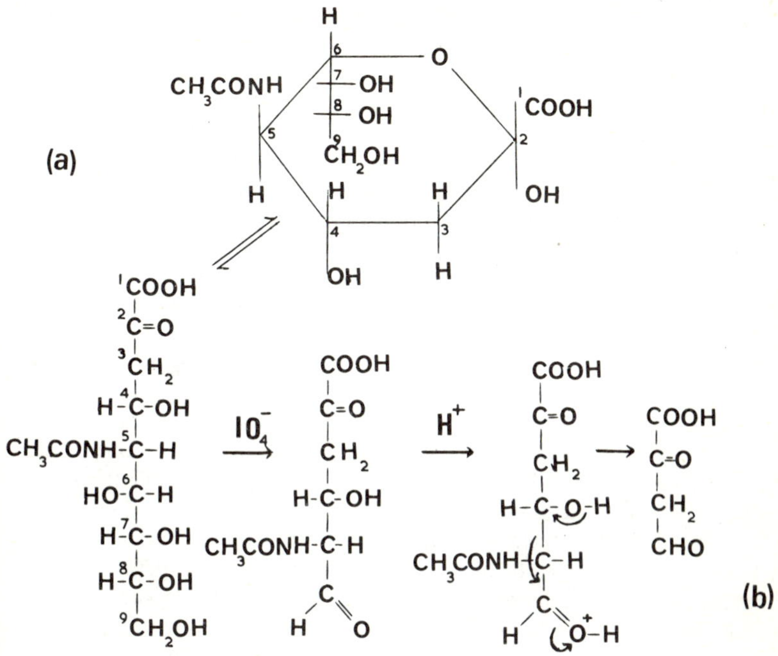

Figure 10. Structure of N-acetylneuraminic acid, and its breakdown in the Warren (1959) assay.
(a) N-acetyl neuraminic acid (5-acetamido-3, 5-dideoxy-D-glycero-D-galacto-nonulosonic acid)
(b) β-formyl pyruvic acid

and again, that the preparation of FSH is a poor substrate for this enzyme.

The results of the experiments with galactose oxidase suggest that either N-acetylneuraminic acid prevents the action of this enzyme or that N-acetylneuraminic acid is linked to galactose. Approximately six galactose units are resistant to periodate oxidation and it may be that these occur near the chain ends, linked to N-acetylneuraminic acid residues.

The results of the attempted partial acid hydrolysis are unusual since oligosaccharides are usually obtained from glycoproteins under conditions described here. Alkaline and alkaline borohydride hydrolyses appear to split the CP 150 molecule into at least two sub-units, but the yield of oligosaccharides was again poor as compared with results from other glycoproteins. Glycopeptide linkages involving serine or threonine are cleaved by alkali since they can undergo elimination of the carbohydrate moiety, but since alkaline hydrolysis failed to give any Morgan-Elson chromogen, it is concluded that the carbohydrate-peptide linkage(s) does not involve 2-amino-2 deoxy-D-glucose or -D-galactose, linked to either serine or threonine.

The amino-acid composition of the product from alkaline borohydride treatment of a glycoprotein often provides information about the carbohydrate-peptide linkage. For example, an O-seryl derivative may be converted into alanine, and an O-threonyl group into α-aminobutyric acid. The results of alkaline borohydride treatment of CP 150 (Table III) present an interesting picture. Increased recoveries of aspartic acid, serine, glycine and alanine were observed while threonine remained constant. The increase in alanine may be due to reduction of an O-serine species, but could also arise from reduction of cystine. Cystine and methionine were lost. A peak 'A' was present on the analyzer trace but it is not possible to distinguish between α-aminobutyric acid and glucosamine in this case. On account of its size, 'A' may represent a mixture of these two compounds. The increase in this peak may represent an O-threonyl link but α-aminobutyric acid may also arise from the reduction of methionine. It is possible therefore that both threonine and serine are involved in the glycopeptide linkage.

Some further conclusions regarding the structure of FSH have been drawn from the results of the periodate oxidation experiments by Kennedy and Butt (1969). For clarity the formulae of the relevant carbohydrates are given in Figure 11.

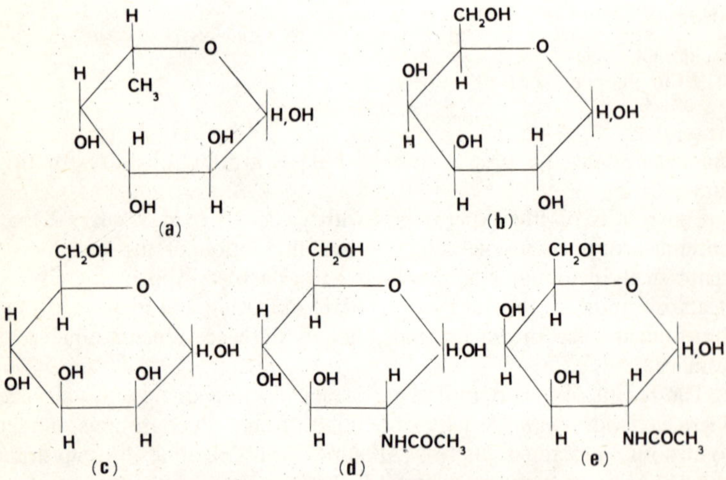

Figure 11. Structures of carbohydrate units.
(a) L-fucopyranose
(b) D-galactopyranose
(c) D-mannopyranose
(d) 2-acetamido-2-deoxy-D-glucopyranose
(e) 2-acetamido-2-deoxy-D-galactopyranose

The fucose units cannot be linked in the 1- and 3- positions since no fucose was recovered from the oxidized and reduced products. Since the amount of propane-1,2-diol recovered after oxidation, reduction and hydrolysis was approximately equivalent to the original content of fucose, all the fucose units must be linked in the 1- and 2- positions or occur as terminal non-reducing units.

Some of the galactose units (41 per cent) were recovered intact and are thus linked in the 1- and 3- positions or are involved in branch points. Since no threitol was identified in the polyol analysis it appears that the susceptible galactose units are linked in the 1- and 2- positions or the 1- and 6- positions, or occur as terminal non-reducing chains ends.

About 80 per cent of the mannose units were oxidised, and since no erythritol was identified these units are not 1- and 3- or 1- and 4- substituted, but are 1- and 2- or 1- and 6- substituted or terminal non-reducing units. It is unlikely that they all exist as terminal units on account of the high proportion of mannose in the preparation. The resistant units would be linked in the 1- and 3- positions or occur at branch points.

Over 80 per cent of the 2-amino-2-deoxy-D-glucose and 2-amino-2-deoxy-D-galactose units were oxidised and, unless some occur as end groups, these appear to be 1- and 6- substituted. Assuming that in the parent substance they are all N-acetylated, then linkage involving C-2 would not occur, and neither would oxidation of the 2- and 3- positions. Hence the units that do remain intact would be 1- and 3- or 1- and 4- substituted, or occur at branch points.

From the results of the amino acid analyses before and after the 28 hr. oxidation experiment it was calculated that 82 per cent of the protein was retained. It is interesting that oxidation of interstitial-cell stimulating hormone under identical conditions proceeds to this stage within 1 hr. and cystine is completely destroyed (Gan *et al.*, 1968).

Photo-oxidation (Table VI) usually results in destruction of histidine, tyrosine, tryptophan, methionine and cysteine residues, although sometimes the effect is quite selective (Koshland *et al.*, 1962; Westhead, 1965; Ingram and Hochester, 1967; Ingram, 1967). Tryptophan in CP 150 was totally destroyed after only two minutes. Both histidine and tyrosine showed a more gradual but significant decrease. Methionine appeared to be affected very little, while cystine showed a decrease only after 20 minutes' exposure. The increase in glycine may be significant; possibly it was a product of the degradation of another amino-acid.

A 'positive halogen' oxidizing agent such as chloramine-T might be expected to cause the destruction of tryptophan, histidine, tyrosine, methionine and cystine. Since all the oxidized CP 150 was recovered on gel filtration, and the peak position coincided with untreated CP 150, it would appear that no fragmentation of the glycoprotein had taken place (Figs. 3 and 4). The amino-acid analysis (Table III) shows that tryptophan appears to have been totally lost, while methionine and cystine were both partially destroyed. Some of the cystine is converted into cysteic acid. Other amino-acids do not appear to have been affected to any significant extent.

When CP 150 which had previously been treated with neuraminidase (CP 150 NF) was oxidized by chloramine-T under similar conditions a different result was obtained. The product could be separated into three fractions of high, medium and low molecular weight. The high molecular weight material represented approximately 6.0 per cent of the total protein recovered, while the medium and low molecular weight fractions represented 33.0 and 61.0 per cent respectively.

The high molecular weight material differs markedly from the other fractions (Table V), and clearly this is not simply unchanged starting material. Most noticeable are the high values for threonine, serine, glycine and alanine, and the small amounts of tyrosine and phenylalanine. The medium and low molecular weight fractions also differ, although not to such a great extent. Thus, the medium molecular weight material contains more proline and tyrosine, and most of the methionine.

However, despite these differences in the individual fractions, the overall composition agrees well with that of the starting material, except that both methionine and cystine have been destroyed. Tryptophan was not determined in these fractions but the change in the ultra-violet spectrum on treatment of CP 150 NF with chloramine-T suggests that tryptophan has been destroyed (Fig. 8).

In terms of the overall compositions, therefore, chloramine-T brings about similar chemical changes in CP 150 and CP 150 NF in that tryptophan, cystine and methionine are oxidized. However, whereas CP 150 remains as high molecular weight material, the neuraminic acid-free preparation undergoes extensive fragmentation. This suggests a role for the neuraminic acid residues in holding the molecule together so that either the molecule is less susceptible to cleavage reactions or the molecule, having been oxidized, does not readily dissociate into fragments.

There is a shoulder in the region of 440 nm in the visible spectrum of CP 150 which has been treated with diazosulphanilic acid. This is compatible with coupling of histidine with this reagent; only a small amount of tyrosine couples since there is little evidence of a shoulder at 500 nm (Pontremoli *et al.*, 1967).

This work suggests a complex structure for the FSH molecule. The preliminary studies with trypsin indicate that the molecule appears to be susceptible to this proteolytic enzyme. Studies of the degradation products obtained and of those by the other methods described should prove helpful in obtaining more information on the structure.

ACKNOWLEDGEMENTS

The authors thank Professor M. Stacey, Dr S. A. Barker, Dr A. C. Crooke and Dr W. R. Butt for their interest and advice, and Mrs C. Law for technical assistance.

The work was supported by a block grant from the Medical Research Council. One of us (M.F.C.) acknowledges receipt of an S.R.C. research studentship.

REFERENCES

Aminoff, D. (1961). *Biochem. J.* **81**, 384.

Amir, S.M., Barker, S.A., Butt, W.R. & Crooke, A.C. (1966). *Nature, Lond.* **209**, 1092.

Bahl, O.P. (1969). *J. biol. Chem.* **244**, 575.

Barker, S.A., Gray, C.J., Kennedy, J.F. & Butt, W.R. (1969a). *J. Endocr.* **45**, 275.

Barker, S.A., Hatt, B.W., Kennedy, J.F. & Somers, P.J. (1969b) *Carbohydrate Res.*, **9**, 327.

Beaven, G.H. & Holiday, E.R. (1952). In *Advances in Protein Chemistry*, Vol. 7, p. 319. Ed. Anson, M.L., Bailey, K. & Edsall, J.T. New York: Academic Press.

Butt, W.R. (1967). In *The Chemistry of the Gonadotrophins*, p. 95. Ed. I.N. Kugelmass, Springfield, Illinois: Thomas.

Chase, M.W. (1938). *J. Bact.* **36**, 383.

Cho Tun, H., Kennedy, J.F., Stacey, M. & Woodbury, R.R. (1969).*Carbohydrate Res.* **11**, 225.

Gan, J., Papoff, H. & Li, C.H. (1968).*Biochim, biophys. Acta,* **170**, 189.

Hartree, A.S. (1966). *Biochem. J.* **100**, 754.

Ingram, J.M. (1967). *Can. J. Biochem. Physiol.* **45**, 1505.

Ingram, J.M. & Hochester, R.M. (1967). *Can. J. Biochem. Physiol.* **45**, 929.

Kennedy, J.F. & Butt, W.R. (1969). *Biochem. J.* **115**, 225.

Klenk, E. & Ullenbruck, G. (1966). *Hoppe-Seyler's. Z. physiol. Chem.* **307**, 266.

Koshland, D.E. Jr., Strumeyer, D.H. & Ray, R.J. (1962). *Brookhaven Symp. Biol.* **15**, 101.

Kuhn, R. & Gauche, A. (1965). *Chem. Ber.* **98**, 395.

Kuhn, R. & Krüger, G. (1956). *Chem. Ber.* **89**, 1473.

Li, Y. (1967). *J. biol. Chem.* **242**, 5474.

Morgan, W.T.J. & Elson, L.A. (1934). *Biochem. J.* **26**, 988.

Noltmann, E.A., Mahowald, T.A. & Kuly, S.A. (1962). *J. biol. Chem.* **237**, 1146.

O'Dell, W.D., Swain, R.W. & Nydick, M. (1964). *J. clin. Endocr. Metab.* **24**, 1266.

Pontremoli, S., Grazi, E. & Accoisi, A. (1967). *J. biol. Chem.* **242**, 61.

Reichert, L.E. Jr., Kathan, R.H. & Ryan, R.J. (1968). *Endocrinology,* **82**, 109.

Reissig, J.L. Strominger, J.L. & Leloir, L.F. (1955). *J. biol. Chem.* **217**, 959.

Roos, P. (1968). *Acta Endocr. Copenh.* Suppl. 131.

Snaith, S.M. & Levy, G.A. (1968). *Biochem. J.* **110**, 663.

Warren, L. (1959). *J. biol. Chem.* **234**, 1971.

Westhead, E.W. (1965). *Biochemistry, Wash.* **4**, 2139.

The Action of a Neuraminic Acid-free Derivative of FSH on Mouse Ovaries Cultured In Vitro

Margaret Ryle, M.F. Chaplin, C. J. Gray and J. F. Kennedy

SUMMARY

Two preparations of follicle stimulating hormone were treated with neuraminidase to remove N-acetyl neuraminic acid. Neither showed any biological activity in the mouse ovarian augmentation assay, though both initially retained full immunological activity. Each was tested twice for *in vitro* biological activity by measuring the uptake of ^{14}C-thymidine in 15 day old mouse ovaries grown in organ culture. The experiments showed that thymidine uptake was enhanced, as it is by normal FSH. The potency of one of the preparations was 45 per cent of that of normal FSH. The possible role of NANA in relation to the structure and function of FSH is discussed.

INTRODUCTION

A procedure for obtaining a reproducible, quantitative *in vitro* response to human pituitary follicle stimulating hormone (FSH), based on a standard organ culture technique, has already been described (Ryle, 1969a) but further modifications have made it more sensitive and less variable. It depends on the stimulation of cell division in the follicles of intact infantile mouse ovaries cultured for several days. Radioactively-labelled thymidine added to the medium is taken up, nearly all being incorporated into deoxyribonucleic acid (DNA), (How *et al.*, 1970). The amount of label present in the tissue at the end of the culture period provides a measure of the mitotic rate and hence, by comparison with suitable controls, of the degree of stimulation by the hormone preparation being tested.

Parallel morphological studies show that FSH markedly stimulates the growth of follicles, particularly of those with less than four layers of cells (Ryle, 1969b). Luteinizing hormone (LH) also stimulates cell division, particularly in larger follicles. It does not usually induce a significant increase in thymidine uptake, however, probably because there are few of these larger follicles in the ovaries of the 15-day old mice used in the standard procedure. It might do so if the culture period were prolonged beyond the usual three to five days. Thus the thymidine uptake response is not truly specific for FSH, though with carefully controlled conditions it appears to be relatively so and the technique is well suited to the study of the biological activity of highly purified FSH and its chemically and enzymatically modified derivatives.

It has previously been shown that the complete enzymatic removal of N-acetyl neuraminic acid (NANA) from human pituitary FSH results in total loss of biological activity, as judged by the mouse ovarian augmentation assay, although its immunological activity is not immediately affected (Butt *et al.*, 1970). This loss of biological activity could conceivably arise from: (1) the failure of the hormone to reach the target site, (2) a failure to recognise and attach to it or (3) having attached, a failure to provoke the normal response. Any particular chemical modification could theoretically affect one or more of these functions. A modified FSH molecule which attached to the target site but did not induce the normal biological response might function as a competitive inhibitor. The present work is an investigation of the effect of NANA-free FSH (FSH-NF) alone and in combination with intact FSH on the *in vitro* ovary.

METHODS

Tissue. The ovaries were taken from mice of the small but prolific CFW strain (M.R.C. Laboratory Animals Centre, Carshalton, Surrey). The stock was systematically random-bred to maintain heterozygosity. The animals intended for experimental use were reared from birth in pooled foster litters of ten females (Smith and Ryle, 1968). One such foster litter was used for each experiment at the age of 15 days. Their weights ranged from 6.0 to 7.5 g, the biggest range within any one foster litter being 6.0 to 7.0 g. The mice were killed abruptly and their ovaries removed rapidly and aseptically.

Culture procedure. Each ovary, weighing 0.25 to 0.5 mg, was placed singly on a stainless steel mesh support in a small petri dish containing 4 ml of fluid. A modified Eagle's medium was used (Vantsis and Wildy, 1962) mixed with 5 per cent serum from a hypophysectomized sheep and the appropriate concentrations of FSH or FSH-NF. The dishes were put in closed jars, gassed with 5 per cent CO_2 in air, and sealed at 2 lb/sq. in. above atmospheric pressure. Before adding the tissue they were incubated at 37°C, overnight in Expts. I and III and for approximately 2 to 4 hours in Expts. II and IV, to allow pH and temperature to stabilize. The final pH was approximately 7.6. After adding the tissue the jars were re-gassed, sealed and incubated as before.

Addition of thymidine. The cultures were incubated overnight before adding the tracer to allow time both for the hormone to diffuse through the tissue and for the effects of the animal's own gonadotrophins on DNA formation to die away. This decline was estimated from earlier work to require about 9 to 12 hours (Ryle, 1969c). Each dish then received 0.05μCi of (2-^{14}C) thymidine (56 mCi/mM; Radiochemical Centre, Amersham) in 0.1 ml Eagle's medium. The jars were re-gassed, sealed and incubated for a further 3 or 4 days.

Sample preparation and counting. At the end of the culture period each stainless steel support was irrigated with about 5 ml of physiological saline. The ovary from it was then dissolved by overnight incubation at 37°C in 0.5 ml hydroxide of hyamine-10X (Packard Instrument Ltd.). The solution was taken up in 6 ml. of the scintillating fluid (0.25 per cent w/v PPO, 0.005 per cent w/v POPOP, in toluene) and transferred to a counting vial. The carbon-14 content of each sample was measured in a Packard Liquid Scintillation Counter and the crude counts corrected for background and, by automatic external standardization, for quenching.

Hormone preparations. Two preparations of FSH-NF were tested. The starting material, the numbers of the corresponding culture experiments and the dates on which media were set up and tissue added are shown in Table I. Preparation A was prepared from CP 1 (Butt *et al.* 1961): Preparation B from CP 150 (Chaplin *et al.* 1970). Neither showed biological activity *in vivo* but their immunological activity was unchanged when assayed shortly after preparation (Butt *et al.* 1970). The concentrations quoted in this work are based on these immunoassays and are referred to in 'international units'. However Sample 1 of Preparation B (Table I) proved to have only 10 per cent of its expected immunological activity when re-checked shortly after the culture experiment was carried out. Preparation A was diluted with Eagle's medium to 50 'i.u.'/ml and each of the two samples of Preparation B to 10 'i.u.'/ml before passage through Millipore filters for sterilization. Each was diluted further as required when setting up the cultures.

In all experiments dishes were also set up with starch-gel purified FSH (Butt *et al.* 1963) to check the ovaries for normal responsiveness. The activity of this preparation was estimated by the mouse ovarian augmentation assay. It was then diluted to 10 i.u./ml before Millipore filtration. Material prepared in this way contained less than 0.012 i.u. LH to 1 i.u. FSH, as estimated by the rat ovarian ascorbic acid depletion assay.

Experiment design. Two designs were used. The first (Expts. I and IV) was intended to check for any toxic effect and, if none was apparent, to test for a competitive inhibitory type of action. In each of these two experiments the mice were divided into two groups. In the first group one ovary of each mouse received no hormone and the other

TABLE I

Details of preparations of neuraminic acid-free follicle stimulating hormone (FSH–NF).

FSH–NF Preparation	Starting material	Used in expt. no.	Date medium set-up	Date ovaries explanted	Duration of culture after ^{14}C–thymidine added
A	CP 1	I	21.3.69	22.3.69	3 days
A	CP 1	II	9.4.69	9.4.69	4 days
B Sample 1	CP 150	III	13.5.69	14.5.69	4 days
B Sample 2	CP 150	IV	11.6.69	11.6.69	4 days

received a standard dose of FSH-NF. In the second group every ovary received a standard dose of normal FSH and one from each mouse received in addition the standard dose of FSH-NF.

The second design (Expts. II and III) was aimed at estimating the relative potency of the FSH-NF in terms of normal FSH. The mice were divided into three groups. In the first, one ovary of each mouse was given the lowest dose of normal FSH and the other the lowest dose of FSH-NF. The ovaries of the second group were similarly allocated to the intermediate doses and the ovaries of the third group to the highest doses of each preparation.

RESULTS

The results of Expts. I and IV are shown in Table II. The modified preparation, FSH-NF, did not inhibit the action of normal FSH and in fact substituted for it or supplemented its activity. In every case where data were available from both ovaries of individual mice (13 mice) the ovary incubated with FSH-NF contained more thymidine than its partner incubated without FSH-NF. The standard errors of the mean counts in Table II indicate that the variation between ovaries of different mice receiving the same treatment was relatively small.

The results of Expts. II and III are shown in Table III. In both experiments the uptake of thymidine increased linearly with increasing concentrations of FSH. The effects of FSH-NF in Expt. II appeared to be nearly maximal but the slope of the concentration/uptake curve for the two lower doses was not significantly different from that of the three concentrations of FSH. The calculated potency of the FSH-NF from this 3:2 assay was 45(20-200) per cent of that of FSH. In Expt. III the effects of the FSH-NF were slight and the slope of the concentration/uptake curve was low, but was not significantly lower than that of the FSH. The calculated potency in this case was only 7.6(1.8-15.8) per cent of intact FSH.

DISCUSSION

It was anticipated that FSH-NF might have an inhibiting effect on the activity of normal FSH. This could have been tested *in vivo* but negative results might have been misleading since absence of inhibiting could have arisen from impaired transport. This possibility was avoided by testing *in vitro* where the tissue was flooded by simple diffusion of FSH-NF from the medium. However, when the ovaries were removed from the culture dishes in the first experiment, although one of each pair was obviously larger than the other, it was, surprisingly, found to be the one treated with FSH-NF. This visual impression suggests that

TABLE II

Mean cpm/ovary and standard errors of means in the treatment groups of Experiments I and IV

(In brackets, number of ovaries in group)

		– FSH–NF	+ FSH–NF (5.0 'i.u.'/ml)
EXPT. I	– FSH	1632 $\pm$ 9 (3)	3164 $\pm$ 324 (3)
	+ FSH (0.5 i.u./ml)	3048 $\pm$ 214 (2)	3577 $\pm$ 314 (3)

		– FSH–NF	+ FSH–NF (2.5 'i.u.'/ml)
EXPT. IV	– FSH	1705 $\pm$ 139 (4)	2216 $\pm$ 76 (5)
	+ FSH (1.0 i.u./ml)	3323 $\pm$ 442 (4)	3915 $\pm$ 296 (5)

TABLE III

Mean cpm/ovary and standard errors of means in the treatment groups of Expts. II and III
(In brackets, numbers of ovaries in group)

	FSH		FSH–NF	
EXPT. II	0.125 i.u./ml.	3278 ± 140 (3)	1.25 'i.u.'/ml	4732 ± 489 (3)
	0.25 i.u./ml	3613 ± 290 (3)	2.5 'i.u.'/ml	5027 ± 271 (3)
	0.5 i.u./ml	5008 ± 277 (3)	5.0 'i.u.'/ml	4994 ± 284 (3)
EXPT. III	0.125 i.u./ml	2499 ± 242 (3)	0.5 'i.u.'/ml	1985 ± 210 (3)
	0.25 i.u./ml	2570 ± 385 (3)	1.0 'i.u.'/ml	2013 ± 360 (2)
	0.5 i.u/ml	3908 ± 146 (3)	2.0 'i.u.'/ml	2243 ± 382 (3)

Index of precision (λ) Expt. II 0.23
Expt. III 0.39

Validity Tests:

No significant corvature in Expt. II for FSH or in Expt. III for FSH or FSH NF

No significant deviation from parallelism between the 3 levels of FSH and 1.25 'i.u.' and 2.5 'i.u.'/ml of FSH–NF in Expt. II, or between FSH and FSH–NF in Expt. III.

the increased uptake of thymidine due to FSH-NF, like that due to normal FSH, was associated with increased DNA formation and cell division. Nevertheless the measurement of thymidine incorporation into DNA is necessary to confirm that the stimulation is of the usual type, and that the ability to accelerate ovarian cell division is not removed along with NANA. Additional studies to establish more precisely the relative potency of the modified material are also required. Meanwhile, on the basis of these results, using two separate preparations of FSH-NF, we will provisionally conclude that this material, totally inactive *in vivo* in the ovarian augmentation assay, successfully activates mouse ovaries *in vitro.*

What then is the function of the NANA? Why is it necessary for effective stimulation *in vivo?* It is not required to provoke the normal response once it reaches the ovary. Moreover, since FSH-NF is effective *in vitro* in concentrations of the same order as the normal hormone, it is unlikely to be of great importance for recognition of and attachment to the target site. It seems probable that it is essential for successful transport from the sub-cutaneous injection site to the ovary. If it is a question of facilitating passage into the blood stream - which does not seem very probable - a positive ovarian augmentation assay should result from intravenous injection. This needs to be tested. Alternatively, the removal of NANA may make the molecule more vulnerable to enzymic degradation *en route,* though in this context it should be remembered that 5 per cent sheep serum was present in the culture medium.

Immunoassays have shown a gradual loss in the activity of FSH-NF over nine weeks, during storage at -20°C, and it has been suggested previously that this may result from progressive changes in the tertiary molecular structure (Butt *et al.* 1970). Presumably any such changes would be accelerated in solution at 37°C.Yet even in Expt. I, where the dishes of medium were set up and put in the incubator the day before the ovaries were explanted to them, a clear-cut positive response was obtained. Moreover, there was also a marked response in Expt. IV six weeks after preparation of the FSH-NF, when its estimated potency by immunoassay was only 7 per cent of the original. Is the loss of immunological activity therefore not necessarily associated with that of *in vitro* biological activity? The unexpectedly low levels of both types of activity in the sample of Preparation B used in Expt. III may have arisen from some different type of deterioration.

The results presented here perhaps raise more questions than they answer, but they narrow slightly the range of possible correlations between structure and function that one may postulate for the FSH molecule. It is hoped that similar studies, in conjuction with *in vivo* assays and with immunological work, using FSH modified in other ways, may help to restrict this range of possibilities yet further.

ACKNOWLEDGEMENTS

We wish to thank Dr A. C. Crooke, Dr W. R. Butt and Dr S. A. Barker for their advice and encouragement in the course of this work and Mr S. S. Lynch for the immunoassays and for helpful discussions during preparation of the paper. Dr P. C. Leglise and Dr Martinet (C.N.R.Z., Jouy-en-Josas, France) kindly supplied the hypophysectomized sheep serum. The work was supported in part by the Ford Foundation.

REFERENCES

Butt, W.R., Crooke, A.C. & Cunningham, F.J. (1961). *Biochem. J.* **81,** 596.

Butt, W.R., Crooke, A.C., Cunningham, F.J. & Wolf, A. (1963). *J. Endocr.* **25,** 541.

Butt, W.R., Lynch, S.S., Chaplin, M.F., Gray, C.J. & Kennedy, J.F. (1970). This volume, page

Chaplin, M.F., Gray, C.J. & Kennedy, J.F. (1970). This volume, page 77.

How, M.J., Chaplin, M.F. & Ryle, M. (1970). *Biochim. Biophys. Acta.* **213,** 226.

Ryle, M. (1969a), *J. Reprod. Fert.* **19,** 87.

Ryle, M. (1969b), *J. Reprod. Fert.* **20,** 307.

Ryle, M. (1969c), *J. Reprod. Fert.* **19,** 349.

Smith, M.J. & Ryle, M. (1968). *J. Inst. anim. Tech.* **19,** 74.

Vantsis, J.T. & Wildy, P. (1962), *Virology,* **17,** 225.

Papain Digestion of FSH from Human Pituitary Glands

P. Rathnam and B. B. Saxena

Highly purified follicle stimulating hormone (FSH) has been prepared from human putuitary glands (Butt *et al.* 1961; McShan *et al.* 1964; Reichert and Parlow, 1964; Roos and Gemzell, 1965; Papkoff *et al.* 1967; Saxena and Rathnam, 1967 and 1968). It has been shown that the biological activity of FSH is resistant to trypsin, chymotrypsin and papain (Amir *et al.* 1966; Butt, 1967; Saxena *et al.* 1968; Reichert, 1969). This paper describes the separation of a biologically active component from FSH by papain digestion.

MATERIALS AND METHODS

Human pituitary FSH was prepared from acetone preserved glands supplied by the National Pituitary Agency, Baltimore, Md., according to the procedures described earlier (Saxena and Rathnam, 1967 and 1968). A recrystallized preparation of papain was obtained from Mann Research Lab. Inc., N.Y. All reagents were of analytical grade and prepared in glass distilled water.

Digestion of FSH with papain. In the light of known non-specificity of proteolytic attack of papain on proteins in general (Bailey, 1967), the experimental conditions were standardized to obtain optimum and reproducible digestion of FSH with papain. In a typical experiment, 10 mg of FSH were weighed in a 10 ml Erlenmeyer flask and dissolved in 2 ml 0.1M phosphate buffer of pH 6.0 containing 1mM disodium EDTA (Fischer Scientific Co., N.J.). To this solution 0.5 mg of papain was added to yield an enzyme to substrate ratio of 1:20. To activate the enzyme, the solution was made 5 mM in cysteine hydrochloride (van Camp Lab., Terminal Island, Calif.) and the solution was re-adjusted to pH 6.0 with 0.1N NaOH. The solution was incubated for 24 hrs. at 37°C in a Dubnoff shaker. At the end of the incubation, aliquots of the enzymatic digest were analyzed for ninhydrin reactive material (Rosen, 1957) using norleucine as standard as well as for the presence of peptides by treatment with 5-dimethyl amino 1-naphthalene sulphonyl chloride (DNS-C1) according to the method of Gray and Hartley (1963). The DNS-derivatives were separated by thin layer chromatography on silica gel. The chromatogram was developed in benzene:

pyridine:acetic acid (1:4:1) and examined for the number of fluorescent spots under UV light. A blank and a control experiment with untreated FSH was performed simultaneously.

Fractionation of the papain digest of FSH by gel-filtration on Sephadex G-25 (40 to 120 μ.). The papain digest of the FSH was applied to a 1 x 100 cm column of Sephadex G-25 (Pharmacia, Piscataway, N.J.). The column was equilibrated with 1 mM phosphate buffer of pH 7.0 and maintained at 4°C. The column was eluted at the rate of 6 ml/hr by the same buffer using a refrigerated fraction-collector and a UV-cord (LKB Inst., Stockholm, Sweden) to monitor the absorption of the eluate at 280 mμ. The fractions 1 to 4 were pooled, as shown in Figure 1, analyzed for total protein (Gornall *et al.* 1949; Lowry *et al.* 1951) and hexoses (Francois *et al.* 1962) as well as assayed for biological and immunological activities.

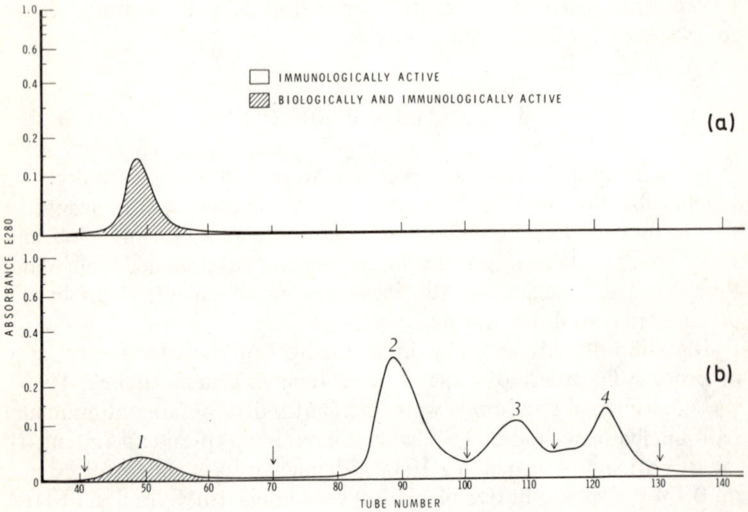

Figure 1. Gel-filtration on Sephadex G-25 of (a) 5 mg FSH; (b) papain digest of 10 mg FSH.

Ion-exchange chromatography of the fraction 1 from Sephadex G-25 column. Dowex 50-X2, cation exchange resin (Biorad, Richmond, Calif.) was washed repeatedly in water to remove fine particles. A 1 x 100 cm column was filled with the resin under 200 psi and maintained at 4°C. The resin was washed with 0.2N NaOH and equilibrated with a 0.2N sodium citrate buffer of pH 5.1. The fraction 1 from Sephadex G-25 was dissolved in 0.5 ml of the same buffer and applied to the column. A linear gradient of pH 5.1 to 10.5 using 0.2N sodium citrate buffer was used to elute the column at a flow rate of 0.5 ml per minute.

Using a stream splitter, one-fifth of the column eluate was conveyed to an Autoanalyzer (Technicon Corp., Ardsley, N.Y.) to determine the ninhydrin reactive material as norleucine and four-fifths of the eluate was collected through a 280 mμ UV-cord, on a refrigerated fraction-collector. The fractions 1, 2 and 3 were pooled as shown in Figure 2, lyophilized and desalted on a Sephadex G-25 column as described earlier. Each fraction was analyzed for total protein content as well as biological and immunological activites.

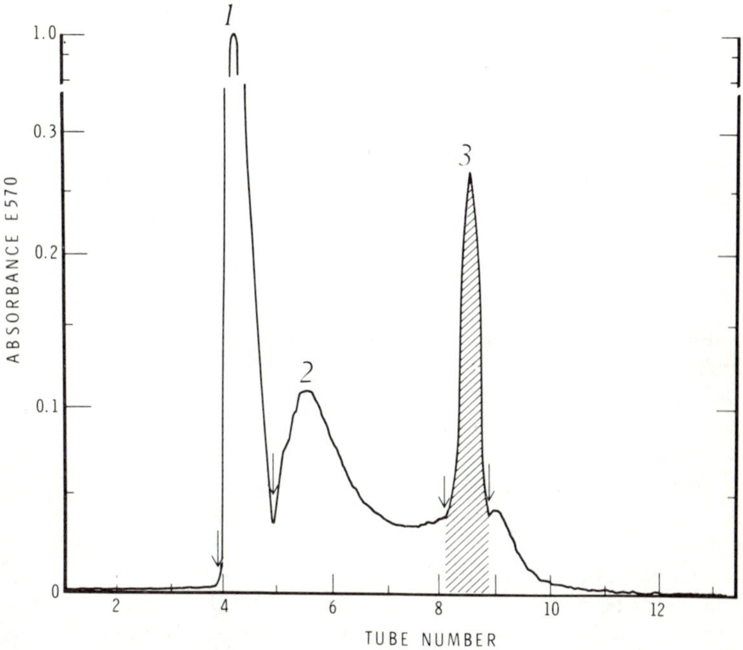

Figure 2. Ion-exchange chromatography of fraction 1 (4.36 mg) from Sephadex G-25 on Dowex 50-X2. The hatched area representing fraction 3 contained FSH activity.

Biological and immunological activities. The follicle stimulating activity of the hormone and its components obtained after enzymatic digestion was determined by the ovarian augmentation method (Steelman and Pohley, 1953) using 21-day old rats obtained from the Holtzman Company, Madison, Wisc. The hormone and its components were also tested in the radioimmunoassay of FSH and LH (Saxena *et al.* 1968). Cross- reaction with antisera to FSH and HCG was examined by immunoelectrophoresis in agar gel (Scheidegger, 1955).

Adsorption of FSH on glass and plastic. There exists a wide range in the biopotency estimates of highly purified preparations of FSH from different laboratories (Amir *et al.* 1966; Butt, 1967; Reichert, 1969;

Saxena and Rathnam, 1968). A similar variation has been encountered in the potency estimates of FSH before and after lyophilization (Amir *et al.* 1966) as well as when lyophilized material was assayed for 6 months at 2 month intervals (Saxena and Rathnam, 1968). The variations among preparations and animal response may account only partly for the discrepancies. Hence, experiments were undertaken to establish the extent of adsorption of FSH on glass and plastic in the presence of various solvents as well as its effect on the biological activity of FSH (Table 1). Ten thousand cpm of a ^{131}I-FSH were allowed to remain for 24 hr. in the vials in a total volume of 500 μl of the solvent. The vials were then emptied, washed three times with respective solvents and counted in an autogamma well; the per cent recovery of the labelled hormone was calculated. The FSH was also dissolved in various solvents and assayed for biological activity.

Amino acid and Hexosamine analyses. Two 1 mg aliquots of fraction 3 from the ion-exchange column were dissolved in 2 ml of 5.7N HC1 in two separate glass tubes; the tubes were evacuated and sealed under nitrogen. The contents of one tube were hydrolyzed for 24 hr. while those of the other were hydrolyzed for 72 hr. at 105°C. The hydrolysates were dried in a desiccator under vacuum, dissolved in water and lyophilized to remove residual HC1. The 24 and 72 hr. hydrolysates were redissolved in 0.5 ml of 0.01N HC1 and analyzed for amino acids and hexosamines on an Autoanalyzer according to the procedures of Spackman *et al.* (1958) and Hamilton (1963), respectively.

Neutral sugar and sialic acid determination. An aliquot of 0.5 mg of fraction 3 from the ion-exchange column was dissolved in 0.5 ml of water adjusted to pH 2.0 with 0.01N HC1. The solution was incubated at 37°C for 24 hr. to liberate sialic acid (Neuberger and Marshall, 1966). The incubate was lyophilised and redissolved in 0.1 ml of water. The residual protein was precipitated by the addition of 5 ml of chilled 95 per cent ethanol. The supernatant was dried under a stream of nitrogen and analyzed for sialic acid by the resorcinol method of Svennerholm (1957). The precipitate was dissolved in 1N HC1 and hydrolyzed at 100°C for 6 hr. The hydrolysate was dried under vacuum and analyzed for neutral sugars (Francois *et al.*, 1962).

RESULTS AND DISCUSSION

Enzymatic digestion. As shown in Table II, the generation of ninhydrin reactive material and DNS-derivatives during papain digestion were indicative of proteolytic action of papain on FSH.

Gel filtration of the 'papain digest of FSH'. The gel filtration of the enzymic digest on Sephadex G-25 is shown in Figure 1. Fraction 1 contained the biological activity and showed 55 per cent immunological

TABLE I

Percent Recovery of FSH in the Presence of Various Solvents

Solvent	Glass vial [1]		Plastic vial [1]		Glass vial [2]
	Untreated	Siliconized	Untreated	Siliconized	
Saline	55	54	45	70	72 ± 2.8
Saline + 0.1% albumin	99	98	99	98	124 ± 15.8
Saline + 1% mannitol	78	51	61	79	100 ± 1.7
Saline + 1% lactose	79	58	52	74	22 ± 0.8
Saline + 1% sucrose	83	69	65	63	–

1 Represents average percent recovery of 10,000 c.p.m. of ^{131}I-FSH (SA 300 uC/ug), from three experiments.

2 Represents percent recovery of 185 (160-213) NIH-FSH-S3 units $\pm$ standard error.

TABLE II

Papain Digestion of Human Pituitary FSH

	Yield mg	FSH u./mg	Total FSH units	% recovery	No. DNS derivatives	Norleucine uM/mg
Untreated FSH	10	185 ± 10	1850	100	none	trace
Papain-treated FSH[2] pH 6.0, 24 hr, 37°C	–	136 ± 1.6	1360	74	5	12.7
Sephadex G-25 (fraction 1)	4.36	99 ± 10.4	432	24	3	10.0
Dowex 50-X2 (fraction 3)	1.54	132 ± 4.2	203	11	1	4.5

[1] NIH-FSH-S3 u./mg ± standard error.

[2] Enzyme to substrate ratio, 1:20

TABLE III

Amino acid and Carbohydrate Analyses of FSH and fraction 3 from the Ion-Exchange column

Amino acid[1]	FSH (24 hr hydrolysate)	Fraction 3 (24 hr hydrolysate)	Fraction 3 (72 hr hydrolysate)
Lys	4.5	2.6	2.3
His	2.6	1.3	1.1
Arg	3.7	2.2	2.2
Asp	7.3	3.2	3.9
Thr	2.0	2.7	3.3
Ser	4.0	2.0	2.1
Glu	10.9	3.7	3.5
Pro	6.2	2.9	2.3
Gly	3.4	1.2	1.1
Ala	4.7	1.6	1.4
1/2 Cys	5.6	3.1	2.4
Val	5.3	2.2	2.0
Met	2.3	0.8	0.6
Ileu	2.9	1.7	1.6
Leu	5.9	2.4	2.0
Tyr	5.3	2.8	2.0
Phe	5.0	2.1	1.6
Try	0.76	–	–
Carbohydrate[1]			
Hexose	6.1	6.8	–
Sialic acid	5.0	3.4	–
Glucosamine	5.3	28.8	–
Fucose	1.2	–	–

[1] As g/100g. Uncorrected for moisture, ash or hydrolysis.

cross-reaction in the radioimmunoassay of FSH. Fractions 2, 3 and 4 were devoid of FSH activity but showed 65, 60 and 24 per cent cross-reaction, respectively, in the FSH radioimmunoassay. Fractions 1, 3 and 4 did not cross-react, whereas fraction 2 did show 15 per cent cross-reaction in the radioimmunoassay of LH. As shown in Figure 3, fraction 1 showed a single precipitin line in immunoelectrophoresis using anti-FSH serum and similar results were obtained with fractions 2, 3 and 4. These results suggest that biological and immunological activities of FSH may reside at different loci of the molecule. Fractions 1, 2, 3 and 4 contained 37, 13, 15 and 35 per cent, respectively, of the total hexoses contained in the papain digest. The presence of carbohydrates in all fractions also suggests that the sugar residues may be linked at more than one point in the protein chain of the FSH molecule.

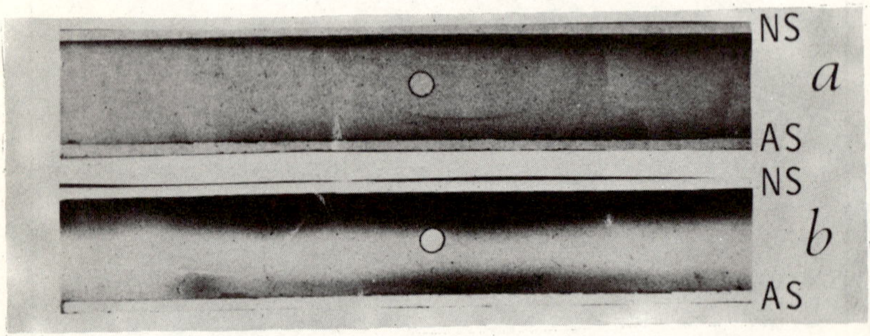

Figure 3. Immunoelectrophoretic patterns of (a) fraction 1 from the Sephadex G-25 column, and (b) fraction 3 from the Dowex 50-X2 column. The wells contain normal rabbit serum (NS) or the antisera to FSH (AS).

Ion-exchange chromatography of fraction 1 from the Sephadex G-25 column. As shown in Figure 2, fractions 1 and 2 were inactive and the FSH activity was contained in fraction 3.

Yield and activity. A fresh solution of FSH, isolated from the polyacrylamide column and subsequently desalted on Sephadex G-25, contained an average of 185 u./mg (Saxena and Rathnam, 1967). When the lyophilized FSH was dissolved in saline, an average of 133 u./mg were recovered (Saxena and Rathnam, 1968), suggesting a loss of approximately 28 per cent of the original activity. However, when the FSH was dissolved in saline containing 0.1 per cent albumin, 124 per cent of the original activity was recovered (Table 1). The recovery of FSH in the presence of mannitol was greater than in saline alone. Similarly, 99 per cent of the labelled FSH was recovered from the vials with phosphate buffer containing 0.01 per cent albumin. In the light of these observations all preparations were assayed in saline containing 0.1 per cent albumin.

The papain digest of 10 mg of FSH contained 136 u./mg, and 1360 total units, representing 74 per cent recovery of the starting units. Subsequent gel filtration on Sephadex G-25, yielded 4.36 mg protein in fraction 1 containing 99 u./mg, representing 432 total units and 24 per cent recovery of the starting units. The FSH activity was recovered in 1.54 mg protein of fraction 3 from the ion-exchange column. The specific activity of this component was 132 u./mg, which is greater than the specific activity of fraction 1 from Sephadex G-25 column and less than the activity of the untreated FSH. However, only 203 total units were recovered representing a yield of only 11 per cent of the starting units. The loss in total FSH units during papain digestion, gel filtration and ion-exchange chromatography may partly be due to the loss of material during experimentation, however, the decrease in specific activity appears to be due to the fragmentation of the molecule.

Comparison of amino acid and carbohydrate analyses of untreated FSH with fraction 3 from the ion-exchange column. (Table III). In both preparations, the quantities of aspartic and glutamic acids were higher than the other amino acids. With the exception of threonine, other amino acids in fraction 3 were approximately half of those in the FSH. There was a noteworthy increase in the hexosamine content of fraction 3 as compared with that of FSH. No significant difference in the neutral sugar content of fraction 3 and FSH was found.

ACKNOWLEDGEMENTS

Thanks are due to Miss Mary Ann Davis for excellent technical assistance. NIH-FSH-S3 and 2nd-IRP-HMG were obtained by the courtesies of the Endocrine Study section of the National Institutes of Health, Bathesda, Md., and Dr D. R. Bangham of Medical Research Council, National Institutes for Medical Research, Mill Hill, London, N.W., respectively.

Professor Saxena is a Career Scientist Awardee of Health Research Council, NYC, Contract 1-621. This study has been supported by USPHS grants AM 11187 and TIAM 5350, Population Council grant M 69-18, and American Cancer Society grant P-440.

REFERENCES

Amir, S.M., Barker, S.A., Butt, W.R. & Crooke, A.C. (1966). *Nature, Lond.* **209,** 1092.

Bailey, J.L., (1967). In *Techniques in Protein Chemistry,* 2nd ed. p. 149, New York: Elsevier.

Butt, W.R., (1967). In *Recent Research on Gonadotrophic Hormones,* p. 167. Ed. Bell, E.T. & Loraine, J.A. Edinburgh: Livingstone.
Butt, W.R., Crooke, A.C. & Cunningham, F.J. (1961). *Biochem. J.* **81,** 596.
Francois, C., Marshall, R.D. & Neuberger, A. (1962). *Biochem. J.* **83,** 335.
Gornall, A.G., Bardawill, C.J. & David, M.M. (1949). *J. biol. Chem.* **177,** 751
Gray, W.R. & Hartley, B.S. (1963). *Biochem. J.,* **89,** 50**P.**
Hamilton, P.B. (1963). *Analyt. Chem.* **35,** 2055.
Lowry, O.H., Rosebrough, N.J., Farr, A.L. & Randall, R.J. (1951). *J. biol. Chem.* **193,** 265.
McShan, W.H., Saxena, B.B. & Creek, R.O. (1964). *Can. J. Biochem. Physiol.* **42,** 841.
Neuberger, A. & Marshall, R.D., (1966). In *Glycoproteins, their Composition. Structure and Function,* p. 197. Ed. Gottschalk, A. New York. Elsevier.
Papkoff, H., Mahlman, L.J. & Li, C.H. (1967), *Biochemistry, Wash.* **6,** 3976.
Reichert, L.E. Jr. (1969). *Endocrinology,* **84,** 9.
Reichert, L.E. Jr. & Parlow, A.F. (1964). *Endocrinology,* **74,** 236.
Roos, P. & Gemzell, C.A. (1965). In *Gonadotrophins: physiocochemical and Immunological Properties,* p. 11, Ciba Fdn. Study Grps. No. 22 London: Churchill.
Rosen, H. (1957). *Archs. Biochem. Biophys.* **67,** 10.
Saxena, B.B., Demura, H., Gandy, H.M. & Peterson, R.E. (1968). *J. clin. Endocr. Metab.* **28,** 519.
Saxena, B.B. & Rathnam, P. (1967). *J. biol. Chem.* **242,** 3769.
Saxena, B.B. & Rathnam, P. (1968). In *Gonadotrophins 1968,* Ed. Rosemberg, E. p.3, Los Altos, California: Geron-X.
Scheidegger, J. (1955). *Int. Archs. Allergy appl. Immun.* **7,** 103.
Spackman, D.H., Stein, W.H. & Moore, S. (1958). *Analyt. Chem.* **30,** 1190.
Steelman, S.L. & Pohley, F.M. (1953). *Endocrinology,* **53,** 604.
Svennerholm, L. (1957). *Biochim. Biophys, Acta,* **24,** 604.

Amino Acid Composition of Some Purified Urinary LH Preparations

David G. Anderson, Piero Donini and Vernon C. Stevens

Donini, Montezemolo and Puzzuoli (1964) have extracted and purified LH from human menopausal gonadotrophin (HMG). Further purification by Donini (1967a, b), Donini *et al* (1968a, b), has yielded LH of high purity. The carbohydrate composition and some physical characterization measurements have been reported on this material (Donini *et al.* to be submitted).

Human pituitary luteinizing hormone (LH) has been purified and the amino acid composition reported by Hartree (1967), Butt (1967) and Kathan *et al* (1967). As yet no amino acid composition has been reported for urinary LH. The present investigation was undertaken to purify urinary LH sufficiently to warrant a study of its amino acid composition.

MATERIALS AND METHODS

Purification methods. Human menopausal gonadotrophin (HMG) was purified by methods described elsewhere (Donini *et al.* 1964). This material was further purified by batchwise separation of FSH and LH on DEAE - cellulose (Roos and Gemzell, 1964)followed by a second batchwise separation on CMC-70 (Reichert, 1967). Subsequently the material is passed over Sephadex G-100 (Donini *et al.* 1968). The fraction of highest specific activity is passed over a DEAE-cellulose column equilibrated with 0.002 M phosphate buffer, pH 6.6. The LH eluted in 0.1 M NaCl from the DEAE-cellulose column is passed through a Sephadex G-200 column equilibrated with 0.01 M phosphate and 0.05 M NaCl buffer, pH 6.9.

In addition to the steps carried out above, the LH material obtained from the Sephadex G-200 was fractionated with SE-Sephadex C-50 and by preparative polyacrylamide electrophoresis. A column 1.2 x 28 cm of SE-Sephadex C-50 was equilibrated with 0.0125 M sodium diethyl barbiturate, 0.0025 M diethyl barbituric acid and 0.0125 M sodium acetate buffer (BA), pH 8.6. A fraction from the Sephadex G-200 fractionation was applied to the column and eluted stepwise with 0.1 M and 0.5 M NaCl in the above buffer.

Preparative gel electrophoresis was done in 10 per cent polyacrylamide gel buffered with 0.05 M triethylamine hydrochloride (TEA), pH 10.4. The electrode buffers were of the same composition. The gel was composed of 10 per cent (w/v) acrylamide, 0.12 percent (w/v) methylenebisacrylamide, 0.03 per cent (v/v) tetramethylethylenediamine, 0.03 per cent (w/v) ammonium persulphate and buffer. After a sample (20 mg) was dissolved in one-eighth the concentration of the TEA buffer containing sucrose, the solution was layered on 2.5 cm of gel and subjected to a current of 60 mA at 70 volts. Fresh buffer was pumped into the electrode vessels to maintain a constant pH during the electrophoresis. All fractionations, collections, and dialysis were done below 10°C. The protein fractions were dialysed and lyophilized.

Analytical polyacrylamide electrophoresis. Two procedures were used. Electrophoresis in tris (hydroxymethyl) aminomethane (THAM) and glycine buffer, pH 8, was done under the conditions described by Davies (1964). Electrophoresis at pH 10.4 was done using 'small pore gel' only of a composition similar to that of Davies except that the buffer was 0.05 M TEA in the gel and the electrode compartments. The samples were applied to the gel in sucrose solution and diluted buffer as described by Hjerten *et al.* (1965).

Assays. Bioassays were determined by ovarian ascorbic acid depletion assay (OAAD), (Parlow, 1961), and uterine weight increase of immature mice (Loraine and Brown, 1954). The radioimmunoassays of the material prior to the SE-Sephadex fractionation were done by the method of Donini *et al.* 1968a. Subsequent radioimmunoassays were done by methods described elsewhere (Stevens 1969). The second IRP was used as a standard.

Immunoelectrophoresis. Immunoelectrophoretic analyses were performed in agar buffered with 0.025 ionic strength sodium diethylbarbiturate, and sodium acetate (BA), pH 8.6 (Little and Donahue, 1968). A potential gradient of approximately 20 V/cm was applied for one hour.

Bovine anti-HCG was used to detect LH in the gel and was obtained by immunizing animals against 3000 i.u./mg human chorionic gonodotrophin (HCG).

Amino acid analysis. Hydrolyzed samples were analysed in a Technicon Amino Acid Analyzer. The system used was essentially that of Piez and Morris (1960) with some modifications of the buffer system. Samples were dried at 106°C for six hours in an evacuated desiccator containing $P_2 0_5$. Sealed tube hydrolysis was done according to the method of Moore and Stein (1963). The HC1 was partially removed from the hydrolysate and a small amount of distilled water added before it was evaporated to dryness in an evacuated vessel. Norleucine was used as an internal standard. Samples from the

SE-Sephadex fractionation were hydrolyzed 48 hours whereas samples from the polyacrylamide electrophoresis were hydrolysed 28, 52, and 100 hours.

RESULTS

Purification. Bioassay of the LH material eluted from batchwise separation on DEAE-cellulose gave 5.6 i.u./mg FSH and 64.8 i.u./mg LH. The biological activity after the batchwise purification on CMC-70 was 2.0 i.u./mg FSH and 107 i.u./mg LH. The fractionation on Sephadex G-100 gave a LH activity of 228 i.u./mg by radioimmunoassay. Urinary LH from the first DEAE-cellulose column (C-1) was found to have a biological activity of 753 i.u./mg but lost about 70 per cent of this activity during the first five months as is shown in Table I. However, the immunological activity (about 450 i.u./mg) determined by radioimmunoassay did not seem to change appreciably.

TABLE I

Bioassay and radioimmunoassay of lyophilized urinary LH, fraction C–1, stored for 5 months

Date	OAAD (a) i.u./mg	Radioimmunoassay (a) i.u./mg
July 3	753	433
July 5	622	464
July 26	383	--
Oct. 25	155	--
Oct. 30	142	--
Nov. 8	220	--
Dec. 20	--	400 (b)

(a) Donini

(b) Stevens

Figure 1 and Table II present results of Sephadex G-200 chromotography of fraction C-1. One major peak was found with a smaller peak of more rapidly eluted material. Fractions were pooled as indicated in Figure 1 and designated S-1 to S-7. The radioimmunological activity determined about a month later (by Stevens) was slightly lower but probably not significantly lower.

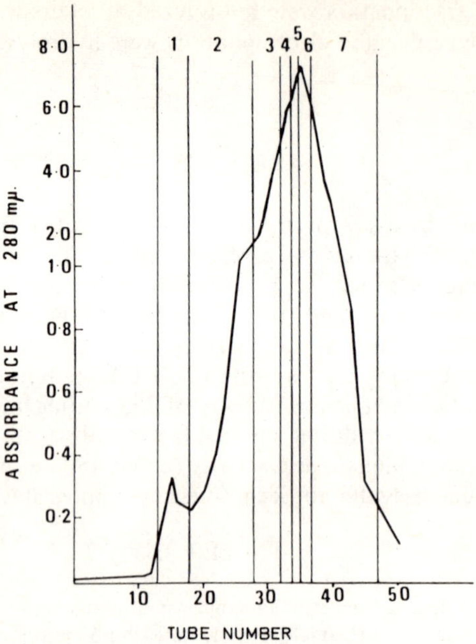

Figure 1, Molecular sieve chromatography of fraction C-1 on Sephadex G-200. Buffer: 0.01 M phosphate and 0.05 M NaCl, pH 6.9.

TABLE II

Bioassay and radioimmunoassay of lyophilized fractions from Sephadex G–200 chromatography

Fraction	OAAD (a) i.u./mg	Radioimmunoassay i.u./mg (a)	Radioimmunoassay i.u./mg (b)
S – 1	––	401	––
S – 2	23	94	––
S – 3	134	841	––
S – 4	155	973	880
S – 5	180	1234	907
S – 6	91	731	700
S – 7	––	196	––

(a) Donini

(b) Stevens

Fractionation of 80 mg of S-6 in SE-Sephadex is illustrated in Figure 2 and pooled in four fractions (SE-A, B, C, and D) as indicated.

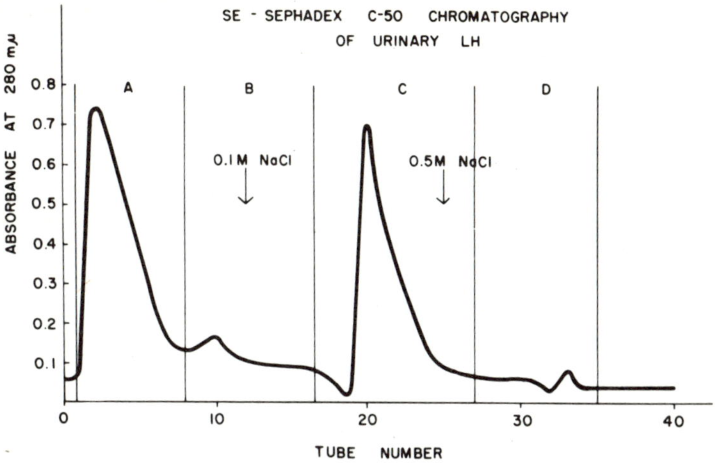

Figure 2, Chromatography of fraction S-6 on SE-sephadex C-50. Buffer: (1) 0.025 ionic strength sodium diethylbarbiturate and sodium acetate (BA), pH 8.6. (2) 0.1 M sodium chloride in BA. (3) 0.5 M sodium chloride in BA.

Fraction SE-D contained less than a milligram and was inactive material. Radioimmunoassay of fractions SE-A, B, and C gave specific activites of 100, 240, and 1100 i.u./mg respectively. Approximately 87 per cent of the material and 61 per cent of the activity was recovered. These results are presented in Table III. The sharp decrease in the absorbance observed just prior to the elution of fraction SE-C and D is probably due to adsorption of barbiturate ions on the SE-Sephadex.

TABLE III

Protein recovery and activity of fractions obtained from SE-Sephadex C-50 chromatography of fraction S–6

Material	Weight mg.	Radioimmunoassay specific activity i.u./mg	Total activity i.u.
Starting material	80.3	700	56,300
SE – A	32.6	100	3,300
SE – B	8.4	240	2,000
SE – C	28.7	1100	28,700
% Recovery	87		61

Each of these three fractions was subjected to Sephadex G-100 chromatography in BA buffer of the same composition used in the fractionation in SE-Sephadex. Elution volumes were approximately the same indicating a molecular weight of approximately 3 x 10^4. When portions of S-3, S-6, and S-7 were subjected to SE-Sephadex chromatography it was found that the ratio of the per cent protein found in the 0.1 M NaCl eluate to the activity in the respective starting material was roughly the same for each of the fractions. Thus, probably the protein found in the 0.1 M eluate was responsible for most of the activity of the starting material. It is interesting to note that urinary LH of high specific activity (810 i.u./mg) purified differently from that described in this report and from another lot of HMG was not adsorbed on SE-Sephadex under the same conditions used in this study.

The results of preparative electrophoresis of fraction SE-C is given in Figure 3. One broad major peak can be seen with material of lower

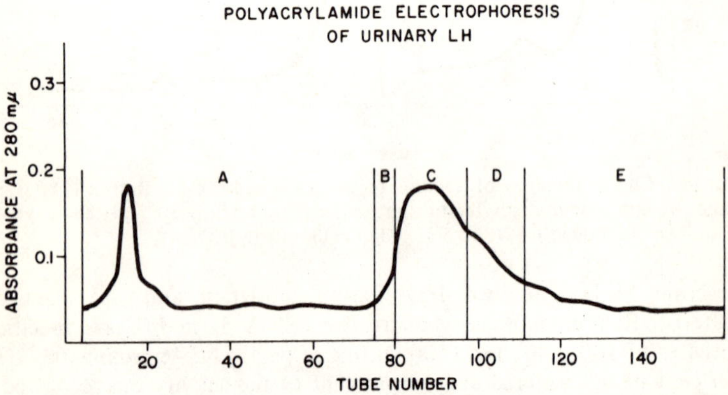

Figure 3, Preparative polyacrylamide electrophoresis of fraction SE-C from the SE-Sephadex fractionation. Gel concentration was 10 per cent. Buffer: 0.05 M triethylamine hydrochloride, pH 10.4.

mobility migrating behind. After the fractions were collected the gel was stained with amido-schwartz. No stainable protein remained in the gel. Five pools were made (E-A, B, C, D, and E). Yields and radioimmunological activities are summarized in Table IV. In another electrophoresis experiment LH active material was found throughout the main peak.

TABLE IV

Protein recovery and activity of fractions obtained from polyacrylamide electrophoresis of fraction (SE–C)

Fraction	Weight mg.	Specific activity i.u./mg
Starting material	20.0	1100
IE – A	1.5	–
E – B	0.5	–
E -- C	6.9	740
E – D	4.5	515
E – E	4.2	160

Analytical polyacrylamide electrophoresis. Figure 4 shows the results of electrophoresis of fraction E-C at pH 8 and 10. In both analyses one broad band can be seen. After electrophoresis of the fraction S-6, the stained gel contained at least two broad bands.

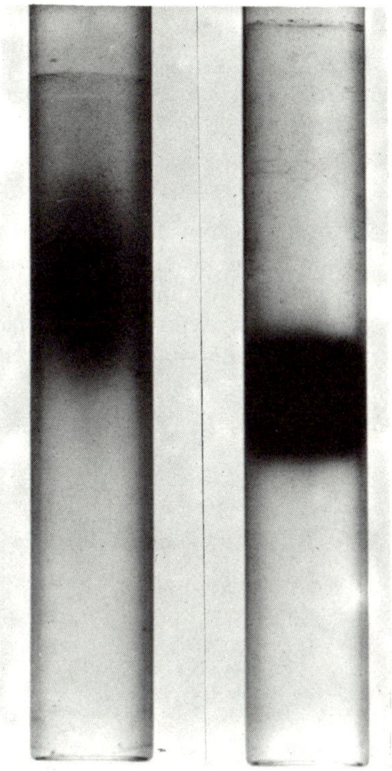

Figure 4 Analytical polyacrylamide electrophoresis of fraction E-C from the electrophoretic fractionation. In both cases a 7.5 per cent separating gel was used. Direction of electrophoresis is from top to bottom. Left: Electrophoresis at pH 8-9 by the method of Davies, (1964). Right: Electrophoresis at pH 10.4 in 0.05 M triethylamine hydrochloride.

Immunoelectrophoresis. The immunoelectrophoretic patterns are shown in Figures 5, 6, and 7. Fractions S-3, 4, 5, 6, and 7 gave essentially identical patterns, two long arcs of cross-reacting components and one non or partially cross-reacting component of lower mobility as can be seen in Figure 5. If a larger amount of antigen is analyzed, the two components of higher mobility merge and appear as one long arc.

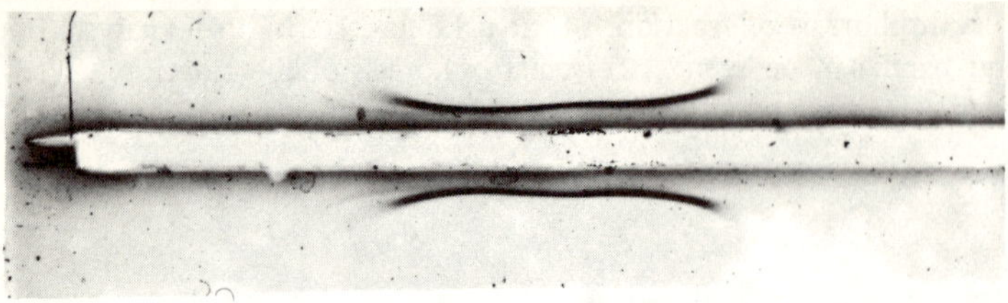

Figure 5, Immunoelectrophoresis of fraction S-6 from the Sephadex G-200 fractionation. Electrophoresis is from left to right. The trough contained bovine anti-HCG and each well contained the antigen.

In Figure 6 are shown the patterns formed by the fractions from the SE-sephadex fractionation. The lower mobility component is present only in fraction SE-B, whereas fractions SE-A and C both contain the two cross-reacting components. After polyacrylamide electrophoresis the two cross-reacting components were still in fractions E-C, D, and E.

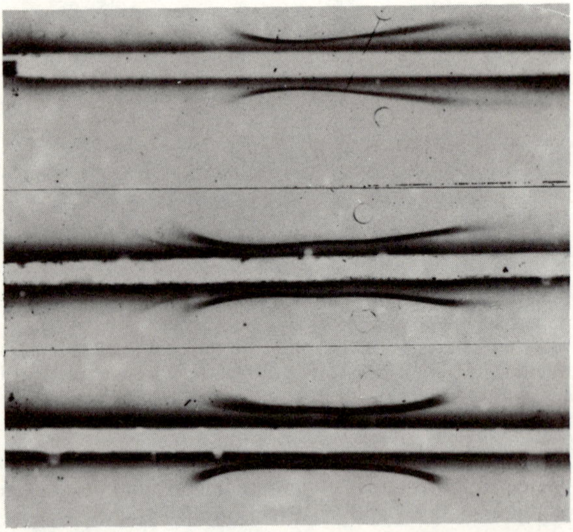

Figure 6, Immunoelectrophoresis of fractions from the SE-Sephadex fractionation. Electrophoresis is from left to right. The trough contained bovine anti-HCG and each well contained the antigen.

Top: fraction SE-A. Centre: fraction SE-B. Bottom: fraction SE-C.

However, fraction E-C (Fig. 7) contained a much smaller amount of the higher mobility component than did the other fractions.

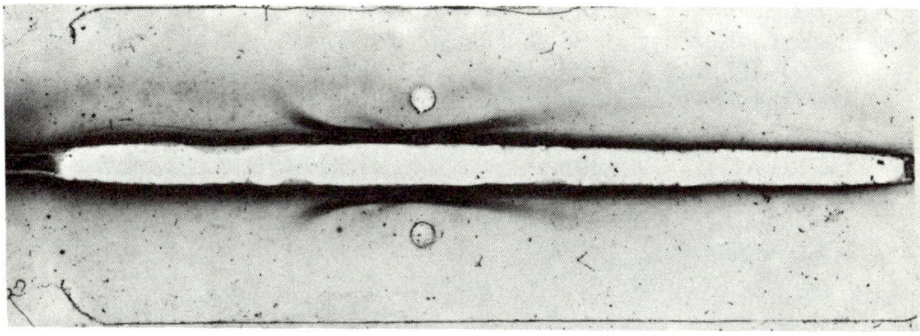

Figure 7. Immunoelectrophoresis of fraction E-C from the polyacrylamide fractionation. Electrophoresis is from left to right. The trough contained bovine anti-HCG and each well contained the antigen.

Amino acid analysis. The amino acid composition of the most highly purified LH preparation (E-C) is presented in Table V. These values were corrected for amino acid decomposition during hydrolysis. First order rate plots of the amino acid values as a function of hydrolysis time were extrapolated to zero time in order to obtain values for those amino acids which were found to undergo decomposition.

Six amino acids, threonine, serine, proline, tyrosine, half-cystine, and methionine, were found to have values which decreased with time. The largest correction was 9 per cent per 24 hours for half-cystine and threonine. The correction for serine and proline was about 7 per cent and about 3 per cent for tyrosine and methionine. No time dependence was found in the values for the remaining amino acids therefore the values reported were obtained from the average of the results from the different hydrolysis times.

The amino acid chromatograms did not show significant amounts of cysteic acid or methionine sulphoxide. Three unknown peaks containing less than 0.010μM amino acid per mg of material hydrolyzed (based on the colour yield of lysine) were found on the chromatograms. Since these peaks increased slightly in size as the hydrolysis time increased, they were assumed to be degradation products of amino acids.

A minimum molecular weight frequently can be calculated from the amino acid composition. Unfortunately the amino acids which are in lowest concentration, histidine and methionine, are among those which are usually determined with the least accuracy (Bender, *et al.* 1959). In spite of this problem the data was analysed by the method described by Roos (1968) and yielded a minimum molecular weight of 7.5 x 10^3.

The amino acid composition of all four preparations analysed is presented in Table VI. For the purpose of comparison, these values are based on 10 lysine residues. Included in the Table are published amino acid values found in pituitary LH preparations.

TABLE V

Amino acid composition of the most highly purified urinary LH fraction, E–C, from the polyacrylamide electrophoresis fractionation. Hydrolysis was done for 28, 52, and 100 hours and the results used to correct the values for decomposition.

AMINO ACID	μmol/mg*	AMINO ACID	μmol/mg*
Aspartic acid	0.674	Isoleucine	0.242
Threonine	0.435	Leucine	0.452
Serine	0.580	Tyrosine	0.241
Glutamic acid	0.618	Phenylalanine	0.242
Proline	0.384	Lysine	0.439
Glycine	0.412	Histidine	0.127
Alanine	0.357	Arginine	0.275
Valine	0.422	Tryptophan	N.D.
Half-cystine	0.316	Amide nitrogen	N.D.
Methionine	0.134		

* Based on the dry weight of the protein preparation.

N.D. – Not Determined

DISCUSSION

The urinary LH prepared in this study represents material with a range of electrophoretic mobilities. This suggests the material contains molecules differing in charge, size, or shape. Since a broad mobility range for LH was also found in preparative polyacrylamide electrophoresis, it is possible that this material is LH which has undergone modification on being excreted and/or through the preparative procedures. The presence of non-LH derived proteins cannot be excluded.

At one stage in the purification of urinary LH, biological activity was found to decrease considerably on storage (Table 1) while the activity by radioimmunoassay remained relatively constant. Apparently, some molecular modification occurred on storage which affected the site of biological but not immunological activity. A decline in biological activity of another highly purified urinary LH preparation has also been observed by Stevens, Anderson and Powell (1969). Throughout this study it is assumed that the modification of the biological activity of the LH did not affect the primary amino acid structure of the protein and that the immunologically active

TABLE VI

Amino acid composition of fraction SE–A, SE–B, and SE–C. For comparison, the values for fraction E–C and some pituitary LH preparations were included. The values were arbitrarily based on 10 lysine residues and do not necessarily represent residues per mole of protein. Tryptophan and nitrogen were not determined.

	PREPARATION					
AMINO ACID	SE–A	SE–B	SE–C	E–C	A	B
Aspartic acid	18	16	19	15	16	14
Threonine	13	12	12	10(9)	17	16
Serine	15	14	15	13(11)	13	13
Glutamic acid	19	18	18	14	19	20
Proline	13	12	11	9(8)	26	24
Glycine	12	10	10	9	15	14
Alanine	9	8	9	8	11	13
Valine	10	8	10	10(9)	20	17
Half-cystine	17	18	14	7(6)	20	–
Methionine	3	2	3	3	5	–
Isoleucine	5	4	6	5	7	7
Leucine	8	8	10	10	14	14
Tyrosine	5	5	8	5	6	5
Phenylalanine	5	4	5	6	6	7
Lysine	10	10	10	10	10	10
Histidine	4	5	4	3	7	5
Arginine	10	7	9	6	16	11

SE–A, B, and C – 48 hour hydrolysis, uncorrected for decomposition.
E–C – Corrected for decomposition. The 48 hour hydrolysis values are included in parenthesis where they differ from the corrected values.
A – Pituitary LH, 48 hour hydrolysis (Hartree, 1967)
B – Pituitary LH, 48 hour hydrolysis (Kathan *et al.* 1967)

material isolated contained the same amino acid composition as the biologically active material.

Fractions S-3 through S-6 (Fig. 1 and Table II) all contained material with high immunological activity. SE-Sephadex chromatography of fractions S-3, 6, and 7 indicated that the amount of LH recovered in the most active fraction from these fractionations was proportional to the magnitude of the activity in the fractions placed on the column. A greater loss in activity than could be explained by the loss in total protein occurred during this fractionation. Sephadex G-100 chromatography showed that although the molecular weights of these fractions were almost half that reported by Donini *et al.* (1968a), they did not represent the subunits reported by Reichert and Ward (1969) to be present in human pituitary LH.

A further purification of urinary LH on polyacrylamide gel did not increase the specific activity of the preparation but did remove material of a lower specific activity and differing mobility. Although the final preparation contained material with a range of mobilities (Fig. 4), it represented a much smaller range than did the starting material and no material separating into two distinct bands as reported by Donini *et al.* (1968a). Bell *et al.* (1969) have shown that HCG preparations of differing electrophoretic mobilities all have essentially the same amino acid composition. A similar phenomena may occur with urinary LH. The immunoelectrophoretic results (Fig. 6, bottom) show that fraction SE-C is composed of at least two cross-reacting components, the non-crossing or partially cross-reacting component present in S-6 (Fig. 5) was removed. Figure 7 shows that fraction E-C still contains two cross-reacting components, although the component of higher mobility represents a much lower fraction of the material. Cross-reacting LH components were not found by Donini (1967b) and Donini *et al.* (1968b). Twice the ionic strength and a slightly lower pH was used in the later study and may account for the difference or the difference may be characteristic of the material used.

The amino acid analysis done on the most highly purified urinary LH (E-C) is compared with the amino acid composition of the SE-Sephadex fractions (Table VI). It can be seen that all these fractions, SE-A, B, C, and E-C, are similar in composition, although some differences occur. The largest difference occurs in glutamic acid, proline, half-cystine and arginine. The amino acids appearing in high concentration or low concentration are present in high concentration or low concentration respectively in all the preparations with the exception of half-cystine and arginine. The same type of general agreement is found when the composition of urinary LH is compared with pituitary LH. The largest difference occurs with the proline content. The pituitary preparations contain more than two times the number of proline residues found in the urinary LH preparations when adjusted to the reference of 10 lysine residues. Although the amino acid compositions of the various preparations show many similarities, the differences are significant. The differences between fraction SE-C and E-C indicate that the broad range in mobilities of LH found in the polyacrylamide electrophoresis results (Fig. 3) must be due, at least in part, to amino acid differences. It is possible that the fractions SE-A, B, and C, and E-C differ in amino acid composition because of different degrees of contamination with proteins not related to LH. Since the amino acid composition of SE-A and SE-C are very similar it seems improbable, however, that the differences in the specific activity of the preparations could be due to protein not derived from LH. Fraction SE-C has a specific activity eleven times that of fraction SE-A. The most probable explanation is that these fractions were derived from LH during the excretion process or during the various procedures required for isolation.

Although the amino acid analysis of urinary LH must be considered preliminary, it suggests several interesting considerations. Urinary LH is perhaps not a single molecular species but represents molecules which have closely related amino acid compositions. Urinary LH is probably a hydrolytic product of pituitary LH. More extensive chemical analysis of urinary LH is to be undertaken to clarify some of these considerations.

REFERENCES

Bell, I.I., Canfield, R.E. & Sciarra, J.J. (1969). *Endocrinology,* **84,** 298.

Bender, A.E., Palgrave, J.A. & Doell, B.H. (1959). *Analyst. Lond.* **84,** 526.

Butt, W.R. (1967). In *Recent Research on Gonadotrophic Hormones,* p. 130. Ed. Bell, E.T. & Loraine, J.A. Edinburgh: Livingstone.

Davies, B.J. (1964). *Ann. N.Y. Acad. Sci.* **121,** 404.

Donini, P. (1967a). In *Recent Research on Gonadotrophic Hormones,* p. 123. Ed. Bell, E.T. & Loraine, J.A. Edinburgh: Livingstone.

Donini, P. (1967b). In *Recent Research on Gonadotrophic Hormones,* p. 153. Ed. Bell, E.T. & Loraine, J.A. Edinburgh: Livingstone.

Donini, P., Montezemolo, R. and Puzzuoli, D. (1964). *Acta Endocr. Copenh.* **45,** 321.

Donini, P., Puzzuoli, D., D'Alessio, I., Bergesi, G. & Donini, S. (1968a). In *Gonadotropins,* p. 37. Ed. Rosemberg, E. Los Altos, California: Geron-X.

Donini, P., Puzzuoli, D., D'Alessio, I., & Donini, S. (1968b). *Il Farmaco,* **23,** 317.

Hartree, A.S. (1967). In *Recent Research on Gonadotrophic Hormones,* p. 131. Ed. Bell, E.T. & Loraine, J.A. Edinburgh: Livingstone.

Hjerten, S., Jerstedt, S. & Tiselius, A. (1965). *Analyt. Biochem.* **11,** 219.

Kathan, R.H., Reichert, L.E. & Ryan, R.J. (1967). *Endocrinology,* **81,** 45.

Little, J.R. and Donahue, H. (1968). In *Methods in Immunology and Immunochemistry,* Vol. **2,** p. 390. Ed. Williams, C.A. & Chase, M.W. New York: Academic Press.

Loraine, J.A. & Brown, J.B. (1954). *J. Endocr.* **18,** 77.

Moore, S. & Stein, W.H. (1963). In *Methods of Enzymology,* Vol. VI, p. 820. Ed. Colowick, S.P. & Kaplan, N.O. New York: Academic Press.

Parlow, A.F. (1961). In *Human Pituitary Gonadotropins,* p. 196. Ed. Albert, A. Springfield, Illinois: Thomas.

Piez, K.A. and Morris, L. (1960). *Analyt. Biochem.* **1,** 187.

Reichert, L.E. (1967). *Endocrinology,* **80,** 319.

Reichert, L.E. & Ward, D.N. (1969). *Fedn Proc. Fedn Am. Socs exp Biol,* **28,** 505.

Roos, P. (1968). *Acta Endocr. Copenh.* Suppl. **131**, 25.
Roos, P. and Gemzell, C.A. (1964). *Biochim. Biophys. Acta.* **93**, 217.
Stevens, V.C. (1969). *J. clin. Endocr. Metab.* **29**, 904.
Stevens, V.C., Anderson, D.G. & Powell, J.E. (1969) unpublished.

Purification and Properties of Human Pituitary Luteinizing Hormone

Mary Braikevitch and Anne Stockell Hartree

In a previous publication (Hartree, 1966) the preparation of human luteinizing hormone (LH) of high biological activity (5.3 times standard NIH-LH-S3) was described. Studies of the homogeneity and amino acid composition of the preparation are presented here and are compared with published data on human LH prepared by other methods. Dissociation into half molecules at acid pH has been demonstrated for both sheep (Li and Starman 1964, Ward and Arnott, 1965) and human LH (Ryan, 1968). Studies of the pH dependence of dissociation of human LH in acid solution and of its reassociation at pH 5 are reported here.

MATERIALS AND METHODS

Preparation of human LH. Human LH, Fraction DEAE-I was prepared from acetone-dried human pituitaries as described previously (Hartree, 1966). It was further purified by chromatography on Amberlite IRC-50 in pH 7.5 phosphate buffer, 5 mM in Na^+ (Hartree *et al.* 1964). Under these conditions LH was adsorbed by the ion exchanger and, after removal of unadsorbed material (fraction IRC-1) by washing with starting buffer, LH (fraction IRC-2) was eluted with 1 M ammonium acetate. Both fractions were precipitated with 5 volumes of 95 per cent ethanol at 4°C centrifuged, washed with ethanol and ether, and dried *in vacuo.* The average yield of fraction IRC-2 from 100 g pituitary powder (1000 human pituitaries) was 54 mg (Hartree, 1967) and the average potency was 5 times standard NIH-LH-S3 (Hartree, 1966). The inert fraction IRC-1 was obtained in a much lower average weight yield (3 mg from 100 gm pituitary powder). After storage in the dry state at room temperature for 18 months, one preparation of IRC-2 (10-12-64) had a potency of 7 times standard NIH-LH-S6 while another preparation of IRC-2 (9-2-65) had decreased in potency to 2 times standard NIH-LH-S6 (Hartree, 1967).

Polyacrylamide gel electrophoresis. This was by the method of Davis (1964) using tris-glycine buffer, pH 8.3, in the reservoirs. The gels were

7 per cent polyacrylamide and samples (approximately 100 μg each) were applied in 40 per cent sucrose solution. Electrophoresis was carried out at 1 mA/tube for 20 min. to allow the sample to enter the gel and then at 3mA/tube for 1.5 to 2.5 hr. The gels were stained for 1 hr. in a solution of 1 per cent amido black in 7 per cent acetic acid, washed with water and destained electrophoretically.

Determination of N-terminal amino acids. For qualitative determination of N-terminal amino acids the 'dansyl' method of Gray and Hartley (1963) was used with modifications described by Oliver and Iartree (1968).

Amino acid analysis. Samples of LH (0.35 mg) were hydrolysed in onstant-boiling HC1 (6N) in sealed evacuated tubes at 105°C in duplicate for 24, 48, 61, 72 and 96 hr. The tubes were opened, HC1 removed by repeated concentration *in vacuo*, and the analyses were performed on the Beckman Unichrom Amino Acid Analyzer, Additional samples were oxidized with performic acid (Hirs, 1956), hydrolysed for 27, 51 and 73 hr. at 105°C and analyzed on the Technicon Auto-Analyzer. Results of these latter analyses were calculated from molar ratios assuming a value of 14.2 residues of valine per mole of hormone.

Gel filtration. These experiments were performed with a column of Sephadex G-100 (2.32 cm. diameter x 34.5 cm). Approximately 10 mg of LH were dissolved in buffer, I = 0.1, and applied to the column which had been previously equilibrated with the same buffer. At pH 8 ammonium acetate buffer was used, at pH 4-5 the buffer was sodium acetate adjusted with acetic acid, and at pH 2.0 and below KC1-HC1 buffers were employed. The column was run at a flow rate of 30 ml/hr/100 ml bed volume and the eluant was collected in weighed tubes as fractions of approximately 4 ml. The volume of each fraction was estimated by weight and its protein concentration by absorption at 280 mμ.

RESULTS AND DISCUSSION

Homogeneity of human LH. After purification of human LH on IRC-50 most of the protein migrates as a single peak on Sephadex G-100 (Hartree, 1966).

The preparation was tested for immunological homogeneity by Dr T. Chard (personal communication) who found a single strong precipitin line between LH and its antiserum on immunoelectrophoresis. There was no demonstrable contamination with normal serum components as evidenced by absence of precipitin lines between human LH and horse antisera to human serum.

Polyacrylamide gel electrophoresis showed four major components in DEAE-I (Fig. 1). After chromatography on IRC-50 these bands were very weak and were replaced by a single strong band of lesser electrophoretic mobility in fraction IRC-2; The small side-fraction (IRC-1) contained

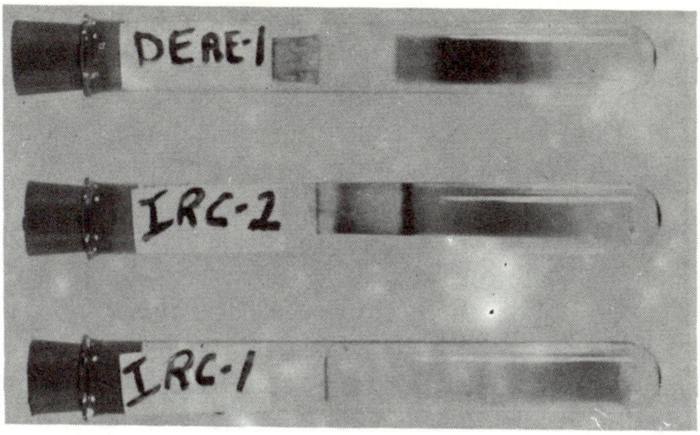

Figure 1. Polyacrylamide gel electrophoresis of human LH fraction DEAE-I and the fractions derived from it by chromatography on IRC-50. Fraction IRC-1 was obtained in a yield of 0.6 per cent by weight and fraction IRC-2, 74 per cent by weight of DEAE-I.

weak bands of greater electrophoretic mobility. The lack of correspondence between the mobilities of the components of DEAE-I and fractions IRC-1 and IRC-2 is not easily explained. It may represent removal of firmly bound low molecular weight material from DEAE-I resulting in a change of charge, or else an alteration in the three-dimensional conformation of the molecule with a concomitant change in electrophoretic mobility. It has previously been suggested that the increases in total LH activity in the final purification of LH from several species may result from dissociation of an LH-inhibitor complex (Reichert, 1961; Hartree, *et al.* 1964; Hartree and Cunningham, 1969). The possibility exists, therefore, that the change in electrophoretic mobility after chromatography on IRC-50 results from dissociation of a low molecular weight inhibitor from LH.

N-terminal amino acids. N-terminal amino acids were determined qualitatively in LH preparations DEAE-I and IRC-2 by the 'dansyl' method. In each case only valine and aspartic acid (or asparagine) were detected. Using the FDNB method Reichert (1966) found these as well as other amino acids with unsubstituted α-amino groups in preparations of human LH (biological activity 2 to 3 times standard NIH-LH-SI) but the amounts of individual amino acids were no greater than 0.36 μmol per μmol (28 mg.) of hormone. Very low or insignificant amounts of N-terminal amino acids have also been reported for sheep, beef and pig LH (Ward *et al.* 1959; Reichert, 1966; Ward *et al.* 1967). A partial explanation for these findings in the case of sheep LH is the demonstration of firm bindings of certain free amino acids by the hormone (Holcomb *et al.* 1968). It is possible, therefore, that the free α-amino groups detected in our studies of human LH represent firmly bound amino acids rather than the amino termini of polypeptide chains.

Amino acid analysis. Analyses on LH preparation IRC-2 (10-12-64) are given in Table I, on the assumption that the molecular weight is 28,000 (Ryan, 1969). Good Agreement on duplicate analyses of protein hydrolysed for 24 hr. and 48 hr. was obtained, but there was a small progressive decrease in recovery of all the amino acids at hydrolysis times greater than 48 hr., presumably because of interaction with the carbohydrate portion of the hormone. For this reason only the results obtained on 24 and 48 hr. hydrolysates are included in the table. Half-cystine was determined as cysteic acid from analyses of performic acid oxidized protein hydrolysed for 27, 51, and 73 hr. Since phenylalanine was not well separated from amino sugars present in the hydrolysates of unoxidized protein, it was estimated from hydrolysates of performic acid oxidized hormone. Tryptophan was determined by the spectrophotometric method of Bencze and Schmid (1957) which also estimates tyrosine. A value of 3.9 tyrosine residues per molecule was found in fair agreement with results obtained on the amino acid analyser. The assumed number of residues of each amino acid per molecule of hormone is also given in Table I.

Analyses for neutral and amino sugars were performed on LH preparation IRC-2 (9-2-65) by Drs J. Pierce and B. Shome (personal communication) using methods described by Kim *et al.* (1967). Based on a molecular weight of 28,000 the number of residues of each sugar determined was: Fucose - 1.7; mannose - 7.0; galactose - 3.0; glucosamine - 9.5; galactosamine - 1.6.

Our analyses are generally in good agreement with preliminary amino acid analyses of human LH reported by Hartree (1967) and with analyses reported for human LH (Biological potency 4 times standard NIH-LH-SI) prepared by a similar method (Shome, Parlow *et al.*, 1968; Shome, Brown *et al.*, 1968). There is fair agreement with the amino acid analyses of the human LH preparations of Kathan *et al.* (1967), but their half-cystine values were much lower than ours. The explanation for the discrepancy may lie in the techniques employed as these investigators did not analyse performic acid oxidized protein, but estimated half-cystine by totalling the concentrations of cystine and cysteic acid in 24hr. hydrolysates of the hormone. Their estimates of hexose were higher and of hexosamine lower than those reported here but there is good agreement in the total hexose plus hexosamine values.

In Table I the amino acid composition of the placental hormone human chorionic gonadotrophin, reported by Bahl (1969), is included for comparison with human LH. Because of their similar biological activities and their strong immunological cross-reaction (Wide *et al.* 1961), similarities in their chemical compositions would be expected. The amino acid compositions of the two human hormones are

TABLE I

Amino acid analysis of human LH (IRC-2)

Except where otherwise indicated the analytical results are the mean and standard deviation of four analyses including duplicates of 24 hr. and 48 hr. hydrolysates. The recoveries of serine and threonine decreased with time of hydrolysis, so these values were extrapolated to zero time. Half-cystine and phenylalanine values are the mean and standard deviation of three analyses of 27, 51 and 73 hr. hydrolysates of performic acid oxidized protein. Tryptophan was determined by the spectrophotometric method of Bencze & Schmid (1957).

Amino Acid	Residues per 28,000 M.W. ± std. dev.	Assumed no. of residues	Assumed no. of residues in human chorionic gonadotrophin (Bahl, 1969)
Lys	6.80 ± 0.14	7	7
His	4.36 ± 0.13	4–5	3
Arg	10.56 ± 0.28	10–11	11
Asp	10.84 ± 0.15	11	12
Thr	12.61	13	12
Ser	12.04	12	16
Glu	15.83 ± 0.36	15–16	13
Pro	17.07 ± 0.50	16–18	20
Gly	10.82 ± 0.29	10–11	9
Ala	8.03 ± 0.25	8	9
Cys	14.93 ± 0.37	14–16	14
Val	14.23 ± 0.13	14–15	13
Met	3.32 ± 0.28	3–4	3
Ile	4.96 ± 0.23	5	4
Leu	10.48 ± 0.25	10–11	10
Tyr	4.76 ± 0.30	4–5	4
Phe	4.75 ± 0.14	4–5	4
Try	1.4	1–2	1

comparable, particularly noteworthy are their high content of proline and half-cystine residues.

Dissociation of LH at acid pH. Because relatively large amounts of material were required for these experiments, preparation DEAE-I was used. At pH 8.0 and at pH 4.95, LH was eluted from Sephadex G-100 as a symmetrical peak, the ratio of the elution volume of the peak (V_e) to the column volume (V_t) being 0.51. At pH 4.0, a slight shoulder representing material of lower molecular weight appeared and at pH 1.3 most of the material was eluted as a single peak with V_e/V_t=0.61 These results are consistent with dissociation into sub-units of approximately half the original molecular weight (Andrews, 1964).

The percentage of native (undissociated) hormone at various pH values below 8 was estimated from the area under the curve of each elution profile (Fig. 2). Most of the dissociation occurred between pH 1.75 and pH 2.0 and the hormone was 50 per cent dissociated at pH 1.9.

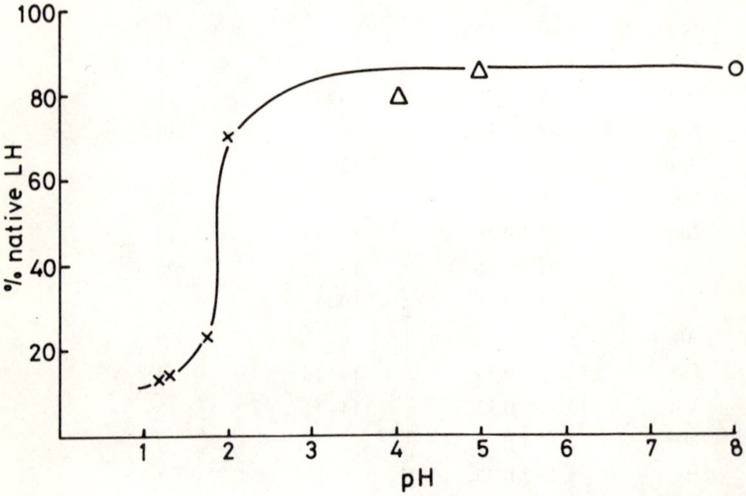

Figure 2. pH dependence of the acid dissociation of human LH. The buffers used at 0.1 ionic strength were: X-KC1-HC1; Δ - sodium acetate; O - ammonium acetate.

The main peak eluted at pH 1.3 was concentrated by rotary evaporation, dialysed against 0.1 M sodium acetate-acetic acid buffer at pH 4.95 and run again on Sephadex G-100. The material was eluted as a single peak with elution volume identical to that of native LH indicating that reassociation had occurred. The pattern obtained on polyacrylamide gel electrophoresis of reassociated protein was not significantly different from that of the native hormone.

ACKNOWLEDGMENTS

We are indebted to Dr T. Chard for carrying out immuno-electrophoretic experiments on highly purified human LH, to Drs J. Pierce and B. Shome for analyses of neutral and amino sugars, and to Miss M.B. Thomas for expert technical assistance. The Medical Research Council provided human pituitaries and a grant in support of this research. We thank Professor F.G. Young, F.R.S., for his encouragement.

REFERENCES

Andrews, P. (1964). *Biochem. J.* **91**, 222.
Bahl, O.P. (1969). *J. biol. Chem.* **244**, 567.
Bencze, W.L. & Schmid, K. (1957). *Analyt. Chem.* **29**, 1193.
Davis, B.J. (1964). *Ann. N.Y. Acad. Sci.* **121**, 404.
Gray, W.R. & Hartley, B.S. (1963). *Biochem. J.* **89**, 59P.
Hartree, A.S. (1966). *Biochem. J.* **100**, 754.
Hartree, A.S. (1967). In *Recent Research on Gonadotrophic Hormones*, p. 124 Ed. Bell, E.T. & Loraine, J.A. Edinburgh: Livingstone.
Hartree, A.S., Butt, W.R. & Kirkham, K.E. (1964). *J. Endocr.* **29**, 61.
Hartree, A.S. & Cunningham, F.J. (1969). *J. Endocr.* **43**, 609.
Hirs, C.H.W. (1956). *J. biol. Chem.* **219**, 611.
Holcomb, G.N., Lamkin, W.M., James, S.A., Wade, J. & Ward, D.N. (1968). *Endocrinology*, **83**, 1293.
Kathan, R.H., Reichert, L.E., Jr. & Ryan, R.J. (1967). *Endocrinology*, **81**, 45.
Kim, J.H., Shome, B., Liao, T-H. & Pierce, J.G. (1967). *Analyt. Biochem.* **20**, 258.
Li, C.H. & Starman, B. (1964). *Nature, Lond.* **202**, 291.
Oliver, L. & Hartree, A.S. (1968). *Biochem. J.* **109**, 19.
Reichert, L.E., Jr. (1961). *Biochim. biophys. Acta*, **53**, 586.
Reichert, L.E., Jr. (1966). *Endocrinology*, **78**, 186.
Ryan, R.J. (1968). *J. clin. Endocr. Metab.* **28**, 886.
Ryan, R.J. (1969). *Biochemistry, Wash.* **8**, 495.
Shome, B., Brown, D.M., Howard, S.M. & Pierce, J.G. (1968). *Archs Biochem. Biophys.* **126**, 456.
Shome, B., Parlow, A.F., Ramirez, V.D., Elrick, H. & Pierce, J.G. (1968). *Archs Biochem. Biophys.* **126**, 444.
Ward, D.N. & Arnott, M.S. (1965). *Analyt. Biochem.* **12**, 296.
Ward, D.N., Adams-Mayne, M., Ray, N., Balke, D.E., Coffey, J. & Showalter. M. (1967). *Gen. comp. Endocr.* **8**. 44.
Ward, D.N., McGregor, R.F. & Griffin, A.C. (1959). *Biochim. biophys. Acta*, **32**, 305.
Wide, L., Roos, P. & Gemzell, C. (1961). *Acta Endocr. Copenh.* **37**, 445.

Studies on the Chemistry of Interstitial Cell-stimulating Hormone

Harold Papkoff and Choh Hao Li

INTRODUCTION

In the past decade much progress has been made in understanding the chemistry of the protein and polypeptide hormones. Increased knowledge of the interstitial cell-stimulating hormone (ICSH, LH) of the anterior pituitary gland has been part of this progress. This brief account will summarize some of the contributions from this laboratory which relate to the chemistry of several species of ICSH. For a full account of the studies on ICSH performed in other laboratories the reader is referred to the published work of W. R. Butt (UK), A. Hartree (UK), M. Jutisz (France), L. E. Reichert, Jr. (USA), R. J. Ryan (USA), B. Saxena (USA) and D. N. Ward (USA), among others.

PURIFICATION STUDIES

A decade ago Squire and Li (1959) and Ward *et al.* (1959) described purification procedures for the preparation of ovine ICSH. These procedures suffered from the disadvantages of being tedious and yielding only low amounts of hormone. The method used in our laboratory has been superceded by the procedure described by Papkoff *et al.* (1965). In this procedure the glands are extracted at pH 4.0 and some of the inert proteins removed by treating the extract with HPO_3. The hormone is then precipitated from the extract with $(NH_4)_2SO_4$ and further purified by chromatography on sulphoethyl-Sephadex C-50 followed by gel filtration on Sephadex G-100. The yield, which averages 300 mg/kg of pituitaries, is considered good. This procedure can also be used for the preparation of bovine ICSH, although the yields are somewhat lower (Papkoff, 1966; Papkoff and Gan, 1969).

Human ICSH has been purified from a side fraction obtained during the preparation of growth hormone (GH). The side fraction was material which precipitated with GH at 0.5 saturated $(NH_4)_2SO_4$ and remained unadsorbed on Amberlite IRC-50 under conditions in which GH was adsorbed. Further purification steps have been described by Squire *et al.* (1962), and include step-wise elution of the material from CM-cellulose and electrophoresis of the most active fraction obtained from CM-cellulose by column zone electrophoresis at pH 3.6. This procedure yields a highly purified and active material. Sulphoethyl-Sephadex C-50 has also been employed for the purification of human ICSH from the above-mentioned side fraction. In this procedure (Papkoff, 1965) the material was chromatographed on the gel at pH 4.0, 0.03μ acetate buffer, and an inert fraction removed with a pH 5.4, 0.03 μ acetate buffer. The hormone was eluted by increasing the ionic strength of the buffer to 0.2. Final purification was achieved on Sephadex G-100.

CHARACTERIZATION STUDIES

Physical studies. The three species of ICSH have been examined by a variety of electrophoretic techniques including free boundary electrophoresis, zone electrophoresis in troughs of starch, columns of cellulose, and strips of cellulose acetate, and disc electrophoresis in columns of polyacrylamide. Free boundary experiments showed ovine and human ICSH to be homogeneous and to have isoelectric points at pH 7.3 and 5.4, respectively (Squire and Li, 1959; Squire *et al.* 1962). Bovine ICSH was observed to have electrophoretic properties identical to ovine ICSH and a pI probably close to that of ovine ICSH (Papkoff and Gan, 1969).

When examined in the analytical ultracentrifuge ovine, bovine and human ICSH were found to be monodisperse in a slightly alkaline buffer (pH 8.4) and to have very nearly the same sedimentation coefficient ($S_{20,w}$, 2.67-2.74). The molecular weight of ovine ICSH calculated from S and D measurements is approximately 30,000 in neutral and slightly alkaline solutions (Squire and Li, 1959). A value that agrees closely with this has also been obtained from sedimentation-equilibrium studies on ovine ICSH (Reinhardt and Squire, 1965). Li and Starman (1964) observed, however, that in strongly acidic solutions (pH 1.3, HCl-KCl) ovine ICSH undergoes a dissociation. The $S_{20,w}$ value drops to 1.64 and the molecular weight calculated from sedimentation-equilibrium data becomes 15,000, or about one-half that observed in neutral solutions. The dissociation of ovine ICSH into units of 15,000 was also shown with the preparation of Papkoff *et al.* (1965). The units, although of nearly the same molecular weight, were shown to be chemically non-identical with respect to carbohydrate and amino acid content. The dissociation is discussed in greater detail below, and some of the physical data is summarized in Table I.

The molecular weight of bovine ICSH as determined by osmometry in 0.05 M NH_4HCO_3 was found to be 25,200, uncorrected for salt and moisture content (Papkoff and Gan, 1969). Bovine ICSH has also been observed to dissociate in strongly acidic solutions (Papkoff, 1966; Papkoff and Gan, 1969).

Human ICSH was determined to have a molecular weight of 26,000 from sedimentation-equilibrium studies at pH 3.6 (Squire *et al.*, 1962).

TABLE I

Properties of Ovine, Bovine, and Human interstitial Cell-stimulating hormone (ICSH, LH)

		Ovine	Bovine	Human
Molecular wt.	pH 8.4	30,000	25,200	26,000
	pH 1.3	15,000	–	–
Sedimentation	(pH 8.4)	2.74	2.67	2.71
coefficient ($S_{20,w}$)		1.64	1.66	–
Isoelectric Pt. (pI)		7.3	–	5.4
Spectrophotometric:				
% tyrosine		3.1	3.0	3.27
% tryptophan		0.44	0.44	0.85
$O.D._{275}$ (1 mg/ml)		0.423	0.460	0.630
N-terminal				
FDNB		Absent	Absent	Absent
Edman PTH		Proline	Proline	–
C-terminal				
carboxypeptidase		Serine Leucine	Serine Leucine	–

Chemical studies. It has long been known that ICSH is a glycoprotein. Sugars which have been qualitatively identified in acid hydrolysates of both ovine and bovine ICSH are mannose, galactose, fucose, glucosamine and galactosamine. The amount of each sugar in ovine ICSH has been specifically determined by Fujino *et al.* (1968). Our analyses, which are summarized in Table II, show that the carbohydrate content varies from 12 to 14 per cent in three species of ICSH. We have found the carbohydrate content of bovine ICSH preparations to be consistently lower than that of ovine ICSH, the bovine ICSH having 5.3 per cent hexose as compared to 6.5 per cent for ovine. Hexosamine and fucose values are also lower in bovine ICSH. Although bovine and ovine ICSH do not contain significant quantities of sialic acid, there is enough sialic acid present in human ICSH (0.68 per cent) to account for one residue per molecule of human ICSH. The sialic acid in human ICSH does not appear to be necessary for biological activity, since reaction of human ICSH with neuraminidase does not affect its hormonal activity (Papkoff, unpublished). The carbohydrate analyses we have obtained are summarized in Table II.

Amino acid analyses (Table III) show that ovine, bovine and human ICSH are very similar in amino acid content, and that ovine and bovine ICSH are nearly identical. These latter two species are characterized by

TABLE II

Carbohydrate content* of Ovine, Bovine, and Human interstitial Cell-stimulating hormone (ICSH, LH)

Sugar	Species		
	Ovine	Bovine	Human
Hexose	6.5	5.3	5.9
Hexosamine	7.8	7.0	5.1
Fucose	1.4	0.9	0.6
Sialic acid	0.37	nil	0.68

*g per 100 g protein

the absence of tryptophan, and a very high content of cystine and proline. Human ICSH, on the other hand, contains tryptophan, has a higher content of tyrosine, and a much lower content of proline. On the basis of the lysine and arginine content one would expect a tryptic digest of performic acid oxidized ovine or bovine ICSH to yield approximately 25 peptides. This is indeed the case (Papkoff and Gan, 1969) as seen in peptide mapping experiments. This result also suggests that the 15,000 molecular weight units are not identical. If the two units were identical only 12 or 13 peptides would be expected.

We recently investigated our ovine ICSH preparation for the presence of terminal residues (Samy *et al.*, 1969a). We found, as have others, that significant quantities of DNP-amino acids could not be obtained from the reaction of fluorodinitrobenzene (FDNB) with ovine or bovine ICSH. This finding implied that proline was at the N-terminus because DNP-proline is known to be very unstable in 6 N HC1 at 110°. Subsequently, we were able to identify proline as the N-terminal amino acid residue of ovine ICSH by means of the Edman reaction. PTH-proline could be obtained from native ICSH, performic acid oxidized ICSH, and the sub- unit preparations, C I and C II. The PTH-proline was identified qualitatively by thin layer chromatography on silica gel-coated glass plates. It was observed that the yield of PTH-proline was always higher in the oxidized preparations. By the Edman procedure proline was also identified as the N-terminal amino acid of bovine ICSH. Again it was observed that a greater yield of PTH-proline could be obtained from oxidized bovine ICSH than from the native molecule.

TABLE III

Amino acid content of Ovine, Bovine, and Human ICSH

Amino acid	Residues/30,000		
	Ovine*	Bovine**	Human ‡
Lysine	13.2	13.4	10.4
Histidine	5.6	5.6	4.4
Arginine	10.0	9.6	7.8
Aspartic acid	12.0	14.0	12.4
Threonine	15.6	14.0	15.4
Serine	13.6	13.0	10.8
Glutamic acid	14.6	14.8	13.6
Proline	25.0	22.2	12.2
Glycine	12.6	12.2	8.4
Alanine	16.0	14.8	9.0
1/2 Cystine	18.0	16.6	15.8
Valine	13.0	11.4	11.0
Methionine	5.0	5.2	3.6
Isoleucine	6.8	5.8	7.4
Leucine	12.0	13.6	9.8
Tyrosine	6.0	6.6	9.6
Phenylalanine	6.6	7.4	6.6
Tryptophan	Absent	Absent	1.4

* Taken from Papkoff *et al.* (1965)

** Taken from Gan *et al.* (1968)

‡ Uncorrected for hydrolytic destruction, moisture and ash

Studies have been made by Reichert (1966) and Ward *et al.* (1967) on the C-terminal amino acids of ICSH. Reichert's data from carboxypeptidase experiments suggested that serine and leucine are the C-terminal amino acids of ovine ICSH. Ward *et al.* (1967) found that carboxypeptidase liberated serine, followed by tyrosine, in stoichiometric amounts, and additionally, small amounts of aspartic acid. From these and other studies (Holcomb *et al.*, 1967) they concluded that serine and aspartic acid are the C-terminal groups in ICSH. However, our kinetic studies (Samy *et al.*, 1969a) on the reaction of ovine ICSH and oxidized ICSH with carboxypeptidase A support Reichert's observations that serine and leucine are the C-terminal residues. We have also shown (as will be discussed below) that serine and leucine can each be associated with an individual subunit.

The subunit nature of ICSH. With the recognition that ovine ICSH is dissociated into dissimilar subunits by strong acid or, as de la Llosa and Jutisz (1969) have shown, by urea or guanidine, it became imperative to find a means of separating the subunits in order to begin structural studies. We achieved separation by means of countercurrent distribution (Papkoff and Samy, 1967). When ovine ICSH was submitted to as few as nine transfers in the system 40 per cent $(NH_4)_2SO_4$-0.2 per cent dichloroacetic acid-n-propanol-ethanol (60:60:27:33), the subunits separated into two fractions: one of very low partition coefficient found in the aqueous phase of the initial tubes and one of very high partition coefficient found in the organic phase of the final tubes. Bovine ICSH was found to behave in an identical fashion (Papkoff and Gan, 1969) and there are indications that human ICSH does as well (Reichert and Ward, 1969). We have designated the subunit having the low partition coefficient as C I and the subunit with the high partition coefficient C II, and propose that this nomenclature be applied to all species of ICSH which display this behavior (the subunit would be preceded by a letter to indicate the species). Nearly equal amounts of the two subunits were obtained by this procedure. Further purification was achieved by gel filtration on columns of Sephadex G-100 equilibrated with 0.05 M NH_4HCO_3 in the case of subunit C I and equilibrated with pH 4.0, 0.05 μ acetate in the case of subunit C II. The purified subunits were shown to be homogeneous by disc electrophoresis.

Terminal group analyses by the methods described previously showed that each subunit and its performic acid oxidized derivitive contained proline at the N-terminus. In kinetic studies with carboxypeptidase A it was shown that serine was the first amino acid released in oxidized C I. A final amount of 0.87 mol of serine was released per mol of C I and, additionally, 1.85 mol of tyrosine per mol were released. The C-terminal sequence of C I was thereby determined to be......Tyr.Tyr.Ser.OH. By the same methods the C-terminal sequence of oxidized C II was found to be......Cys.Ile.Leu.OH.

In Table IV the amino acid compositions of the subunits are compared with the composition of native ICSH. Subunit C I contains much more lysine (3 to 4 times) than C II and greater amounts of aspartic acid, threonine, glutamic acid, tyrosine and phenylalanine. C II, on the other hand, contains more arginine, proline (2 times) and leucine (3 times) than C I. With respect to amino acid composition, C II is the more hydrophobic of the two subunits as would be expected from its higher partition coefficient. The bovine ICSH subunits have amino acid compositions much like those of ovine ICSH (Papkoff and Gan, 1969).

TABLE IV

Amino acid composition of the Subunits of Ovine ICSH*

Amino acid	Ovine ICSH	Subunit, CI	Subunit, CII
	μmol/mg protein**		
Lysine	0.372	0.593	0.171
Histidine	0.181	0.177	0.152
Arginine	0.324	0.216	0.344
Aspartic	0.367	0.396	0.282
Threonine	0.410	0.555	0.352
Serine	0.357	0.386	0.445
Glutamic	0.395	0.519	0.370
Proline	0.780	0.538	1.120
Glycine	0.312	0.287	0.412
Alanine	0.418	0.455	0.422
1/2 Cystine	0.505	0.486	0.464
Valine	0.415	0.309	0.352
Methionine	0.150	0.208	0.111
Isoleucine	0.212	0.137	0.201
Leucine	0.402	0.203	0.581
Tyrosine	0.234	0.297	0.099
Phenylalanine	0.248	0.320	0.185

* Taken from Papkoff and Samy (1967)

** Uncorrected for hydrolytic destruction, moisture and ash

The carbohydrate compositions of the ovine and bovine subunits are shown in Table V. The C I subunit contains about twice as much carbohydrate as C II, and this difference in carbohydrate content contributes to the greater hydrophilic character of C I.

Although this survey is primarily concerned with chemical aspects of ICSH, a few words are appropriate here regarding the biological activity of the subunits. In the most highly purified preparations it has been found that the biological activity of the individual subunits is only 0.5 to 2 per cent of native ICSH. Activity can be restored, however, by recombination of the subunits (incubated at room temperature for 20 hours at a concentration of 10 mg per ml of pH 6.0 phosphate buffer before diluting for assay). As seen in Table VI, upon recombination of

TABLE V

Carbohydrate Content* of the Subunits of Bovine and Ovine ICSH

Sugar	Bovine			Ovine		
	Native	CI	CII	Native	CI	CII
Hexose	5.3	7.3	3.7	6.5	8.3	4.3
Hexosamine	7.0	8.7	5.5	7.8	8.6	5.8
Fucose	0.9	1.6	0.9	1.4	1.3	1.0

*g per 100 g glycoprotein

TABLE VI

Biological activity of Ovine ICSH, the subunits, and reconstituted ICSH

	Sp. activity*	95% C.L.	λ	Assays
Native ICSH	2.00	1.11–4.71	0.214	6
Subunit, CI	0.15	0.13–0.19	0.078	2
Subunit, CII	0.17	0.14–0.22	0.084	2
CI + CII **	2.47	1.75–3.94	0.130	3

* Assayed by the ovarian ascorbic acid depletion method (Parlow). Activity expressed in terms of NIH-LH-SI.

** 10 mg each per ml pH 6.0 phosphate; r.t. 20 hr. before dilution for assay.

the subunits a specific activity of 2.47 x NIH-LH-S1 was obtained. This represents complete restoration of the biological activity of the hormone.

Structural studies on ovine ICSH subunit C II. Determination of the amino acid sequence of the C II subunit has been under way and part of the sequence has been published (Samy *et al.*, 1969b). To investigate the structure advantage was taken of the fact that C II possessed only two methionine residues. Cleavage of the molecule at these residues was achieved by reacting C II with cyanogen bromide (CNBr), which produced three peptides, two in which the C-terminal methionine had been converted to homoserine lactone. One of the peptides (CNBr-2) was isolated from the reaction mixture by gel filtration on Sephadex G-25, on which it was retarded. The unretarded material was oxidized with performic acid to break disulphide bonds and this reaction mixture was filtered through Sephadex G-25 once again to separate the two remaining peptides, CNBr-1A and CNBr-1B. The peptides were further purified by additional gel filtration and chromatography on CM-cellulose. Their amino acid compositions were then determined and the results are shown in Table VII. Peptide CNBr-1A contains 62 residues,

TABLE VII

Composition of CNBr Peptides of Ovine ICSH subunit CII

	CNBr–2	CNBr–1B	CNBr–1A
N-terminal	Lys	Proline	Val
Homoserine lactone	+ (1)	+ (1)	–
Carbohydrate	–	–	+
Amino acids			
Lys	1	0	1
His	0	1	3
Arg	1	3	3
$CySO_3H$	0	2	8
Asp	0	1	4
Thr	0	1	4
Ser	0	1	5
Glu	0	2	3
Pro	4	5	10
Gly	0	2	3
Ala	0	1	4
Val	2	3	4
Leu	2	2	6
Ileu	1	0	1
Tyr	0	1	1
Phe	0	1	2
	12	27	62

+ present

– absent

has an N-terminal valine, possesses carbohydrate and, because of the absence of homoserine at the C-terminus, must be the C-terminus of the C II polypeptide chain. The peptide CNBr-1B possesses 27 amino acids and by virtue of its N-terminal proline must be the N-terminus of C II. The remaining peptide, CNBr-2, has 12 amino acids, an N-terminal lysine, and must be located between the other two peptides. The sequence is thus: (CNBr-1B)→(CNBr-2)→(CNBr-1A).

Our work on the sequence of the first two CNBr peptides (Samy *et al.*, 1969b), which was accomplished by the usual amino acid sequence techniques, allows us to postulate the following structure for the C II subunit of ovine ICSH:

$\underset{1}{\text{Pro}}$-Gln-Arg-Val-CyA-Thr-Tyr-His-Gln-$\underset{10}{\text{Leu}}$-Arg-Phe-Ala-Ser-Val-Arg-

Leu-Pro-Gly-$\underset{20}{\text{Pro}}$-CyA-Pro-Val-Asp-Pro-Gly-Met-Lys-Arg-$\underset{30}{\text{Val}}$-Leu-Pro-

Val-Pro-Pro-Leu-Ile-Pro-Met-Val-[$\text{Lys}_1\text{His}_3\text{Arg}_3\text{CyA}_7\text{Asp}_4\text{Thr}_4\text{Ser}_5$

$\text{Glu}_3\text{Pro}_{10}\text{Gly}_3\ \text{Ala}_4\ \text{Val}_4\ \text{Leu}_5\ \text{Tyr}_1\ \text{Phe}_2$]-CyA-Ile-$\underset{101}{\text{Leu}}$.

We hope that the entire amino acid sequence of the bovine subunit C II will be completed shortly and that the structure of subunit C I will also be known in the not too distant future. Knowledge of the nature of the carbohydrate sequence and linkage will also be necessary before the structure of the ICSH molecule can be correlated with its biological activity. To our knowledge ICSH is the first protein hormone which has been shown to have a subunit structure. Many areas of additional study have been opened up by the demonstration of this subunit structure and by the successful separation and characterization of the subunits. Already we have learned that the subunits have very dissimilar immunochemical properties (to be published), and one area of interest will be to correlate immunochemical properties with biological activity. We look forward to further increasing our understanding of the chemistry of the ICSH molecule.

ACKNOWLEDGEMENT

We thank the many associates, professional and technical, who have contributed to these studies. This work was supported in part by a grant from the National Institute of Arthritis and Metabolic Diseases, NIH, USPHS, A 6097. One of us (H.P.) is a Career Development Awardee of the NIH.

REFERENCES

Fujino, M., Lamkin, W.M., Mayfield, J.D. & Ward, D.N. (1968). *Gen. comp. Endocr.* **11**, 444.

Gan, J., Papkoff, H. & Li, C.H. (1968). *Biochim. biophys. Acta,* **170**, 189.

Holcomb, G.N., James, S.A. & Ward, D.N. (1967). *Biochemistry, Wash.* **7**, 1291.

Li, C.H., and Starman, B. (1964). *Nature* **202**, 291.

Llosa, P. de la, & Jutisz, M. (1969). *Excerpte Med. Int. Congr. Ser. No.* **161**, 229.

Papkoff, H. (1966). *Excerpta Med. Int. Congr. Ser. No.* **112**, 334.

Papkoff, H. & Gan, J. (1969). Submitted for publication.

Papkoff, H., Gospodarowicz, D., Candiotti, A. & Li, C.H. (1965). *Archs Biochem. Biophys.* **111**, 431.

Papkoff, H. & Samy, T.S.A. (1967). *Biochim. biophys. Acta,* **147**, 175.

Reichert, L. E., Jr. (1966). *Endocrinology,* **78**, 186.

Reichert, L.E., Jr. & Ward, D.N. (1969). *Fedn Proc. Fedn Am. Socs exp. Biol.* **28**, 505.

Reinhardt, W.P. & Squire, P.G. (1965). *Biochim. biophys. Acta,* **94**, 566.

Samy, T.S.A., Papkoff, H. & Li, C.H. (1969a). *Archs Biochem. Biophys.* **130**, 674.

Samy, T.S.A., Papkoff, H. & Li, C.H. (1969b). *Archs Biochem. Biophys.* **132**, 315.

Squire, P.G. & Li, C.H. (1959). *J. biol. Chem.* **234**, 520.

Squire, P.G., Li, C.H. & Andersen, R.N. (1962). *Biochemistry, Wash.* **1**, 412.

Ward, D.N., Adams-Mayne, M., Ray, N., Balke, D.E., Coffey, J. & Showalter, M. (1967). *Gen. comp. Endocr.* **8**, 44.

Ward, D.N., McGregor, R.F. & Griffin, A.C. (1959). *Biochim. biophys. Acta,* **32**, 305.

On the Isolation and Characterization of Subunits of Human Pituitary Luteinizing Hormone

Leo E. Reichert, Jr., Darrell N. Ward, Gordon D. Niswender and A. Rees Midgley, Jr.

The possibility that ovine luteinizing hormone (LH) may be composed of subunits was first indicated by the studies of Li and Starman (1964) who, on the basis of sedimentation experiments, concluded that the hormone was a dimer at neutral pH's and could be dissociated into monomeric components under acidic conditions. Later studies on the nature of the N–terminal and C-terminal amino acids of ovine, bovine, porcine and human LH (Reichert, 1966) gave results which were compatible with the concept that LH consisted of more than one chain (Greep, 1968) and suggested that this may be a ubiquitous feature of heterologous LH molecules. The first direct evidence for the existence of two dissimilar subunits in LH was obtained by Ward *et al.* (1966) who proposed a model for ovine LH consisting of two chains of differing composition, but similar molecular dimension. Papkoff and Samy (1967) succeeded in isolating the subunits of ovine LH through use of counter-current distribution, opening the way for a detailed study on the primary structure of the hormone.

We have previously shown that incubation of human LH with 8M urea caused a marked decrease in its molecular dimensions, as measured by gel filtration (Reichert and Midgley, 1968). By analogy with the studies of De La Llosa *et al.* (1967) we felt it likely that this reflected a dissociation of human LH into subunits. In this report, we wish to describe our initial attempts to achieve a separation and isolation of these presumed subunits, as well as present some preliminary information on their physical properties.

Biological activity of LH and/or subunits was measured using the ovarian ascorbic acid depletion assay (Parlow, 1961) with results expressed in terms of NIH-LH-S1. One milligram of this reference preparation was considered to contain one unit of activity. Immunological activity was measured by radioimmunoassay using methods reported previously (Midgley, 1966; 1967). The standard in these assays was LER-907, a partially purified fraction derived from acetone-dried glands: 1 mg of LER-907 contained LH activity equivalent to 0.067 mg of NIH-LH-S1, as detrmined by bioassay. The radioimmunoassay values reported here were obtained through use of these conversion factors.

The human LH fraction utilized in these experiments was prepared by procedures reported elsewhere (Reichert, 1966). It had a biological activity of 2.54 u./mg, an immunological activity of 3.64 u./mg and gave no evidence of heterogeneity when examined by polyacrylamide gel electrophoresis (Reichert *et al.*, 1968). Fractions prepared in identical fashion have proven to be of immunochemical grade purity (Albert *et al.*, 1968).

The counter-current distribution (CCD) system employed by Papkoff and Samy (1967) with ovine LH, was used in our studies on the dissociation and separation of human LH subunits. This system employs 40 per cent ammonium sulphate-0.2 per cent dichloracetic acid-n-propanol-ethanol in proportions of 60:60:27:33. When applied to human LH, however, two modifications were found essential. First, sufficient dichloracetic acid was added to the lower phase to neutralize all basic groups on the hormone. This was necessary to insure adequate solubility of the LH, as well as proper distribution characteristics during partition. Second, the human hormone was incubated in the lower phase for 45 minutes at 37°C prior to partition. If this is not done, incomplete dissociation occurs, resulting in contamination of the subunit fractions with intact hormone.

In a typical experiment, 10 mg of human LH were carried through 11 transfers in the system described above, with each phase consisting of one ml. When absorbance at 275 mμ was plotted against tube number, two components were observed, one with a partition coefficient (K) of less then 0.1, and the other with a K of 2.14. Tubes 1-4 (low K fraction) and tubes 5-11 (high K fraction) were pooled, diluted with water, and then alcohol and salts removed by filtration through a Diaflo UM-2 membrane. Protein recovery was 6 mg in the low K fraction, and 1.25 mg in the high K fraction, for a total of 73.5 per cent. For convenience, the low K fraction has been designated as C-1, and the high K fraction as C-2.

Both subunit fractions showed definite biological and immunological activity. The biological activity of C-1 was 0.19 u./mg, and immunological activity was 0.55 u./mg. The biological activity of C-2 was 0.60 u./mg, and its immunological activity was 2.49 u./mg.

At this point in the development of the problem, we were required to make a decision as to how most profitably to expend the limited amounts of subunit material available. The chemical purity of C-1 and C-2 was uncertain, and our feelings that these fractions were or contained human LH subunits were based on an assumed analogy between our results and those obtained in similar experiments with ovine and bovine LH. We opted to utilize biological and immunological techniques in an effort to demonstrate the presence of subunits. Our approach was to attempt their recombination. Presumably, the reconstituted molecule would have biological and immunological activity significantly greater than could be predicted on the basis of the activity in either subunit fraction alone, and would approach the levels found in the intact precursor molecule.

The conditions described by Papkoff and Samy (1967) for recombination of ovine LH subunits required amounts of material which were prohibitive in the present study. Therefore, an alternate system was developed. 500 μg of Fraction C-1 were dissolved in 0.25 ml of 0.01M phosphate buffer, pH 7.0. 500 μg of Fraction C-2 were similarly dissolved. The two solutions were combined and allowed to incubate for 16 hours at 38°C. Initial assays were performed immediately following the incubation, and aliquots were then frozen until thawed for subsequent analysis. The biological activity of the incubated mixture of subunit fractions C-1 and C-2 was 1.67 u./mg and its immunological activity was 3.30 u./mg. This represented a 4.2-fold augmentation of biological activity and a 2.2-fold augmentation of immunological activity over levels which would be predicted assuming no recombination of subunits (Table I). Recovery of biological activity was 67 per cent and of immunological activity, 91 per cent. The results were interpreted as indicating that fractions C-1 and fraction C-2 were, or contained, subunits of human LH.

Fractions C-1 and C-2 were next examined by gel filtration through Sephadex G-100. Column size was 2.5 cm x 90.0 cm, the solvent was 0.007M phosphate-0.003M borate, pH 8.0, + 0.1M NaCl and the system was operated in the cold at a flow rate of 15 ml per hour. Ultra-violet absorbancy was monitored at 215 μ. Column charge was 1 mg in 1 ml of buffer. Each fraction contained a minor component (about 25 per cent) emerging from the column with an average distribution coefficient *Kav, of 0.32. Fraction C-1 had a major component with a Kav = 0.49, and fraction C-2 had a major component with a Kav = 0.46. Since highly purified human LH, when run in the identical system, had a Kav = 0.32, it was felt the minor component present in fractions C-1 and C-2 probably represented undissociated LH. In earlier studies (Reichert and Midgley, 1968) urea-dissociated human LH was shown to emerge from a similar gel filtration system with a Kav value of 0.46. Therefore, it seemed reasonable to conclude that the larger Kav component in each fraction represented an isolated subunit. We did not feel the possible presence of undissociated hormone in C-1 and/or C-2 necessitated a reinterpretation of the biological and immunological augmentation data presented earlier and summarized in Table I.

*The average distribution coefficient, Kav = Ve-Vo/Vt-Vo, where Ve = elution volume of the protein; Vo = outer volume of column, and Vt = total volume of column. For a discussion of this value see Siegel and Monty, 1965.

TABLE I

Description	Biologic activity[a] u/mg	Immunologic activity[a] u/mg
Intact human LH	2.54	3.64
Fraction C–1 (low K subunit)[b]	0.19	0.55
Fraction C–2 (high K subunit)	0.60	2.49
Fraction C–1 + C–2[c]	1.67	3.30
Predicted potency[d]	0.40	1.52
Augmentation	4.2	2.2

a. Biologic activity determined by OAAD bioassay, immunologic activity by radioimmunoassay. See text for details.

b. Fractions obtained after CCD but before gel filtration. Note that these values may reflect, in part, the presence of undissociated LH.

c. Values obtained after incubation as described in text.

d. Predicted on the basis of a 1:1 mixture (by weight) with no augmentation.

A number of physical constants were calculated based on the assumption that the Kav = 0.42 and Kav = 0.46 components represented subunits of human LH. The approach used has been described in detail in recent studies with bovine LH subunits (Reichert *et al.* 1969). In brief, the Kav for a number of standard proteins are experimentally determined using the identical system employed for gel filtration of the subunit fractions. From this, a plot is made of the $Kav^{1/3}$ vs the Stokes radius of these proteins, values for which are obtained from the literature. The Kav for the unknown protein is then determined and the Stokes radius obtained from the reference plot. Theoretical justification for this approach has been given by Siegal and Monty (1966) and Laurent and Killander (1964). Using the Stokes radius value thus determined, accepting values for the sedimentation coefficient for human LH in neutral and acid solutions as reported by Ryan (1969) and employing standard equations (see Rogers *et al.* 1965 and Siegal and Monty, 1965), we obtained the results summarized in Table II.

Adequate amounts of human C-1 (low K) subunit were prepared by gel filtration to permit preliminary determinations of its amino acid composition. In comparison with bovine LH C-1 subunit (low K), the human C-1 subunit had considerably less lysine and methionine, but

TABLE II

Description	Native LH	Subunit C–1	Subunit C–2
Stokes radius (mμ)	2.82	2.24	2.32
Molecular weight	26,100	14,600	15,300
Diffusion coefficient ($cm^2 sec^{-1}$)	7.62	9.59	9.25
Frictional ratio	1.45	1.44	1.49

significantly more serine and glutamic acid. The relative composition of the other amino acids was essentially the same. Differences in composition between the human and bovine LH C-2 subunits (high K) may be inferred on the basis of their partition coefficients in the counter-current distribution system. The K value for human C-2 was 2.14, while that for the bovine C-2 subunit was greater than 10 (Reichert *et al.* 1969). Despite these marked differences, it has been possible to form a hybrid molecule consisting of the human C-1 (low K) and bovine C-2 (high K) subunits which possesses high biological and immunological activity (Reichert *et al.* 1969).

In summary, highly purified human LH has been dissociated in subunits and these have been separated and isolated by counter-current distribution and gel filtration. It has been possible to recombine the subunits to form a molecule having biological and immunological properties similar to those of the native hormone. The Stokes radius, molecular weight, diffusion coefficient, fractional ratio of the human LH and subunits have been calculated. Differences in amino acid composition between human and bovine C-1 (low K) subunits have been noted by direct analysis, and differences between the human and bovine C-2 (high K) subunits have been inferred on the basis of varying partition coefficients.

ACKNOWLEDGEMENTS

This is publication #929 from the Division of Basic Health Sciences, Emory University, Atlanta, Georgia. One of us (ARM) is a Career Development Awardee of the National Institute of Child Health and Human Development. This study was supported, in part, by U.S. Public Health Service Grants AM-3598 (LER), HD-04105 (ARM) and AM-09801 (DNW).

REFERENCES

Albert, A., Rosemberg, E., Ross, G.T., Paulsen, C.A. & Ryan, R.J. (1968). *J. clin. Endocr. Metab.* **28,** 1214.

De La Llosa, O., Courts, C. & Jutisz, M. (1967). *Biochem. biophys. Res. Commun.* **26,** 411.

Greep, R.A. (1968). *Perspect. Biol. Med.* **12,** 124.

Laurent, T.C. & Killander, J., (1964). *J. Chromat.* **14,** 317.

Li, C.H. & Starman, B. (1964). *Nature, Lond.* **202,** 291.

Midgley, A.R., Jr. (1966). *Endocrinology,* **79,** 10.

Midgley, A.R., Jr. (1967). *J. clin. Endocr. Metab.* **27,** 295.

Papkoff, H. and Samy, T. (1967). *Biochim. biophys. Acta,* **147,** 175.

Parlow, A. (1961). In *Human Pituitary Gonadotropins,* p. 301. Ed. Albert, A. Springfield, Illinois: Thomas.

Reichert, L.E., Jr. (1966). *Endocrinology,* **78,** 186.

Reichert, L.E., Jr., Kathan, R. & Ryan, R. (1968). *Endocrinology,* **82,** 109.

Reichert, L.E., Jr. & Midgley, A.R., Jr. (1968). In *Gonadotropins, 1968,* p. 25. Ed. Rosemberg, E. Los Altos, California: Geron-X.

Reichert, L.E., Jr., Niswender, G.D., Midgley, A.R. & Ward, D.N. (1969). In *Program of Fifty-first Meeting, Endocrine Society,* p. 51. New York, N.Y.

Reichert, L.E., Jr., Rasco, M.A., Ward, D.N., Niswender, G.D. & Midgley, A.R., Jr. (1969). *J. biol. Chem.* **244,** 5110.

Rogers, K.S., Hellerman, L. & Thompson, T.E. (1965). *J. biol. Chem.* **240,** 198.

Ryan, R.J., (1969). *Biochemistry, Wash.* **8,** 495.

Siegel, L.M. & Monty, K.J. (1965). *Biochem. biophys. Res. Commun.* **19,** 494.

Siegel, L.M. & Monty, K.J. (1966). *Biochim. biophys. Acta,* **112,** 346.

Ward, D.N., Fujino, M. & Lamkin, W.M. (1966). *Fedn Proc. Fedn Am. Socs exp. Biol.* **25,** 348.

Amino- and Carboxy-terminal Sequences of Human Chorionic Gonadotrophins

Om P. Bahl

Previous communications from this laboratory reported a method for the purification of human chorionic gonadotrophin (HCG) and its physicochemical properties (Bahl, 1969a). The number and size of the carbohydrate units and their tentative monosaccharide sequence was also reported (Bahl, 1969b). The molecular weight of this preparation of HCG, determined by the sedimentation equilibrium method was 47,000 ± 3,000 and that determined by gel filtration was 59,000 ± 4,000. The reduction and carboxamidomethylation yielded a product with a molecular weight of approximately half of the intact molecule. Bell *et al.* (1969) obtained results similar to ours on the molecular weight of HCG as well as its reduced-alkylated derivative and on the chemical composition. On the basis of our preliminary observations we postulated the presence of two identical chains in HCG. Further evidence in support of this based on a careful examination of the termini of the polypeptide chains, particularly the carboxy-termini, is presented below.

Before any studies on the sequence of amino acids in HCG can be initiated, a knowledge of the number of polypeptide chains or the subunit structure of HCG is absolutely essential. Therefore, the present investigation to determine the carboxy- and amino-terminal amino acids was undertaken. Furthermore, since HCG is not readily available, it is advisable that its homogeneity be ascertained by as many criteria as possible before subjecting it to detailed sequence studies. Consequently, another correlated objective of these studies was to gain more information on the purity of the starting material.

EXPERIMENTAL AND RESULTS

HCG employed in the present work was obtained as previously described (Akabori *et al.* 1952). The carboxamidomethyl HCG was prepared by the reduction of HCG with dithiothreitol followed by alkylation of the reduced material with a twenty-fold excess of freshly crystallized iodacetamide. The conditions used for the reduction and alkylation were essentially similar to those described by Bell *et al.*

(1969). The enzymes carboxypeptidase A and B (DFP-treated) and leucine aminopeptidase were purchased from Worthington, Freehold, New Jersey. Leucine aminopeptidase was treated with di-isopropylphosphofluoridate (DFP) (5 μl of DFP to 5 mg of the enzyme in 0.3 ml solution) before use. Prior to the digestion with carboxypeptidase, the HCG was denatured with alcohol. The digestion was performed in 0.1 M ammonium bicarbonate solution, pH 8.5 (Potts, Berger, Cooke and Anfinsen, 1962). The digestion with leucine aminopeptidase was carried out in 0.1 M triethylamine acetate buffer, pH 8.0, in the presence of Mg^{++} and Mn^{++} ions (Potts, 1967). Hydrazinolysis was carried out by heating 1 to 1.5 μmol of HCG with 1 ml of anhydrous hydrazine at 105°C in an evacuated sealed tube for 10 hours. After the removal of the excess reagent *in vacuo* over P_2O_5, the residue was dissolved in sodium citrate buffer (0.2 M Na^+), pH 2.2. The solution was then analyzed using a Spinco Automatic Amino Acid Analyzer (Spackman *et al.* 1958). The anhydrous hydrazine was prepared by refluxing 95 per cent hydrazine with CaO and toluene followed by distillation of the hydrazine-toluene azeotrope at 93 to 94°C (Akabori *et al.* 1952).

When intact HCG was subjected to treatment with DFP-carboxypeptidase A as well as with a mixture of DFP-carboxypeptidase A and B, one mole of serine per 27,000 g of HCG was released. The amounts of other amino acids such as tyrosine, valine, alanine, leucine and threonine were less than 10 per cent. The hydrazinolysis of HCG, however, showed only serine with a low recovery of 0.6 to 0.7 mol per 27,000 g of HCG. The hydrazinolysis procedure has been noted for low yield of the C-terminal amino acid (Braun and Schroder, 1967). Table I indicates the kinetics of the release of serine with time from HCG by carboxypeptidase. When desialyzed HCG obtained by the hydrolysis of HCG with 0.025 N HCl at 80°C, for 1 hour (Bahl, 1969b) was treated with DFP-carboxypeptidase A, two amino acids, serine and tyrosine, were liberated. After 24 hours approximately 1 mol of each amino acid per 24,000 g of HCG was liberated. The amounts of the other amino acids mentioned above were negligible in this case also. The rate of release of amino acids with time from desialyzed HCG is given in Table II. The hydrazinolysis of the desialyzed HCG yielded only serine, the recovery being 0.6 to 0.7 mol per 24,000 g of desialyzed HCG. Finally, when the reduced and carboxamidomethylated HCG was digested with carboxypeptidase A and B, several amino acids, indicated in Table III, were liberated. The kinetics of their release is shown in Table III.

Leucine aminopeptidase (LAP) did not release any amino acid from the intact or the desialyzed HCG. However, when the reduced and carboxamidomethylated HCG was digested with LAP, several amino acids were liberated (Table IV). After 60 hours 1 mol each of serine and lysine was released per 26,000 g of HCG. The amounts of other amino acids such as aspartic acid, threonine, glycine, alanine, leucine and tyrosine varied between 20 and 25 per cent.

TABLE I

Action of carboxypeptidase A on HCG*

Amino acid	Moles of amino acid per 27,000 g			
	4 hr.	13 hr.	24 hr.	42 hr.
Serine	0.56	0.77	0.95	0.95

* A solution of 7.71 mg of HCG in 0.2 ml water was treated with 5 ml of ethanol. After evaporating the reaction mixture to dryness, the residue was dissolved in 0.9 ml water. A 0.1 ml solution of DFP-carboxypeptidase A and B (0.25 mg protein) in 2 M ammonium bicarbonate, pH 8.5, was added and the digest was incubated at 37°C. Aliquots of 100µl at different time intervals were analyzed using an Automatic Amino Acid Analyzer.

TABLE II

Action of carboxypeptidase A on desialyzed HCG*

Amino acid	Moles of amino acid per 25,000 g				
	1 hr.	2 hr.	5 hr.	13 hr.	24 hr.
Serine	0.57	0.91	1.08	1.00	0.95
Tyrosine	0.38	0.56	0.60	0.75	0.90

* A sample of 5.38 mg protein in 0.5 ml water was treated with 100 µl of DFP-carboxypeptidase A (0.25 mg protein) in 2 M ammonium bicarbonate, pH 8.5. Aliquots of 100 µl at different time intervals were analyzed using an Automatic Amino Acid Analyzer.

TABLE III

Action of carboxypeptidase A on carboxamidomethyl HCG*

Amino acid	Moles of amino acid per 25,000 g		
	4 hr.	17 hr.	24 hr.
Serine	0.61	0.98	1.05
Tyrosine	0.53	0.92	1.10
Valine	0.46	1.30	1.44
Alanine	0.27	0.90	1.10
Threonine	0.26	0.57	0.63
Leucine	0.18	0.58	0.77
Methionine	0.09	0.37	0.42

* A sample of 5.91 mg of carboxamidomethyl HCG in 0.9 ml of water was treated with 0.1 ml of carboxypeptidase A and B in 2 M ammonium bicarbonate, pH 8.5. Aliquots of 100 μl were analyzed using an Automatic Amino Acid Analyzer at different time intervals.

TABLE IV

Digestion of carboxamidomethyl HCG with leucine aminopeptidase*

Amino acid	Moles of amino acid per 26,000 g		
	12 hr.	36 hr.	60 hr.
Serine	0.33	0.80	1.10
Lysine	0.33	0.74	0.80

* A sample of 8.71 mg of HCG in 500 μl of triethylamine acetate buffer and 0.002 M $MgCl_2$, pH 8.1, was incubated with 400 μl of the enzyme solution (0.5 mg) previously dialyzed against the same buffer. Aliquots of 200 μl taken at different time intervals were analyzed using an Automatic Amino Acid Analyzer.

DISCUSSION

Since intact HCG treated with carboxypeptidase A and B as well as with hydrazine yielded only 2 mol of serine per 54,000 (Table I), it suggests that HCG probably has two chains, each being terminated at the carboxy-terminus by serine. Denaturation of HCG with alcohol or urea prior to digestion with carboxypeptidase still did not yield any amino acid other than serine. The desialyzed HCG, which would be expected to be quite denatured, with carboxypeptidase A gave 2 mol of tyrosine in addition to serine per 48,000 g of HCG. The rate of release of these amino acids with time (Table II) suggests that the sequence at the C-terminus is probably -Tyr-Ser. The hydrazinolysis of the desialyzed HCG again showed only serine, thus confirming the above sequence. Finally, the treatment of carboxamidomethyl HCG with carboxypeptidase A yielded several amino acids. The kinetics of their release with time (Table III) would indicate the following possible sequences at the C-terminus of each of the chains of HCG:

-Val-Ala-Val-Tyr-Ser or -Val-Val-Ala-Tyr-Ser.

Intact HCG and desialyzed HCG were not attacked by leucine aminopeptidase. However, the enzyme released two moles of serine and 2 mol of lysine per 52.000 g of carboxamidomethyl HCG. Since the rates of their release at different time intervals were quite comparable, it is difficult to say which one of these is present at the amino-terminus. These preliminary observations, therefore, require further substantiation by careful chemical studies. It may be noted that the release of the above amino acids with leucine aminopeptidase was quite slow, which is consistent with the specificity of the enzyme (Smith and Hill, 1960).

The above data, particularly that on the carboxy-terminus, suggest the presence of two identical chains in HCG*; however, the final proof of this preliminary observation would come from the actual dissociation of the molecule and subsequent separation of the chains if they are different. The size of each chain, on the basis of quantitative determination of the terminal amino acids, is approximately 27,000 or 54,000 for both the chains. This is in agreement with the molecular weight of HCG and carboxamidomethyl HCG, determined by centrifugation and gel filtration (Bahl, 1969a). The mode of attachment

* *Addendum in proof.* Since the symposium, we have been able to dissociate native HCG into two subunits and subsequently separate them by chromatography on DEAE-Sephadex A-50. The two subunits are definitely nonidentical as evidenced by their peptide maps, amino acid and carbohydrate composition. The individual subunits show low biological activity but the reconstituted molecule has most of the biological activity of the native HCG. A preliminary report of this work has been published in *Federation Proceedings. Federation of American Societies for Experimental Biology* (1970), **29**, 599.

of these two chains still remains to be investigated. Preliminary attempts to dissociate HCG into two polypeptide chains by urea or guanidine hydrochloride have not been successful.

Finally, the preparation of HCG, obtained by the method previously described (Bahl, 1969a), appears to be quite homogeneous on the basis of the above studies. The carboxypeptidase treatment and hydrazinolysis of HCG or desialyzed HCG yielded serine or serine and tyrosine respectively in molar proportions and no other amino acid was formed in appreciable amounts, strongly suggesting that the preparation was homogeneous. This is further supported by the fact that leucine aminopeptidase failed to release any amino acids from the intact or desialyzed HCG. Whereas the above findings on the homogeneity of HCG are encouraging for undertaking more detailed investigation of the complete primary structure of the hormone, the problem of the subunit structure of HCG still remains to be resolved, particularly in view of the data presented by Canfield *et al.* (1970).

REFERENCES

Akabori, S., Ohno, K. & Narita, K. (1952). *Bull. chem. Soc., Japan.* **25**, 214.

Bahl, O.P. (1969a). *J. biol. Chem.* **244**, 507.

Bahl, O.P. (1969b). *J. biol. Chem.* **244**, 575.

Bahl, O.P. & Smith, E.L. (1965). *J. biol. Chem.* **240**, 3585.

Bell, J.J., Canfield, E. & Sciarra, J.J. (1969). *Endocrinology,* **84**, 298.

Braun, V. & Schroder, W.A. (1967). *Archs Biochem. Biophys.* **118**, 241.

Canfield, R.E., Agosto, G.M. & Bell, J.J. (1970). This volume, page 161.

Potts, J.T. (1967). In *Methods in Enzymology,* **Vol XI**, p. 659. Ed. Hirs, C.H.W. Potts, New York; Academic Press.

Potts, J.T., Berger, A., Cooke, J. & Anfinsen, C.B. (1962). *J. biol. Chem.* **237**, 1851.

Smith, E.L. & Hill, R.L. (1960). In *The Enzymes,* **Vol. IV**, p. 37. Ed. Boyer. P.D., Lardy, H.A. & Myrback, K. New York: Academic Press.

Spackman, D.H., Stein, W.H. & Moore, S. (1958). *Analyt. Chem.* **30**, 1190.

Studies of the Chemistry of Human Chorionic Gonadotrophin

Robert E. Canfield, Gladys M. Agosto and Jennifer J. Bell

The data outlined here represent a brief progress report of studies of the chemistry of human chorionic gonadotrophin (HCG) that are being conducted in our laboratory. We have recently published data for the amino acid composition of pregnancy HCG (Bell *et al.* 1969) and Bahl simultaneously published data that are in excellent agreement with ours (Bahl, 1969). The intrinsic heterogeneity in HCG preparations had been a matter of concern, and it has been re-assuring to find such excellent agreement in these reports on the composition of the hormone. The sedimentation characteristics of reduced, alkylated HCG suggested that the native molecule was composed of two subunits of identical molecular weight. Professors Papkoff, Reichert and Ward, inferring from their independent studies of LH, have insisted privately that these HCG subunits, while identical in molecular weight, should in fact not be identical in their primary structure. The data that follows suggest that this is correct.

NATURE OF THE HCG SUBUNIT

Samples of purified HCG, comparable to those described earlier were reduced with Cleland's reagent (dithio-threitol) in 10 M urea and alkylated with ^{14}C-iodacetic acid. Amino acid analyses confirmed that all of the cystine had been converted to S-carboxymethylcysteine. This material was dialyzed and stored as a lyophilized powder.

The resulting carboxymethyl HCG derivative is readily soluble at alkaline pH. When aliquots of such a preparation are subjected to polyacrylamide disc gel electrophoresis (with 5 per cent cross-linked gels at a running pH of 9.3), the material staining with amido black appears to be approximately equally divided into two relatively fast moving bands (henceforth referred to as A, the faster, and B, the slower migrating band). Upon careful inspection the A band actually appears to contain two regions migrating very close together. This pattern has been reproduced with four different preparations of reduced, alkylated HCG. Figure 1 illustrates diagramatically the type of pattern that has been seen.

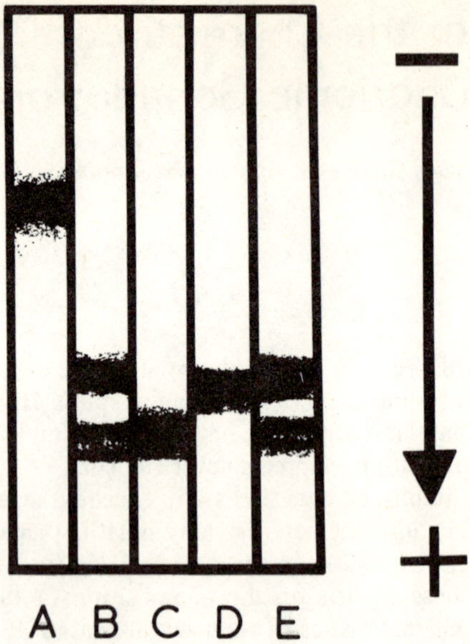

Figure 1. Diagram of gel electrophoresis results for various HCG preparations (see text). A. native HCG; B. Carboxymethyl HCG; C. peak A; D. peak B; E. Recombined peaks A and B.

As one might expect, the introduction of negatively charged groups by alkylation with iodoacetic acid increases the mobility of the peptide chains in this electrophoresis system. The multiple bands that were observed could be accounted for by either amide differences or the known heterogeneity in sialic acid content, but the division into two major regions also raised the possibility of non-identical subunits. Accordingly, preparative gel electrophoresis was performed employing the same buffer system, and the eluted material was found to be distributed into two radioactive peaks that also coincided with the peaks of optical density at 280 mμ (Fig. 2). When aliquots of the two peak fractions were subjected to analytical disc gel electrophoresis, each appeared as a broad band with distinctly different migration characteristics (Fig. 1C and D). When these two were recombined, a pattern identical to that of the original carboxymethyl HCG was reproduced (Fig. 1E). Aliquots of the fractions from each peak were dialyzed against water, lyophilized and then subjected to amino acid analysis. The results are listed in Table I.

The samples from peak A contain twice as much lysine as arginine while the reverse is true of the material from peak B. The peptide chains in peak A had more S-carboxymethyl cysteine than those of

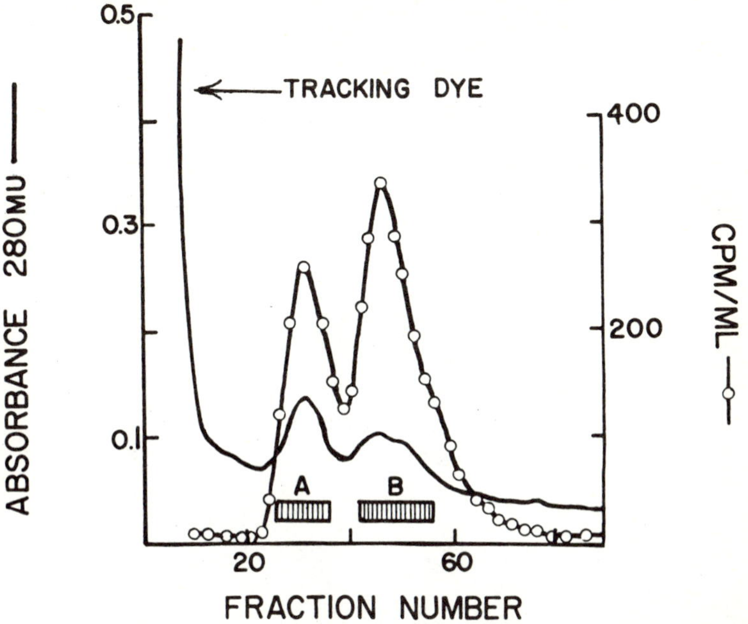

Figure 2. Results of the preparative gel electrophoretic separation of the subunits of carboxymethyl HCG. The fractions indicated by the shaded boxes (A and B) were pooled and subsequently used for the peptide map shown in Figure 3.

peak B and this probably accounts for the greater electrophoretic mobility of peak A. The total radioactivity in the A region is, nevertheless, less than that of the B region. A small amount of precipitate was noted on top of the gel, and this may have accounted for a lower yield of material in the A region. The increased content of tyrosine in the A peptide chains could account for the higher optical density at 280 mμ. Tryptophan analyses have not been performed. The remainder of the amino acid differences are less striking except for the proline difference with the peak B material containing nearly twice as much proline as that of peak A. The distribution of the basic and aromatic amino acids and proline between these two HCG subunits resembles the results for LH with the peak A peptide chain bearing similarities to the CI fragment of LH and peak B chains to the CII fragment (Papkoff and Samy, 1967).

A note of explanation must be made concerning the glycine values. The preparative electrophoresis was performed with a glycine buffer and, despite subsequent dialysis, it is apparent that the final product represented a glycine salt of carboxymethyl HCG. Accordingly, in carrying out the percentage calculation we have made the assumption

TABLE I

Amino acid analyses of subunits isolated from pregnancy HCG†

	Peak A		Peak B		$\frac{A+B}{2}$	Pregnancy ‡ HCG
	μm	(%)	μm	(%)		
Lys	.183	6.63	.125	3.42	5.02	4.3
His	.076	2.75	.027	0.73	1.74	1.8
Arg	.088	3.19	.310	8.49	5.84	6.5
Smc	.298	10.80	.262	7.17	8.98	7.6
Asp	.186	6.74	.278	7.61	7.17	8.4
Thr	.220	7.97	.240	6.57	7.27	7.5
Ser	.242	8.77	.292	8.00	8.38	8.4
Glu	.300	10.87	.250	6.84	8.85	8.6
Pro	.250	9.06	.623	17.06	13.06	12.6
Gly*	.372 (.156)	6.00	.391 (.207)	6.00	6.00	6.0
Ala	.130	4.71	.201	5.50	5.10	5.8
Val	.205	7.43	.284	7.78	7.60	7.3
Met	.055	1.99	.026	0.71	1.35	0.7
Ile	.036	1.30	.117	3.20	2.25	2.6
Leu	.120	4.35	.287	7.86	6.10	6.7
Tyr	.102	3.69	.070	1.91	2.85	2.7
Phe	.111	4.02	.051	1.39	2.70	2.6
Trp**	–	–	–	–	–	–

† Corrections have not been made for destruction during hydrolysis.

‡ Taken from Bell *et al.* (1969)

* The values in parentheses are arbitrary and are equivalent to a content of 6%. (See Text)

** Tryptophan analyses have not been performed.

that the glycine is 6 per cent in each peptide chain, which is the content of glycine in native HCG.

Since our previous data suggested that the subunits are of approximately equal molecular weight, we have averaged the values for moles per mole and listed these next to the data for native HCG. (Table I) While there is some discrepancy between the values in these latter two columns, the average data do bear a distinct resemblance to the

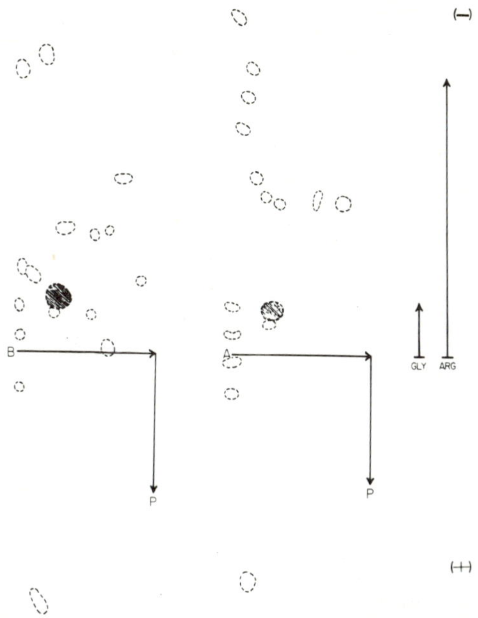

Figure 3. Peptide map of the A and B subunits of HCG following digestion with trypsin. The material was applied in the locations marked 'A' and 'B'. The first dimension (horizontal) represents descending chromatography in an n-butanol-acetic acid-water system (4:1:5) while the second dimension (vertical) reflects electrophoresis in a pyridine acetate system (pH 3.6). Phenol red was applied at the origin as a reference marker and its migration during chromatography is indicated by the horizontal arrow and its migration during electrophoresis by the vertical arrow. Both samples were known to be contaminated with glycine (see text) which appeared as a dark spot in each peptide map and these areas have been shaded. The migration of glycine and arginine standards applied prior to electrophoresis, are also indicated on the figure.

composition of native HCG. At present we are attributing these differences to analytical or purification errors.

It is theroetically possible that these non-identical subunits are produced as an artifact by proteolytic cleavage of native HCG while it is in the pooled urine specimens. If this had happened, the altered subunit might remain attached to the native molecule through a disulphide bond until the process of reduction released it. If this were the case, a 'fingerprint' pattern of the isolated subunits should reveal a number of common peptides. Figure 3 is a tracing of the peptide maps of tryptic digests of the isolated A and B subunits. It is apparent, that many, if not all, of the peptide locations differ in A and B and thus provides us with the definitive evidence that the primary structures of these subunits are not identical. The shaded areas represent the glycine that was present in both preparations as noted above. Since

^{14}C-iodoacetic acid was employed in the alkylation of the HCG subunits, it was possible to locate the S-carboxymethyl cysteine-containing regions of the peptide maps by autoradiography. While these results are not shown in Figure 3, it can be reported that the radioactive regions also have distinctly different patterns of distribution in A and B.

Our peptide mapping conclusions are regrettably in disagreement with those reported by Bahl (1969). However, we believe that all the results noted here conclusively establish that native HCG is composed of two non-identical subunits, a situation analagous to that which has been found in LH (Papkoff and Samy, 1967). This conclusion doubles the amount of work and material that will be necessary for a determination of the amino acid sequence of HCG.

SPECULATIONS

Whenever nonidentical subunits are found in a cysteine-rich protein, one is faced with a question of how the newly synthesized subunits achieved their native conformation. Specifically I am raising the question of whether or not there is a 'proHCG' or a 'proLH' analagous to the proinsulin found by Steiner and Oyer (1967). Reichert has shown that under certain conditions subunits of LH can be dissociated and then will spontaneously reaggregate to form biologically active hormones (Reichert *et al.* 1969) so it would seem that a 'proLH' is not required to achieve the proper alignment of the two chains. Since HCG appears to be quite homologous with LH in its structure, it would be my guess that these subunits reflect (in both hormones) two distinctly different genes in a fashion analagous to the alpha and beta chains of haemoglobin. When the amino acid sequence of the two subunits for both LH and HCG are known, it will be possible to search for homology to determine whether or not the two chains have evolved from the same ancestral gene.

The non-identity of these subunits leads me to speculate about another possibility, namely that native HCG, while normally composed of an A-B alignment, can occur in other forms. Accidents of nature or artifacts produced during purification could also yield A-A and B-B alignments as well as single A and B units. This scrambling of subunits might account for some of the variations seen among bioassays for different batches of HCG (Hamashige *et al.* 1967). In particular, I would like to call attention to a problem that could arise in the process of eluting HCG from kaolin with M NH_4OH. Treating a cystine-rich protein in this fashion in the presence of many urinary contaminants is inviting the process of disulphide interchange, which could significantly alter the three dimensional structure of the hormone.

PRELIMINARY CHARACTERIZATION OF MALE CHORIOCARCINOMA HCG

The methods of purification of HCG in our laboratory continue to change, and I would like to record here our present technique for the isolation of HCG and to illustrate this with data concerning the composition of HCG isolated from the urine of male patients with choriocarcinoma. We are indebted to Dr Griff Ross at the National Institutes of Health and to Dr John Lewis at the Sloan Kettering Institute for assistance in obtaining the choriocarcinoma material.

The present method of purification of HCG is as follows:

Step 1. Crude commercial pregnancy HCG powder (approximately 2500 i.u./mg) is dissolved in 0.05 M sodium phosphate with pH adjusted to 5.0 and then batch adsorbed to sulphoethyl Sephadex previously equilibrated with the phosphate buffer in the cold. This step was suggested to us by Dr Papkoff, and despite the fact that HCG is an acidic protein, it adsorbs quite completely. When high titre choriocarcinoma urine is employed as a starting material, it is diluted with an equal volume of distilled water and adjusted to pH 5.0 and then similarly batch adsorbed to sulphoethyl Sephadex. After washing the ion exchanger with multiple volumes of buffer to remove the chromagens, the protein is eluted with 0.1 M sodium phosphate - 0.5 M sodium chloride (pH 5.0), dialyzed and lyophilized.

When the HCG titer of urine is low, the sulphoethyl Sephadex adsorption step is inefficient and the usual benzoic acid adsorption method is employed.

Step II. The powder that is derived from either sulphoethyl Sephadex adsorption or benzoic acid precipitation is dissolved in 0.05 M sodium phosphate (pH 5.0) and passed over a 2.5 x 200 cm column of Bio-Gel P150 (100-200 mesh) previously equilibrated with phosphate buffer.

Figure 4 illustrates the elution pattern for a sample of pregnancy HCG (top) and a sample of HCG isolated from pooled urine of male patients with choriocarcinoma (bottom). The black circles represent fractions that were taken for amino acid analysis, sialic acid analysis, disc gel electrophoresis and immunoassay. A radioactive marker (^{14}C-glucose) was added to each sample prior to application to locate the salt peak in the effluent. The major peak in the top pattern represents pregnancy HCG. A symmetrical protein peak appears in the same region of eluate when a sample of choriocarcinoma HCG is applied to the Bio-Gel column, and as shown by the dashed lined (open circles) a protein elutes in this region that cross-reacts with pregnancy HCG in the immuno-assay. Although they are not plotted, the proline values have been determined and they are symmetrical with the optical density at this region (280 mμ.) suggesting that this peak represents

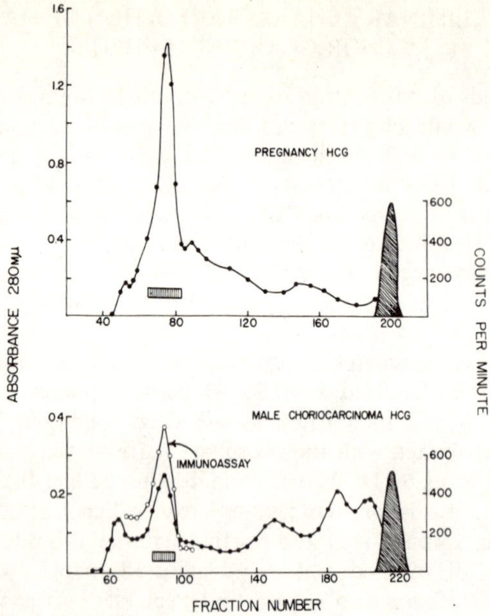

Figure 4. Gel filtration of HCG preparations on Bio-Gel P-150. The top portion of the figure depicts the elution pattern for a sample of pregnancy HCG. The lower portion depicts the elution pattern for a sample of HCG prepared from a pool of urine obtained from male patients with choriocarcinoma. Radioactive glucose was added as a marker and is represented by the shaded area. The fractions indicated by the black dots were saved for characterization as described in the text. The results of radioimmunoassay as shown in the lower figure indicate the point of emergence of protein that is immunologically similar to pregnancy HCG.

choriocarcinoma HCG with little if any contamination. Disc gel electrophoresis of samples from this region reveals broad bands of approximately identical electrophoretic mobility. Table II contains amino acid analyses for the three fractions at the top of the peak – expressed as moles per mole of HCG. These analyses are in fairly good agreement lending further support to the notion that a single species of protein molecules have been eluted in this region.

Clearly there is more work to be done to characterize this male choriocarcinoma HCG and to search for chemical differences from pregnancy HCG. However, I have presented these preliminary data here because the average of the amino acid analyses of the choriocarcinoma HCG fraction is quite similar to that of the composition of pregnancy HCG. The proline content is a particularly good index, and the 3 per cent difference is within the error of the technique of accelerated amino acid analysis. The disagreement in methionine and cystine values may reflect differences in destruction during acid

TABLE II

Amino acid analyses (Mol/Mol HCG) of Choriocarcinoma HCG fractions shown in Figure 4†

	#86	#89	#92	Average	Pregnancy ‡ HCG
Lys	4.91	4.94	4.66	4.83	4.3
His	1.96	1.82	1.76	1:84	1.8
Arg	6.85	6.91	6.46	6.74	6.5
Asp	7.44	7.19	7.64	7.42	8.4
Thr	7.47	7.38	7.61	7.48	7.5
Ser	8.62	8.32	8.58	8.50	8.4
Glu	8.17	8.00	8.02	8.06	8.6
Pro	12.70	13.55	12.74	12.99	12.6
Gly	4.62	4.58	4.68	4.62	6.0
Ala	5.15	2.18	5.29	5.20	5.8
Cys	8.64	8.19	8.48	8.77	7.6
Val	7.59	7.66	7.73	7.66	7.3
Met	1.71	1.60	1.86	1.72	0.7
Ile	2.62	2.68	2.61	2.63	2.6
Leu	6.36	6.53	6.56	6.48	6.7
Tyr	2.67	2.77	2.73	2.72	2.7
Phe	2.46	2.62	2.51	2.53	2.6
Trp*	–	–	–	–	–

† No corrections have been made for destruction during hydrolysis.

‡ Taken from Bell *et al.* (1969).

* Tryptophan analyses have not been performed.

hydrolysis. Thus we have tentatively concluded that the choriocarcinoma amino acid composition is identical to that of pregnancy HCG.

The HCG containing fractions eluted from these Bio-Gel columns are pooled, dialyzed against distilled water and lyophilized.

Step III. The Bio-Gel product is dissolved in 0.02 M NH_4HCO_3 buffer (adjusted to pH 8.5 with NH_4OH) and applied to a column of DEAE Sephadex (40 x 2.5 cm) and developed with a three chamber varigrad with the first chamber containing 0.02 M NH_4HCO_3, the second chamber containing 0.08 M NH_4HCO_3 and the third chamber containing 0.3 M NH_4HCO_3 – each adjusted to pH 8.5 with NH_4OH.

Figure 5 illustrates the pattern of optical density that emerges from such a column when pregnancy HCG (obtained in step II) is applied. The broad peak reflects the heterogeneity of charge among these molecules, but the pooled material has the same amino acid composition as that which was described earlier (Bell *et al.* 1969).

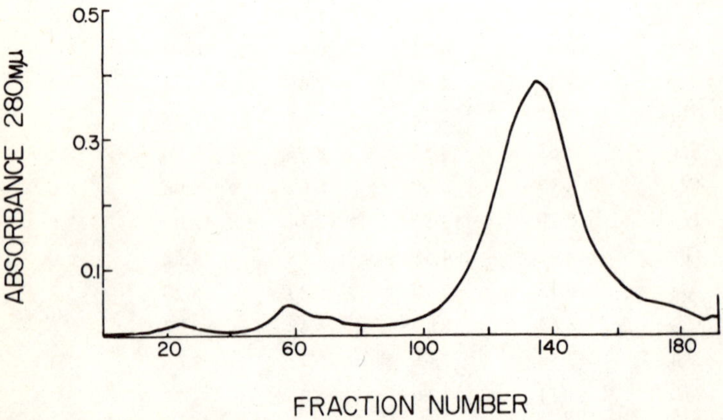

Figure 5. Chromatography of pregnancy HCG on a 2.5 x 40 cm column of DEAE Sephadex. The conditions are described above.

ACKNOWLEDGEMENTS

This study has been supported by Research Grant AM 09579 (NIH). Professor Canfield is a Career Investigator of the Health Research Council of New York.

REFERENCES

Bahl, Om P. (1969). *J. biol. Chem.* **244,** 567.

Bell, J.J., Canfield, R.E. & Sciarra, J.J. (1969). *Endocrinology,* **84,** 298.

Hamashige, S., Astor, M.A., Arquilla, E.R. & Van Theil, D.H. (1967). *J. clin. Endocr. Metab.* **27,** 1690.

Papkoff, H. & Samy, T.S.A. (1967). *Biochim. biophys. Acta,* **147,** 175.

Reichert, L.E. Jr., Midgley R.A. Jr., Niswender, G.D. & Ward, D.N. (1969). *Fifty-first Meeting of the Endocrine Society,* Americana, New York, N.Y.

Steiner, D.F. & Oyer, P.E. (1967). *Proc. natn. Acad. Sci. U.S.A.* **57,** 473.

The Effects of Chemical Modifications on the Biological and Radioimmunological Activity of Pituitary FSH

W. R. Butt, S. S. Lynch, M. F. Chaplin, C. J. Gray and J. F. Kennedy

SUMMARY

A preparation of pituitary follicle stimulating hormone was treated with neuraminidase, various glycosidases, galactose oxidase, proteolytic enzymes, periodate, alkaline borohydride, chloramine-T, diazo-sulphanilic acid and was also subjected to photo-oxidation. The effect of the reactions on biological and radioimmunological activities was investigated and the results were related to the structural alterations which are known to have occurred. Several of these reactions destroyed the biological, but not the immunological activity: the sites on the molecule which are responsible for the two activities are therefore not identical. Certain carbohydrate constituents are essential for the biological, but not the immunological activity. Histidine, methionine, and possibly tyrosine and cystine also appear to be essential for biological but not immunological activity, while tryptophan is not essential for either.

INTRODUCTION

The antigenic groups of protein hormones may be distinct from those parts of the molecule which are responsible for the biological activity. Biological and immunological assays may therefore measure different functions of the same molecule. The present investigations were designed to give information on some of those groups responsible for these activities in a preparation of pituitary follicle stimulating hormone (FSH).

MATERIALS AND METHODS

The FSH (CP1 fraction) was prepared from acetone-preserved human pituitary glands by the method of Butt *et al.* (1961) and further purified by fractionation on Bio-Gel P-150, (CP 150 fraction, Barker *et al.*, 1969).

Details of the chemical and biochemical modifications are described eleswhere (Kennedy and Butt, 1969; Chaplin *et al.*, 1970).

The biological activity of FSH was measured by the ovarian augmentation method in mice (Brown, 1955) and the immunological activity by the solid-phase radioimmunoassay of Butt and Lynch (1968).

RESULTS

Stability.

The average biological activity of four different preparations of CP 150 was 1,725 i.u./mg; the average immunological activity was 2,000 i.u./mg.

One of these preparations was stored at -10°C at pH 7.0 for 16 weeks; there was no change in the immunological activity and the final biological activity was 78 (48-127) per cent of the initial potency, which is probably not a significant change. Dialysis and freeze-drying had no effect on the immunological activity of this preparation and 74 (40-140) per cent of the biological activity was retained. Almost 70 per cent of the biological activity remained after drying the product at 60°C. However, freeze-drying of the undialysed solution caused a loss of about 50 per cent of the immunological and 71 (50-83) per cent of the biological activity.

Modifications

Neuraminidase. Incubation of CP 150 with neuraminidase removed all the N-acetylneuraminic acid residues (Chaplin *et al.* 1970) and the resulting material (CP 150 NF) contained less than 1 per cent of the initial biological activity. Initially there was only slight loss of immunological activity but after two weeks at -10°C only 44.5 per cent of the activity remained and this decreased progressively to 2.7 per cent by nine weeks.

Other glycosidases. Incubation of CP 150 NF with yeast β-D-galactosidase, Jack bean α-D-mannosidase, glycosidases from ram testes and *Coccobacillus* enzymes produced no change in the immunological activity. The only chemical change observed in these experiments was the release of about 10 per cent of the L-fucose after digestion with the *Coccobacillus* enzymes, (Chaplin *et al.*, 1970).

Galactose oxidase. Treatment of CP 150 NF with galactose oxidase caused an apparent increase of at least 100 per cent in immunological activity in two experiments. Chaplin *et al.* (1970) have shown that the enzyme was effective in oxidizing the galactose in these experiments.

Chymotrypsin and trypsin. Digestion of CP 150 with chymotrypsin caused no detectable chemical change or loss of biological or immunological activity.

Limited digestion with trypsin also appeared to cause no essential alteration of biological or immunological activities. Gel filtration, however, showed that some degradation had occurred and material of smaller molecular weight than the parent substance retained biological activity. This material was unstable, however, and the activity was lost over a period of some days but the immunological activity was retained.

Periodate oxidation and reduction. Periodate oxidation and reduction of CP1 (Kennedy and Butt, 1969) destroyed at least 98 per cent of the biological activity but more than 70 per cent of the immunological activity was retained. Most of the carbohydrate moiety was destroyed but the protein appeared to be largely intact.

Alkaline borohydride. Treatment of CP 150 with alkaline borohydride destroyed both biological and immunological activities.

Photo-oxidation. Photo-oxidation of CP 150 in the presence of methylene blue at pH 7.0 caused progressive loss of biological activity with exposure time; the immunological activity was essentially unaffected (Table I). Tryptophan was destroyed within two minutes

TABLE I

Photo-oxidation of CP 150

Time of illumination (min)	Biological activity % of original with 95% fid. lim.	Immunological activity % of original
0	100	100
2	45(20–110)	120
5	12(5–30)	145
10	7(2–15)	105

while histidine and tyrosine decreased progressively with time. Methionine was largely unchanged and cystine was only affected after prolonged exposure (Chaplin *et al.*, 1970).

Chloramine-T. Oxidation of CP 150 with a limited amount of chloramine-T (5 moles per mole of tryptophan) destroyed 39 per cent of the biological activity without affecting the immunological activity. All the tryptophan, and some of the cystine and methionine, were destroyed but histidine and tyrosine were not affected (Chaplin *et al.*, 1970). Some of the cystine oxidized could be detected as cysteic acid (Table II).

TABLE II

Effect of chloramine-T on the biological and immunological activities and oxidisable amino acids of preparations of FSH

	Initial material	Mild treatment	Drastic treatment
Biological activity %	100	61	<9
Immunological activity %	100	100	150
*Cystine	10.9	1.7	0.1
Cysteic acid	0.0	4.0	4.8
Methionine	2.0	1.0	0.1
Tyrosine	5.9	5.5	5.5
Histidine	4.9	5.1	5.7
Tryptophan	5.5	0.0	0.0

*Amino acid as number of residues adjusted to lysine = 10.

Mild treatment refers to use of chloramine-T only (5 mol per mol of tryptophan).

Drastic treatment refers to use of excess chloramine-T followed by excess metabisulphite, as in iodination of proteins for radioimmunoassay.

Oxidation of CP 150 with a large excess of chloramine-T and subsequent reduction with sodium metabisulphite did not alter the immunological activity, but all the biological activity was destroyed. Again, this treatment did not affect histidine or tyrosine, but all the tryptophan, methionine and cystine residues were destroyed (Table II).

The effect of chloramine-T on CP 150 NF was more marked and even the limited oxidation caused degradation to small molecular weight components (Chaplin *et al.*, 1970). There was, however, no change in the immunological activity and the slopes of the inhibition curves in the radioimmunoassay were not different from those of the parent material.

Diazosulphanilic acid. Treatment of CP 150 with diazosulphanilic acid reduced the biological activity by about 50 per cent but did not appreciably affect the immunological activity.

DISCUSSION

Treatment of the preparation CP 150 with alkaline borohydride destroyed both biological and immunological activities. However, many of the modifications to the chemical structure of the preparation destroyed or reduced the biological activity without simultaneously affecting the radioimmunological activity. It seems certain therefore that the sites on the molecule responsible for the two activities are not identical. Since loss of N-acetyl neuraminic acid, or fucose, or destruction of part of the carbohydrate by periodate oxidation leads to loss of biological but not immunological activity, it appears that the carbohydrate portion of the glycoprotein is essential for the biological activity and that the immunological determinants reacting with the present antiserum are on the protein portion.

The role of the N-acetyl neuraminic acid in the biological activity of FSH is not known with certainty. It may be an integral part of the biologically active site on the molecule, or be responsible for the transport of the hormone to the target organ (Ryle *et al.*, 1970). On the other hand, it may contribute to the maintenance of the tertiary structure of the molecule; the progressive loss of immunological activity on storage of the neuraminic-free derivative indicates that its structure is less stable than that of the complete molecule which remained stable on storage over 16 weeks.

The reason for the apparent increase in the immunological activity of CP 150 NF after treatment with galactose oxidase is not known. A non-specific factor may have interfered in the solid-phase system or the chemical change may have increased the affinity of the molecule for the antibody. The reason for the resistance of CP 150 NF to the action of several of the glycosidases is not known: possible explanations were discussed in the paper of Chaplin *et al.* (1970).

Destruction of tryptophan by photo-oxidation or by reaction with chloramine-T does not completely destroy biological activity. The activity seems to depend rather on histidine or tyrosine which are progressively affected by photo-oxidation; diazosulphanilic acid also affects these two residues but the degree of reaction with CP 150 is not yet known. The biological activity also depends on methionine or possibly cystine which are destroyed by reaction with chloramine-T. The immunological activity, on the other hand, is not affected by the changes in any of these amino acids.

The immunological identity of the parent material with the fragments produced by the action of trypsin on CP 150 and the reaction of chloramine-T with CP 150 NF suggests that the molecule is possibly composed of sub-units each of which carry antigenic sites.

ACKNOWLEDGEMENTS

The authors thank Professor M. Stacey, Dr S. A. Barker and Dr A. C. Crooke for their advice and interest in this work, and Mrs C. Law, Miss J. Stacey and Miss J. Clare for their technical assistance. The work was supported by a block grant from the Medical Research Council and the Ford Foundation. One of us (M.F.C.) acknowledges receipt of a research studentship from the Medical Research Council.

REFERENCES

Barker, S.A., Gray, C.J., Kennedy, J.F. & Butt, W.R. (1969). *J. Endocr.* **45**, 275.

Brown, P.S. (1955). *J. Endocr.* **13**, 59.

Butt, W.R., Crooke, A.C. & Cunningham, F.J. (1961). *Biochem. J.* **81**, 596.

Butt, W.R., Lynch, S.S. (1968). *Clinica Chim. Acta,* **22**, 79.

Chaplin, M.F., Gray, C.J. & Kennedy, J.F. (1970). This volume, page 77.

Kennedy, C.J. and Butt, W.R. (1970). *Biochem. J.* **115**, 225.

Ryle, M., Chaplin, M.F., Gray, C.J. and Kennedy, J.F. (1970). This volume, page 98.

Evidence for the Presence of Fragments of FSH in Plasma

W. M. Hunter and H. K. Amin

Present criteria for the specificity of radioimmunoassays for FSH are: absence of any effect when plasma from hypophysectomised subjects are included in the incubates, absence of cross reaction with other pituitary hormones - in particular TSH and LH – and, finally good agreement with bioassay measurements in pituitary material which should include fractions having a wide variety of different FSH:LH ratios. With these criteria satisfied, a number of radioimmunoassays have been established and applied to the measurement of FSH in plasma and urine (Franchimont, 1966, 1968; Faiman and Ryan, 1967). For measurements in urine agreement between bio- and immunoactivity is not essential since urine does not impinge upon target tissues. The value of an immunoassay for urinary measurements would be established if it could be shown that a constant relationship between secreted and excreted FSH existed. The specificity requirements for an immunoassay on blood are quite different however. Here good agreement between immunoactivity and biological activity would appear to be crucial. Unfortunately it is precisely here that the limitations of sensitivity of bioassays, combined with the difficulty of concentrating the gonadotrophins in blood, make such a comparison technically formidable. We have investigated some of these relationships by assaying with two antisera having different specificities, the blood and urine of subjects after i.v. administration of FSH.

There were two assay svstems.

(A) An FSH-anti-FSH system. ^{125}I-FSH was prepared as previously described (Hunter, 1967) from fraction CP_1 cellosolve kindly supplied by Dr W. R. Butt of Birmingham Hospital for Women. The antiserum, anti -FSH-R-22, prepared by Dr B. A. L. Hurn of Burroughs Wellcome was not completely specific for FSH. However, it did not cross react with plasma proteins or with TSH. The preparations used for further specificity studies were FSH, fraction CP_1 cellosolve, which had a biological activity of 800 i.u./mg and which was assumed to be 40 per cent FSH, and LH fraction IRC-50 having a biological activity of 7,750 i.u./mg kindly supplied by Dr Anne Stockell Hartree, Department of Biochemistry, University of Cambridge, which was assumed to represent the pure hormone. Antiserum R-22 showed a complete cross reaction with LH, inhibition curves for the addition of unlabelled FSH and LH were parallel and both were able completely to inhibit binding of ^{125}I-FSH. However, eight times as much LH as FSH by weight was required for equal inhibition. Ten pituitary fractions whose FSH and LH contents were known from bioassay data were assayed in this system and good agreement was found with the sum of the biological activities of all of them provided the LH content was assumed to be only one-eighth of its bioassay value. This suggested (1) that there was little difference between biological and immunological activity for the FSH content of these fractions when this antiserum was used, and (2) that the antiserum did not cross react with other non-hormonal proteins which might be present in the pituitary fractions.

(B) The second immunological system used ^{125}I-LH prepared by iodination of IRC-50 and a subsequent purification procedure involving the adsorption of undamaged fraction to a cellulose column and its elution by means of horse serum. The antiserum, rabbit anti-LH-R6 showed superimposable inhibition curves for LH and FSH when the two hormones were compared on a weight basis. This system did not cross react with plasma proteins. The pituitary mixtures of LH and FSH were also assayed in this system and very poor agreement was found between the values obtained and those expected from the sum of the two biologically determined gonadotrophin contents. This indicated that biologic and immunological activities of the gonadotrophins present were not parallel and/or that the system was measuring other non-hormonal proteins in addition.

The material for i.v. injection was human pituitary fraction CP_1 (batch 60) from Dr Butt. This is the fraction which has been used in Birmingham (by the intramuscular route) for the treatment of infertility in women (Crooke *et al.*, 1963). It was dissolved in saline at 1 mg/ml, filtered through a millipore filter, ampouled and stored at -20°C. The ampouled material was kindly assayed for us by the rat ovarian augmentation assay by Mr S. F. Lunn of the Clinical Endocrinology Unit in Edinburgh and was found to have a potency of 227 i.u./mg (180 - 280 fiducial limits). Using as standard fraction CS-10 which was ascribed a potency of 800 i.u./mg from its bioassay value the ampouled CP_1 60 had an immunoassay value of 244 i.u./mg in immunoassay system A. Dr Butt was kind enough to check the concentration of the ampouled solution by E_{280}. E_{220}, and biuret reaction all of which showed that losses in filtration etc. were negligible at <10 per cent. An LH content of 52 i.u./mg (i.e. 0.7 per cent LH) was found in a specific LH radioimmunoassay, a value which is in agreement with bioassay estimations on other batches of the same fraction.

Following the i.v. injection of 1 mg of this material into three healthy men, an initial rise in plasma 'FSH' during the first 3 to 7 minutes was followed by a rapid fall, presumably due to extravascular mixing. Subsequently, after 15 minutes from the time of the injection the fall in plasma levels conformed to a simple exponential plot which became linear when log 'FSH' was plotted against time. Pre-injection values were regained by 2 to 3 hours without any apparent discontinuity in the curves and plasma half lives of 24, 27 and 38 min. were found for the three subjects.

Using the LH-anti-LH system B a very different apparent disappearance rate was found. After the same pattern during the first minutes the portion of time which had been log linear in system A now showed a complex flattening curve. The semi-logarithmic plot was not linear, pre-injection levels were not regained until 6 hours after the injection and no half life could be calculated. These data suggest that fragments of FSH produced by degradation of the injected material, possibly by the liver or kidney, were present in the plasma after its i.v. injection. An alternative, though less likely, possibility is that non-hormonal components of the injected material might have become altered *in vivo* and that these might then be detected in the non-specific immunoassay system. These altered fragments clearly were not detected by the more specific system A.

Of these two pictures of the plasma disappearance of FSH, that found with system A would seem more likely to approach the true biological half life. Firstly the assay itself is more specific as judged by its performance on pituitary material. Secondly it gave an exponential curve and this has been found also for disappearance rate studies in other specific radioimmunoassays for HGH (Parker *et al.*, 1962) and insulin (Stimmler, 1967). However, in no case yet has the definitive experiment, viz. a comparison of plasma disappearance rates by bioassay and immunoassay been reported.

Turning now to the urinary recovery of the injected FSH our data show completely identical time courses and overall urinary recoveries for the two immunoassay systems. Urine was dialysed against the buffer used in the immunoassay diluent and carrier horse serum was added to mimic the assay diluent. The high and (more important) constant recovery in urine which was 37,38 and 42 per cent of the injected dose by 24 hours after injection suggests that immunoassay of FSH in urine may provide a good measure of secretion rates. After 5 hours 29, 28 and 32 per cent of the total dose was present in the urine. The rapidity with which the injected hormone appeared in the urine also indicates some value for urinary measurements in following the fine detail of the changes in secretion.

The possibility of repeating these assays on plasma and urine following the injection of pituitary FSH, when more completely specific anti-FSH sera become available is awaited with interest.

REFERENCES

Crooke, A.C., Butt, W.R., Palmer, R.F., Morris, R., Edwards, L.R. & Anson, C.T. (1963). *J. Obstet, Gynaec. Br. Commonw.* **70**, 604.

Faiman, C. & Ryan, R.J. (1967). *Proc. Soc. exp. Biol. Med.* **125**, 1130.

Franchimont, P. (1966). *J. label led Compounds* **2**, 303.

Franchimont, P. (1968). In *Protein and Polypeptide Hormones.* Part 1, p. 99, Ed, Margoulies, M. International Congress Series 161. Amsterdam, Excerpta Medica.

Hunter, W.M. (1967). In *Recent Research on Gonadotrophic Hormones.* p.91, Ed. Bell, E.T. & Loraine, J.A. Edinburgh, Livingstone.

Parker, M.L., Utiger, R.D. & Daughaday, W.H. (1962). *J. clin. Invest.* **41**, 262.

Stimmler, L. (1967). *Diabetes,* **16,** 52.

The Cross Reaction between Human Chorionic Gonadotrophin and Hypophyseal Luteinizing Hormone

P. Franchimont

Many authors have demonstrated the existence of a cross reaction between HCG and LH by means of the classical immunological techniques: double diffusion in agar gel, immunoelectrophoresis, haemagglutination and inhibition of the biological activity of the two preparations (Wide *et al.*, 1961; Mougdal and Li, 1961; Goss and Taymor, 1962; Franchimont, 1963; Robyn *et al.*, 1963; Shahani and Rao, 1964; Midgley, 1966; Bagshawe *et al.*, 1966). The overall results generally indicate that there is a complete antigenic identity between LH and HCG. Goss and Lewis (1964), however, believe that the cross reaction is only partial. This article will report the results of radioimmunological studies of the behaviour of LH and HCG.

MATERIALS AND METHODS

1. *Hormone preparations.* FSH (provided by Dr Ryan) with a biological activity of 103.3 i.u. NIH-FSH-S1 per mg.
LH (provided by Dr Hartree) with a biological activity of 7,500 i.u. 2nd IRP-HMG per mg.
HCG (provided by Organon laboratories) with a biological activity of 14,700 i.u. per mg.

2. *Antisera.* An examination was made of 43 anti-HCG sera which were capable of inhibiting the biological activity of HCG. They were incubated with serum from normal children before use.

3. Hormones were labelled with ^{131}I according to Greenwood *et al.* (1963).

4. Radioimmunoassays were performed as previously described (Franchimont, 1966; 1968) by means of the double antibody technique.

5. Labelled HCG (HCG*) and labelled LH (LH*) were examined by starch gel electrophoresis.

RESULTS

1. LH* migrates into the slow α_2-globulin and lipoprotein zones whereas HCG* migrates nearer the anode and is localized in the β-globulin and fast α_2-globulin zones.

2. With many anti-HCG sera the cross reaction between LH and HCG is complete as shown by the similar inhibition of binding to antibody obtained with both hormones. With certain antisera, FSH behaves in the same manner.

3. With a few anti-HCG sera, LH and HCG are dissimilar in their inhibition of HCG* bound to antibody. The percentage bound is maintained at a higher level with unlabelled LH than with unlabelled HCG. On the contrary, with the same antisera the decrease in LH* bound to antibody is identical with both unlabelled HCG and LH (Fig. 1).

4. With one antiserum, the percentage of HCG* bound to antibody was decreased by unlabelled HCG but only slightly reduced by unlabelled LH and not at all modified by unlabelled FSH. A 1/18,000 dilution of this antiserum bound 75 per cent of HCG* and only 16 percent of LH* (Fig. 2).

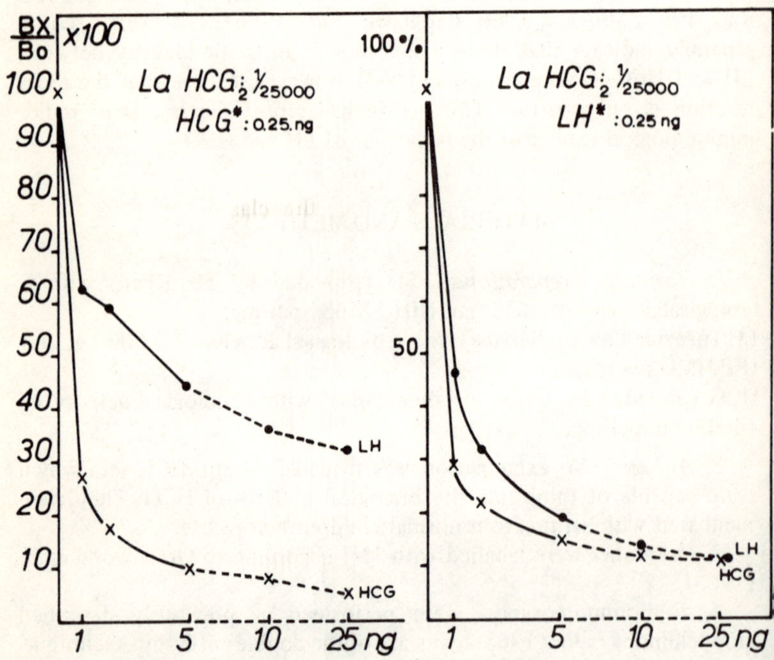

Figure 1. Inhibition curves for LH and HCG with anti-HCG serum (La-HCG) and (1) HCG*, (2) LH*

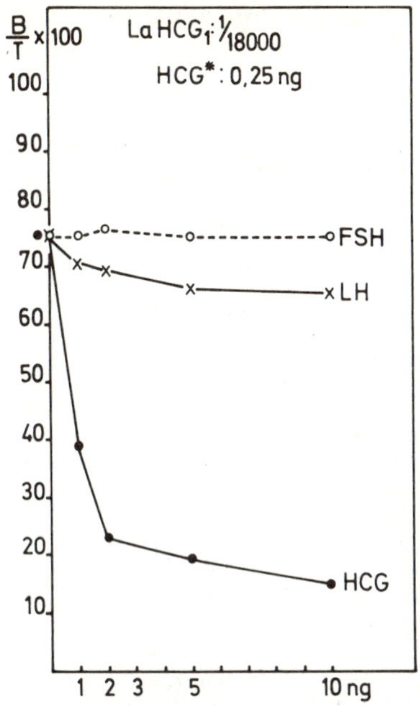

Figure 2. Inhibition curves for FSH, LH and HCG with anti-HCG serum (La-HCG) and HCG*

DISCUSSION

These experiments indicate that the cross reaction between HCG and LH is incomplete. Indeed, there are some anti-HCG sera containing specific antibodies to HCG which are unable to react with LH. With these antisera, unlabelled LH and HCG behave similarly in their inhibition of LH* bound to antibody but differently in their inhibition of HCG* bound to antibody. Moreover, with one antiserum, LH* bound poorly to antibody and unlabelled LH produced only a slight decrease in the percentage of HCG* bound to antibody.

HCG and LH differ with respect to their electrophoretic behaviour on starch gel. HCG migrates to a more anodal zone that LH.

Our results confirm the cross reaction between HCG, LH and FSH. Several of the anti-HCG sera contain antibodies directed only against the common antigenic sites of the three hormones, while others contain antibodies to the two luteinizing hormones but cannot react with FSH. These antisera can therefore be utilized in the radioimmunoassay of LH and HCG. If, however, any of the antibodies in serum are directed specifically against HCG, the radioimmunoassay of LH would have to make use of labelled and unlabelled LH.

CONCLUSIONS

1. The cross reaction between HCG and LH is only partial;

2. The electrophoretic behaviour of these two hormones on starch gel is different;

3. One of the 43 anti-HCG sera tested contained specific antibodies directed against HCG.

REFERENCES

Bagshawe, K., Wilde, C. & Orr, A. (1966). *Lancet,* **1,** 1118.

Franchimont, P. (1963). *Annls Endocr.* **24,** 112.

Franchimont, P. (1966). In *Le Dosage des Hormones Hypophysaires Somatotrope et Gonadotropes et son Application Clinique.* Arscia Bruxelles.

Franchimont, P. (1968). In *Protein and Polypeptide hormones,* p. 99. Ed. Margoulies, M. Int. Congr. Series, No. **161,** Amsterdam: Excerpta Medica.

Goss, D. & Lewis, J. (1964). *Endocrinology,* **74,** 83.

Goss, D. & Taymor, M. (1962). *Endocrinology,* **71,** 321.

Greenwood, F., Hunter, W. & Glover, J. (1963). *Biochem. J.* **89,** 114.

Midgley, A.R. (1966). *Endocrinology,* **79,** 10.

Mougdal, N. and Li, C. (1961). *Nature, Lond.* **191,** 192.

Robyn, C., Limbosch, J. and Hubinont, P. (1963). *Gen. Comp. Endocr.* **3,** 88.

Shahani, S. and Rao, S. (1964). *Acta Endocr.* **46,** 317.

Wide, L., Roos, P. and Gemzell, C. (1961). *Acta Endocr.* **37,** 445.

Production Rate of Follicle Stimulating Hormone in Adult Human Males

Salvatore Raiti, Robert M. Blizzard, Robert Penny and Claude J. Migeon

SUMMARY

The production rate of immunological FSH was determined in six adult males by injecting ^{125}I-FSH intravenously and recovering immunological ^{125}I-FSH in the urine over a three day period. Radioimmunoassay (^{131}I system) was used to measure the stable FSH excreted during this period. Between 20 and 50 i.u. (2nd IRP-HMG.) per day was the calculated production rate in the subjects studied. Labelled and non-labelled FSH were both found to have a disappearance rate (t ½) of 6 hours. FSH renal clearance was found to be 0.64 ml/min. in the one subject studied.

INTRODUCTION

In 1957, Pearlman described a method for measuring the production rate of progesterone in women. Following the injection of a known amount of tritium labelled steroid, the percent excretion of the tritium as a specific urinary metabolite was determined and the specific activity of this metabolite was determined in the urine. From these data the production rate was determined. This methodology is now used extensively to study the production and secretion rate of steroid hormones.

We have adopted this technique to determine the production rate of a protein hormone, follicle- stimulating hormone (HFSH) in man.

In Table I the criteria required for measuring production rates are listed.

METHODOLOGY

Preparation of ^{125}I-FSH for injection.

The experiments were performed on two separate occassions:

1. In the first group, immunochemical FSH (LER-780) was

TABLE I

Criteria required for determining production rates

1. Pure labelled hormone injected in trace amounts.
2. Metabolic behaviour identical to endogenous hormone.
3. Excretion as intact hormone or as a specific metabolite.
4. Urine collections until excretion is completed.
5. Methodology to measure specific activity of excreted hormone or its metabolite.

$$\text{Production rate} = \frac{\text{Dose of immunologically – reactive FSH injected}}{\text{Specific activity of urinary product x Time in days}}$$

iodinated with ^{125}I (Hunter and Greenwood, 1962). The specific activity of the ^{125}I-FSH was 87 μCi/μg. The least-damaged component was separated from the other fractions by G75 Sephadex chromatography (25 cm column) as previously described (Raiti and Blizzard, 1968). Throughout the experiments, human serum albumin (HSA) was used whenever added protein was required. This ^{125}I-FSH could be bound to the extent of 74.5 per cent when reacted with anti-FSH antibody absorbed with HCG and used in the radioimmunoassay system.

2. In the second experiment (performed 12 months later) subject R.S. was re-studied. On this occasion, the pituitary FSH used for labelling and for injection was LER-869. The specific activity of the ^{125}I-FSH was 90 μCi/μg. Separation of the iodinated hormone was carried out by disc electrophoresis (Midgley, 1967); 90.6 per cent of this material could be bound by excess anti-FSH antibody absorbed with HCG.

Mode of injection of ^{125}I-FSH.

Six normal male volunteers aged 22 to 43 years, received an injection of ^{125}I-FSH in amounts of 0.75 - 2.0 μCi. Subject R.S. was studied twice. Twelve hours prior to the injection, each individual began ingesting 350 mg of potassium iodide tablets three times a day and this was continued for 72 hours after injection.

Just prior to injection, the ^{125}I-FSH was precipitated with alcohol to remove any further free ^{125}I, and the precipitate was re-dissolved. It was then passed through a Swinnex disposable millipore filter to obtain sterility. Then 0.75-2.0 μCi of ^{125}I-FSH was injected intravenously. The needle was attached to a three-way stopcock. One syringe contained the ^{125}I-FSH while the second contained sterile normal saline. When the ^{125}I-FSH from the first syringe was completely injected, saline from the second syringe was used to wash repeatedly the first syringe. Via the three-way stopcock, each washing was injected into the patient, thus eliminating losses on equipment. After completion, the first syringe contained no detectable ^{125}I. An aliquot of the ^{125}I-FSH used for injection was saved for repeat antibody- binding studies and also for counting. In this way, the amount of injected ^{125}I was known both in terms of counts per minute (c.p.m.) and in terms of microcuries (by comparison with a ^{125}I standard).

Urine collections.

Urine was collected for 72 hours in 24-hour aliquots. Glacial acetic acid (5-10 ml) was used as preservative. Aliquots of each 24-hour urine specimen were counted in a Nuclear Chicago well counter with a 2in. crystal. The total excreted radioactivity for each day and for the three days was determined and this was related to the amount injected.

Characterization of the ^{125}I excreted in the urine.

The methods listed in Table II were used.

TABLE II

Characterisation of Excreted ^{125}I

1. Protein precipitation.
2. Gel filtration using G – 75 Sephadex.
3. Antibody – Binding properties of excreted protein-bound ^{125}I

Protein precipitation. This technique was used to separate ^{125}I bound to protein from free ^{125}I. Five volumes of absolute ethanol were added to an aliquot of urine which was at pH 5.5. The amount of radioactivity in the precipitate represented the excreted ^{125}I-FSH or its protein metabolites, provided that the iodide had not been split off the FSH and become incorporated into another protein.

To eliminate this possibility, ^{131}I was ingested by two adult male volunteers. A 24-hour urine specimen was collected. No radioactivity was found in the protein precipitate obtained as described above.

Gel filtration with Sephadex G75. This was used as a second method for separating ^{125}I-FSH from free ^{125}I. A glass bead was placed at the bottom of a 10 ml pipette. Sephadex was added to the 10 ml mark. Elution was carried out with 0.1M sodium barbital at pH 8.6. Prior to use this column was primed with 0.1 ml of 30 per cent bovine serum albumin (BSA). The characteristics of the eluted peaks were determined beforehand by applying first ^{131}I-FSH in 2 ml of buffer to the column and identifying the fractions in which the radioactive protein peak was eluted and secondly by applying free ^{131}I in 2 ml buffer to the column and again identifying the fractions in which this was eluted.

To 2 ml of urine to be characterised, 1000 c.p.m. of ^{131}I-FSH were added to serve as an internal control for the ^{125}I-FSH which was present in the urine. This urine was applied to a similar column. The above buffer was used for elution and fractions each of 4 drops were collected. The protein peak was identified firstly by the character of the eluted peaks as described above, and, secondly, by counting the internal standard on the ^{131}I setting on the Nuclear Chicago automatic well counter. All fractions were then counted for ^{125}I, and the ratio of ^{125}I-FSH to free ^{125}I was determined.

Although the ethanol precipitation and column separation permit one to conclude that the protein- bound ^{125}I in the urine excreted behaves like pure ^{131}I-FSH, these techniques do not indicate if the protein-bound ^{125}I represents ^{125}I-FSH.

Disc electrophoresis was one method used to attempt to further characterise the excreted protein bound ^{125}I. However, it was not possible to concentrate the alcohol precipitate of urine sufficiently to obtain adequate quantitative radiographs following disc electrophoresis.

Antibody-binding properties of the excreted protein-bound ^{125}I. These were determined using radioimmunoassay. The alcohol protein precipitate from 10 ml of the pooled urine (from the three days) was obtained and this was re-dissolved in 0.5 to 1 ml of phosphate buffered saline buffer (PBS) so as to obtain adequate ^{125}I counts for quantitation. The radioimmunoassay system was set up so that in a tube there was 0.1 ml of PBS-EDTA, 0.1 ml of the above precipitate and 0.5 ml of concentrated absorbed anti-FSH antibody. Controls (using 0.5 ml of PBS in place of antibody) were set up to determine non-specific precipitation. The tests (each in duplicate or triplicate) were incubated at 4°C for 4 days at which time sheep antirabbit antibody was added to each tube (the volume and concentration required for our antibody having been previously determined). Incubation proceeded for a further 3 days, after which time, the test tubes were centrifuged at 2000g for 20 minutes. The precipitates were washed twice with distilled water (to remove non-specific trapping) and recentrifuged. The washings were added to the supernatants. The percent to total counts bound in each test was corrected by subtracting the values obtained for the controls (which represent non-specific precipitation).

The disappearance rate of ^{125}I-FSH from plasma.

This was measured during the production rate studies on subject R.S.2. Blood was drawn at 15 and 30 minutes and at 1, 1½, 2, 3, 4, 5, 6, 7, 8, 12, 16, 20 and 24 hours. The total radioactivity in 5 ml of plasma from each sample was counted and plotted (A in Figure 1). In a subsequent study on the same subject, performed in an identical manner, 15 ml of blood were drawn on each occasion. The total radioactivity in 5 ml plasma from each sample was measured (plotted as B in Figure 1). Ten ml of acetone (i.e. 2 volumes) were then added to

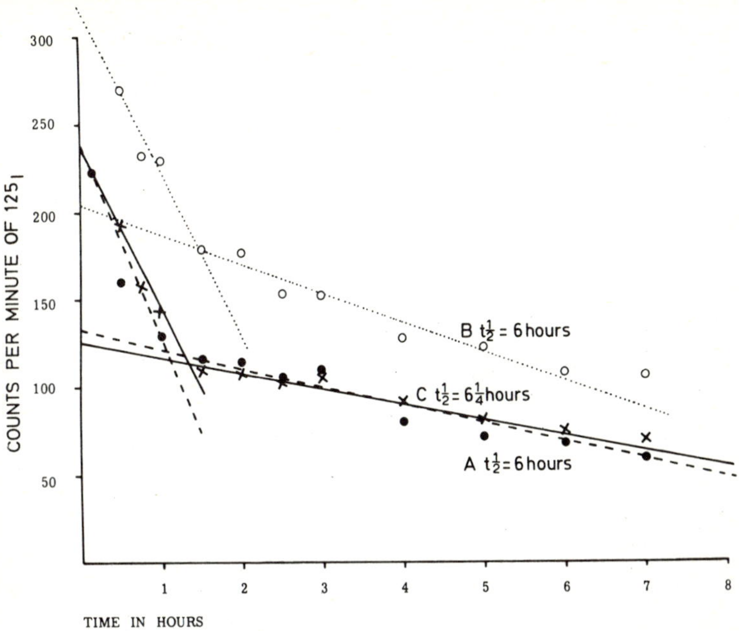

Figure 1

each of the above 5 ml samples. Each was centrifuged. The supernatant was carefully removed. The radioactivity in the protein precipitate (which was redissolved to produce a total volume of 5 ml) of each sample was counted (plotted as C in Figure 1).

An organic hypopituitary patient was injected with labelled (^{131}I-LER 869) FSH and non-labelled FSH simultaneously (Figure 2) (Human FSH kindly donated by Dr W. R. Butt was Batch No. CP/78 and has a bioassay potency of 801 (346-1652) i.u.FSH per mg-mouse ovarian augmentation). The fraction of labelled hormone selected for injection showed no free ^{131}I by alcohol precipitation and was 90.4 per cent precipitated in a double antibody system with excess absorbed antibody on the day of iodination. An interval of two days occured between iodination and injection. Sterility was obtained by passing the injection mixture through a millipore filter. The specific activity of the ^{131}I-FSH was 88.8 μCi/μg. Separation of the iodinated hormone was carried out using Sephadex G-75 column. Blood samples, post injection were obtained every 15 minutes for 1 hour than every 30 minutes for 2 hours and finally every hour for 7 hours.

Measurement of ^{131}I-FSH was done by treating 1 ml aliquots of the patient's serum with excess absorbed antibody at 21°C (room temperature) for 16 hours. A sheep anti-rabbit globulin was added in excess and the mixture was incubated for an additional 4 hours at 37°C. The samples were centrifuged and the supernatant removed by suction.

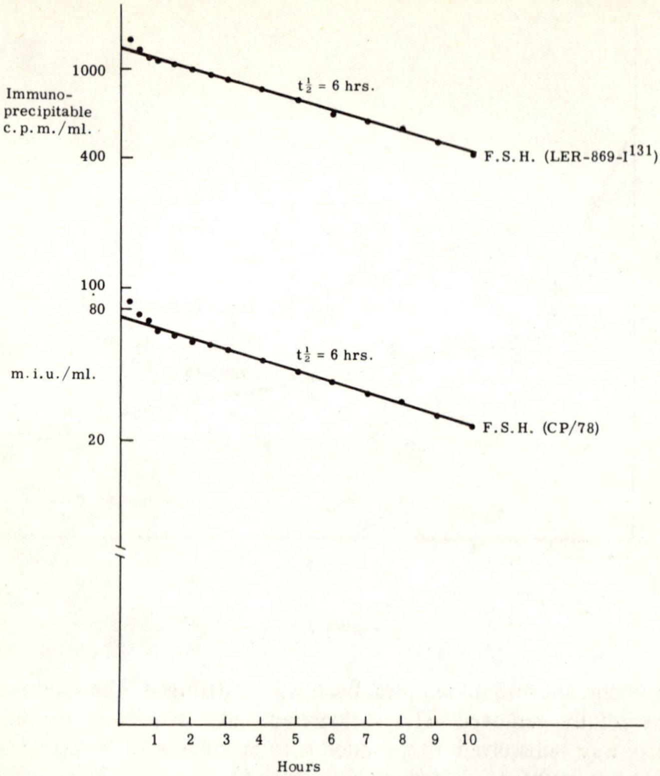

Figure 2

Immuno-precipitated radioactivity was determined in an automatic gamma counter. Counts per minute per millilitre were then plotted on semilog paper against the time post injection that the sample was obtained. Non-labelled FSH was measured by radioimmunoassay (^{125}I system) and m.i.u./ml were plotted on semilog paper against the time post-injection that the sample was obtained.

The renal clearance of FSH.

During the production rate studies on subject R.S.2, the plasma level of FSH was measured by radio- immunoassay at 4-hourly intervals for 24 hours. The results did not show a marked diurnal variation for this subject. The values, in m.i.u./ml of 2nd IRP-HMG were 13.0 at 30 minutes, 14.5 at 1 hour, 12.5 at 2½ hours, 11.0 at 3 hours, 12.5 at 4 hours, 14.0 at 8 hours, 12.0 at 16 hours, 13.0 at 20 hours and 14.5 at 24 hours (each specimen having been run in duplicate or triplicate.) The average value over the 24 hour period is 13.0 m.i.u./ml. If such a value is accepted for the representative plasma level in this subject, the renal clearance of FSH can be calculated using the formula:

$$C = \frac{UV}{P} \text{ where } C = \text{renal clearance}$$

U = urine concentration of FSH in m.i.u./ml
V = urine volume in ml per minute
P = plasma concentration of FSH in m.i.u./ml.

Assay of immunologically-reactive FSH in urine.

The method described by Raiti and Blizzard (1968) was used. This is summarised in Table III.

TABLE III

Measurement of FSH in urine

1. Take 5 ml of the 24hr urine volume
2. Add 0.1ml of 30% bovine albumin
3. Reduce pH to 5.5 with glacial acetic acid
4. Add 25 ml of absolute alcohol
5. Allow precipitation overnight at 4°C
6. Centrifuge at 400 g for 5–10 minutes
7. Discard all supernatant
8. Resuspend precipitate in phosphate buffered saline
9. Bring pH to 7.0 with IN NH_4 OH

RESULTS

Recovery of injected radioactivity.

The percentage of ^{125}I recovered from urine is given in Table IV. Total recoveries ranged from 77 to 92.3 per cent in six of the seven experiments. The recovery of 57.9 per cent in the seventh study is low compared to the others, and is unexplained. Creatinine studies were done on all urines and the creatinine excretion for subject M.W. was comparable to that for the other volunteers. As anticipated, the greatest amount of activity was excreted on day 1 and ranged from 53.2 to 71.4 per cent. Significant amounts were excreted on day 2 and also on day 3 for most subjects. The activity that is unaccounted for may have been excreted in the stools, perspiration, saliva or expired air. Because the thyroid glands were blocked with inorganic iodide, it is unlikely that any iodide was trapped by this gland.

Identification of excreted radioactivity (Table V). Of the excreted radioactivity, 29 to 36.5 per cent could be precipated by alcohol. Presumably, this represented the protein-bound fraction. By gel

TABLE IV

Recoveries of radioactivity in urine
(% of injected c.p.m.)

Patient	Day 1	Day 2	Day 3	Total
P.G.	65.6	15.9	11.1	86.0
R.S.$_1$	68.1	13.8	3.5	85.4
Z.D.	71.4	8.6	2.0	82.0
H.J.	53.2	23.0	8.5	84.7
B.R.M.	68.7	18.4	5.2	92.3
R.S.$_2$	56.5	13.7	6.8	77.0
M.W.	42.3	11.1	4.6	57.9

TABLE V

Characterisation of Excreted ^{125}I

Patient	% of total as protein precipitate	% of total as protein on Sephadex	% of protein precipitate as immunoreactive
P.J.	30.0	32.0	83.0
R.S.$_1$	36.5	29.0	89.0
Z.D.	30.2	27.1	86.2
H.J.	29.0	27.3	84.9
B.R.M.	34.0	30.9	73.3
R.S.$_2$	32.5	--	88.5
M.W.	34.0	28.1	81.4

filtration, a similar percentage (27.1 to 32 per cent) of the total radioactivity appeared in the protein peak.

When the protein precipitates of each urine were dissolved and mixed with excess anti-FSH antibody, 73.3 to 89.2 per cent of this radioactivity appeared in the bound fraction. The antibody-binding properties of aliquots of the injected material were 74.5 and 90.6 per cent for the first and second experiment respectively. Although these figures might appear lower than those for the urine precipitates, the total amount of immunologically reactive ^{125}I-FSH injected was quantitatively at least three times as great as that excreted in the urine.

Excretion of FSH in urine. The daily urinary excretion of immunologically reactive non- radioactive FSH is shown in Table VI. The total three-day excretion varied from 13.5 to 34.7 i.u. of FSH.

TABLE VI

Excretion of FSH in urine
(i.u. of 2nd IRP/24 Hours)

Patient	Day 1	Day 2	Day 3	Total	Average
P.G.	12.2	10.9	10.1	33.2	11.1
R.S.$_1$	11.6	10.7	11.0	33.3	11.1
Z.D.	9.7	8.6	7.3	25.6	8.5
H.J.	6.7	7.7	8.3	22.7	7.6
B.R.M.	7.4	9.5	8.4	25.3	8.4
R.S.$_2$	12.0	11.0	11.7	34.7	11.6
M.W.	5.1	4.6	3.8	13.5	4.5

Calculation of Production rate. This can be calculated from the formula:

$$\text{production} = \frac{\text{dose injected}}{\text{proportion of dose excreted}} \times \text{stable FSH excreted per day.}$$

The amount of immunologically-reactive FSH in the urinary product (of the injected dose) can be calculated by multiplying the percentage of total radioactivity excreted by the percentage that was alcohol precipitable by the percentage of this which was immunologically reactive. Since these results are expressed as percentage of immunologically-intact hormone that was excreted, then the dose injected can also be expressed as percentage of immunologically- intact hormone that was injected. The total excretion of stable FSH over the three- day period is divided by three. Thus for the subject R.S.2, the production rate becomes:

$$\frac{90.6\%}{77\% \times 32.5\% \times 88.5\%} \times \frac{34.8}{3} \text{ i.u. per 24 hours.}$$

or 47.6 i.u. per day.

The production rates for each of the other subjects are given in Table VII.

TABLE VII

PRODUCTION RATES

(i.u. of 2nd IRP HMG/24 HOURS)

P.G.	38.8
R.S.$_1$	29.7
Z.D.	29.7
H.J.	27.4
B.M.R.	27.2
R.S.$_2$	47.6
M.W.	21.1

Disappearance rate of labelled and non-labelled FSH from plasma. The plot of the immuno-precipitated counts per minute per millilitre (^{131}I-FSH) showed a mono-exponential curve with a $t_{1/2}$ of 6 hours (Figure 2 and Table VIII). Non-labelled FSH also exhibited a mono-exponential curve with a $t_{1/2}$ of 6 hours and the specific activity (CPM/mIU) of the points on the curve ranged from 15.8 to 17.4 with a mean of 17.1. In the normal adult male (Figure 1) a similar disappearance rate with a $t_{1/2}$ of 6 hours was found.

Renal clearance of FSH. This can be calculated for subject R.S.2 using the renal clearance formula. The urine volume for day 1 was 860 ml, the total urine excretion of FSH was 12.04 i.u. and so the concentration of immunological FSH was 14 m.i.u./ml. The average plasma level (as previously described) was 13.0 m.i.u./ml. Therefore the renal clearance of FSH for R.S. was 0.64 ml/minute.

TABLE VIII

Disappearance of labelled and non-labelled FSH administered to a hypopituitary patient

Time	Counts per min/ml.	m.i.u./ml 2nd IRP HMG	Specific activity (c.p.m./m.i.u.)
15′	1376	86.0	15.8
30′	1227	76.0	16.1
45′	1136	70.0	16.4
60′	1100	63.0	17.4
90′	1047	60.0	17.2
120′	982	56.0	17.4
150′	927	54.3	17.0
180′	884	50.0	17.4
4 hours	800	46.0	17.4
5 hours	713	40.4	17.4
6 hours	622	36.0	17.4
7 hours	567	32.0	17.4
8 hours	531	30.0	17.4
9 hours	462	26.0	17.4
10 hours	400	23.2	17.4

Mean S.A. = 17.1 (15.8 – 17.4) c.p.m./m.i.u.

DISCUSSION

The criteria to be met in determining the production rate are set out in Table I.

Purity of the ^{125}I-FSH injected. The preparation of human pituitary hormone used for the first experiments (LER-780) was shown to contain 61.1 NIH-FSH-S_1 units per mg (ovarian weight increase) and

0.28 NIH-LH-S_1 units per mg (ovarian ascorbic acid depletion) (Midgley, 1967). By radioimmunological studies, this material could bind to excess absorbed anti-FSH to the extent of 74.5 per cent in our studies. In subsequent experiments in which we reacted this labelled FSH preparation with a mixture of both anti-FSH together with anti-HCG antibodies, a total binding of 90 per cent was achieved. Presumably therefore this FSH preparation contains about 15 per cent of immunological LH.

For the second study, the human pituitary FSH preparation (LER-869) was used. This contains 116.0 NIH-FSH-S_1 units per mg (ovarian augmentation bioassay) and 0.17 NIH-LH-S_1 units per mg (OAAD bioassay) (Midgley and Reichert, 1968). This material could be bound to the extent of 90.6 per cent by excess absorbed anti-FSH antibody and to a total of 95 per cent when anti-FSH and anti-HCG antibodies were used together. Presumably, therefore, it contains 4.5 per cent of immunological LH.

Metabolism of labelled and non-labelled FSH. The comparable half life and specific activity of points on the curve (Figure 2 and Table VIII) exhibited by labelled and non-labelled FSH indicate that they are metabolized in the same manner. Since the injected hormone was of pituitary origin there is reason to believe that it was metabolized and excreted like endogenously produced hormone. The disappearance rate ($t_{1/2}$) of 6 hours found by us is longer than the 3-hour result reported by bioassay (Parlow, 1965). The excretion of ^{125}I-FSH over a 3-day period is consistent with this finding of a relatively long half life of disappearance.

Urinary excretion of ^{125}I-FSH. The ^{125}I-protein excreted was similar in every respect tested, including antibody binding, to the ^{125}I-FSH injected. Therefore, it was only necessary to determine the percent of immunologically reactive ^{125}I-FSH in relation to the amount injected and to relate this to the total amount of stable immunologically reactive FSH excreted. There was no need to search for a specific metabolite of FSH.

Duration of urine collections. The amount of ^{125}I excreted during the third day following injection was small. Although minute quantities of ^{125}I could have been excreted on the fourth day, for practical purposes, the 3-day urine collections contained almost all the radioactivity to be excreted as ^{125}I-FSH.

The use of standards. Throughout the experiments, the 2nd IRP-HMG was used as the standard. The FSH in this preparation is of urinary origin and this standard has been assigned an arbitrary value of 40 i.u. per vial. Although the pituitary production rate values are reported in terms of these units, it is recognised that pituitary and urinary FSH may not be identical. Regardless, about one-third to one-quarter of the pituitary FSH injected was excreted in the urine as immunologically intact FSH, indicating that any changes in this portion of the FSH could not be recognised by immmunoassay studies.

The daily production rate of FSH. The results for production rates for adult males in the two experiments range from 20 to 50 i.u. of 2nd IRP-HMG. However, in the second experiment, the pituitary preparation used could be bound by anti FSH antibody to 90 per cent. Subject R.S. was re-studied using this material. His daily excretion of stable FSH was not significantly different in the two experiments. But his calculated production rate of FSH was about 33 per cent higher than in the first experiment. We believe that these latter results are closer to the true production rate for this patient. Moreover, all the production rate results in the first experiment are probably underestimated by this order of magnitude. Nevertheless, we believe that the basic principles used and the method outlined for determining the production rate of this protein hormone are valid and are applicable for use for this and for other protein hormones.

Renal clearance. The renal clearance value obtained in this study (0.64 ml/min.) is very similar to those reported for FSH by bioassay studies, viz. 0.16 to 0.7 with mean of 0.58 ml/min. (Keller, 1966).

ACKNOWLEDGEMENTS

Dr A. Rees Midgley Jr. prepared the FSH antiserum, Dr Leo Reichert prepared the immunochemical hormone and Dr W. R. Butt supplied the FSH for the half-life studies.

Drs Raiti and Penny are supported by traineeship grant No. TI-AM-5219-08 of the U.S.P.H.S.

The National Pituitary Agency, Endocrinology Study Section and National Institutes of Arthritis and Metabolic Disease made this work possible and it was supported by Research Grants HD-01852 and AM-00180-16.

REFERENCES

Hunter, W.M. & Greenwood, F.C. (1962). Preparation of Iodine-131 labelled human growth hormone of high specific radioactivity. *Nature, Lond.* **194,** 495.

Keller, P.J. (1966). The renal clearance of follicle stimulating and luteinising hormone in post-menopausal women. *Acta Endocr. Copenh.* **53,** 225.

Midgley, Jr. A.R. (1967). Radioimmunoassay of human follicle stimulating hormone. *J. clin. Endocr. Metab.* **27,** 295.

Midgley, Jr. A.R. & Reichert, Jr. L.E. (1968). Specificity studies on a radioimmunoassay for human follicle stimulating hormone. In *Protein and Polypeptide Hormones,* p. 117. Ed. Margoulies, M. International Conference Series No. 161. Amsterdam: Excerpta Medica.

Parlow, A. (1966). The pituitary content of growth and other hormones during fetal and later life. In *Human Pituitary Growth Hormone*, p. 94. Ed. Blizzard, R.M. Columbus: Ross Laboratories.

Pearlman, W.H. (1957). A comparative study of (16-^{3}H) progesterone metabolism in pregnant and non-pregnant women. *Biochem. J.* **65**, *7P*.

Raiti, S. & Blizzard, R.M. (1968). Measurement of immunologically reactive follicle stimulating hormone in human urine by radioimmunoassay. *J. clin. Endocr. Metab.* **28**, 1719.

Since this meeting a legitimate question has been raised regarding the validity of some of the calculations and further experiments are required to verify the production rates presented .

Summary of Discussion

Address of welcome: Professor M. Stacey, F.R.S.

Session I	:	Methods of purification	Chairman:	Dr W.R. Butt
Session II	:	Quantitative analysis		Dr A.S. Hartree
Session III	:	Structure and activity		Dr R.O. Greep
Session IV	:	Standards		Dr D.R. Bangham
Session V	:	Immunological studies		Prof. W.D. Odell
Session VI	:	General Discussion		Dr A.C. Crooke

Professor Stacey welcomed members to Birmingham. He considered the value of this kind of meeting was that specialists in this field were gathered together and they would stimulate one another in preparing for future research.

Methods of Preparation

Methods for the preparation of gonadotrophins for clinical, biological, chemical and immunological studies were discussed at previous conferences and meetings of the 'Gonadotrophin Club'. Some have become standard procedures: these include the use of salts for extraction and fractional precipitation, acetate-ethanolic buffers, ion-exchange celluloses, gel filtration and electrophoresis. Illustrations of their use appears in all the papers presented here.

New techniques discussed included the use of various agents for the purification of HMG, QAE-Sephadex for pituitary FSH and iso-electric focusing. Théoleyre and Jutisz considered that urinary FSH is possibly complexed with other proteins or constituents of urine and that the complexes may be dissociated or denatured by metaphosphoric acid, sulphosalicylic acid and n-butanol. A 25-fold purification was observed by the use of n-butanol followed by DEAE-cellulose, and a 20-fold purification following sulphosalicylic acid and DEAE-cellulose, Sephadex G-200 or SE-Sephadex. Another reagent that has been proposed however, sodium dodecyl sulphate, resulted in the loss of some 50 to 70 per cent of the biological activity (Stevens). Furthermore it was noted that Donini and other workers have obtained much higher specific activities using only classical techniques. One important factor is the potency of the starting material and the methods of Théoleyre and Jutisz have not been applied to highly purified material.

New work was presented on the use of electrofocusing for urinary gonadotrophins (van Hell). Two HMG-preparations having FSH potencies of 700 and 3000 i.u. per mg were examined using a carrier ampholyte with a pH range of 3 to 5. Pooled fractions were dialyzed to remove most of the sugar and carrier ampholyte and then concentrated by lyophilization. The last traces of low molecular weight material were removed by gel filtration on Sephadex G-50.

The results of these experiments are shown in Tables I and II. In both experiments LH activity was found mainly in fractions having a pH above 4 while FSH was in all fractions from pH 2 to pH 8. Possible explanations for the behaviour of FSH are: (1) FSH as isolated from menopausal urine exists in several forms, (2) FSH interacts with different macro-molecular components in urine: this is not thought to be so unlikely in a solution of low ionic strength and (3) FSH is partly degraded during iso-electric focusing.

Donini has found it difficult to remove the last traces of ampholyte from fractions of HMG and agreed that the best method was gel filtration on Sephadex G-50. He obtained four or five fractions from HMG of 500 i.u. per mg, but there was no increase in specific activity in any one fraction. The technique is probably of most value for the fractionation of purer material (van Hell).

TABLE I

Isoelectric focusing of HMG (HTF 13)

Fraction	B	C	D	E	F
pH of eluate	2.0 – 3.2	3.2 – 3.7	3.7– 3.9	4.0 – 4.2	4.2 – 9.3
IU FSH/mg*	250	390 (310–480)	550 (400–680)	1060 (900–1260)	197 (169–228)
IU LH/mg**	5 – 8	5 – 8	5 – 10	20 – 30	160 – 240
mg	3.2	3.4	1.9	1.7	16.6
weight %	13	14	8	7	66
FSH %	5	8	6	10	19
LH %	0.4	0.4	0.2	1	83
FSH/LH	39	60	73	42	1

25 mg starting material
FSH potency: 700(580–910) i.u./mg
LH potency: 120–180 i.u./mg

* augmentation test in rats
** haemagglutination inhibition test

The contents of the tubes obtained after draining the column were pooled using the u.v. density recording of the eluate as a guide. the pH values of the first and the last tube of each fraction are recorded.

TABLE II

Isoelectric focusing of HMG (HTF 17)

Fractions	A	B	C
pH of eluate	2.0 – 3.7	3.7 – 4.0	4.0 – 7.8
IU FSH/mg*	790(610-970)	1160(950-1470)	1200(1030-1390)
IU LH/mg**	15	6	1300
mg	1.7	0.9	0.7
weight %	49	26	20
FSH %	12	9.0	7
LH %	2.9	0.6	100
FSH/LH	53	194	0.9

3.5 mg starting material
FSH potency about 3000 i.u./mg*
LH potency about 250 i.u./mg***

* augmentation test in rats
** radioimmunoassay for LH on the eluates
*** haemagglutination inhibition test

Pooling of the tubes was made by using the results of radioimmunoassay for LH as a guide.

The u.v. density measurement of the eluate did not yield useful information for pooling.

The variations in the concentration of ampholyte and the range of pH used in electrofocusing may lead to some precipitation occurring on the column (Saxena). Bahl cautioned against the use of extremes of pH because of the risk of removing N-acetyl-neuraminic acid residues and the effect on disulphide bonds and alkaline-labile bonds.

A new method for the preparation of pituitary LH of very high specific activity was described by Roos. The procedure employed ammonium sulphate fractionation, DEAE-cellulose, Sephadex G-100 and SE-Sephadex C-50 chromatography and finally, preparative polyacrylamide electrophoresis in a glycine buffer ar pH 8.8.

The yield of LH prior to electrophoresis was about 1.3 mg per 100g fresh pituitary tissue and the biological activity was 2200 i.u. per mg. This preparation was assayed by the seminal vesicle method in immature intact male rats using HCG standard: the result corresponds to 14,000 i.u. LH per mg. The potency was estimated to be about 17,000 i.u. per mg by radioimmunoassay (Dr L. Wide).

The preparative electrophoresis gave four components. The two major components were assayed by radioimmunoassay and appeared to have potencies approximately equal to their immediate precursors.

Some FSH-like material has been separated from preparations of HCG by using the same buffers and chromatographic steps as in the purification of FSH from HMG (Donini). Schuurs estimated the FSH in various preparations of HCG by the ovarian augmentation method in intact rats and found that it decreased relative to the HCG activity as purification proceeded (Table III).

TABLE III

Biological 'FSH' activity found in HCG preparations

Prep.	HCG potency (IU/mg)	FSH potency	
		IU/mg	i.u. per 1000 i.u.HCG
6669–6710	1,000 (800 – 1,220)	1.7	1.7
470A	2,580 (2,280 – 2,910)	4.6	1.8
404	2,940 (2,620 – 3,300)	2.6	0.9
C_3N_6	10,000 (8,600 – 11,700)	<8	<0.8
$C_{21}C_{36}$	10,800 (8,900 – 13,100)	9.1	0.8

Stability and specific activity of gonadotrophins

Urinary FSH, 1255 i.u./mg, has been lyophilized and remained stable for one year at 4°C (Donini). Its residual moisture and that of its precursors was between 0.3-0.4 per cent after drying *in vacuo* over P_2O_5 for 30 hr. Its $A^{1\%}_{1cm}$ at 280 nm of 6.85 did not alter after lyophilization. The colour intensity in the Lowry reaction and the ultraviolet absorption however changed at each purification stage (Table IV). It is possible that the changes in optical density may indicate alterations in the protein: carbohydrate ratio as would be expected in the purification of a glycoprotein (Bahl).

TABLE IV

Changes in ultraviolet absorption and Lowry colour on purification of urinary F.S.H.

i.u.FSH/mg	FSH LH ratio	Lowry colour as % of albumin standard	$A^{1\%}_{280}$	UVmin nm	UVmax nm
61.3	0.87	85	8.67	254	277
121.5	3.2	85	8.29	254	277
671.0	119.0	75	7.36	252	277
1255.0	385.1	70	6.85	251	277

Preparations of purified urinary LH have not been so stable (Donini and Stevens). One preparation was stable for two years but when it was fractionated on CM-cellulose its initial activity of 500 i.u./mg dropped gradually until by 5 months it was only 80 i.u./mg. The immunological activity however was constant and it remained a perfectly satisfactory preparation for raising antibodies. Since in general glycoproteins are relatively stable, contamination with proteolytic enzymes should be considered when losses occur (Bahl). Papkoff and Li have included a heating stage early in their preparative methods to destroy such enzymes but Stevens has not been able to use this method satisfactorily. Bahl suggested that on CM-cellulose the conditions of pH etc. may have encouraged the formations of sub-units but since the loss occurred in the dry state, dissociation was not thought likely to have taken place. Canfield however pointed out that in salt-free lyophilized preparations there may be 8 to 10 per cent water bound and that complete drying of proteins would lead to denaturation. Gray lyophilized pituitary FSH of potency 1500 to 2000 i.u./mg without loss of activity but Rathnam and Saxena observed some decrease in potency in pituitary FSH after lyophilization over a 6 months period. Their work on iodinated preparations indicated that some loss may occur by adsorption on glass or plastic: this could be prevented by the addition of carrier protein (0.1 per cent albumin) or mannitol (1 per cent), the latter being subsequently removable by dialysis. Experiments with iodinated proteins may give variable results. Thus Hunter found that iodinated FSH adsorbed less to glassware than any other protein hormone examined except gastrin, whereas LH adsorbed more to glassware than any of the others.

By analogy with the behaviour of enzymes, loss on lyophilization may be caused by working with protein concentrations which are too low (van Hell). Since some glycosidases cannot be lyophilized because of damage to their tertiary structure however, Bahl has preferred to use pressure dialysis through membranes (Amicon) for concentration. Suitable grades are available for direct application to urine and the process is much quicker than gel filtration. Roos also avoided lyophilization because of possible changes in tertiary structure. Less pure material of up to 300 i.u. FSH/mg can be lyophilized without loss, possibly because it is protected by protein contaminants.

Precipitation with ethanol was used to prepare pituitary LH of 7,500 i.u./mg which remained stable for two years (Hartree). However, a more recent preparation appears to have developed a high molecular weight component over 6 months: similar behaviour has been noted in a human growth hormone preparation of Roos.

Criteria of purity

No single test is adequate to demonstrate the homogeneity of a gonadotrophin. Thus contamination of a preparation of TSH with 30 per cent by weight of LH could not be detected immunologically (Odell). Highly purified HCG preparations which appeared immunochemically identical have been shown to be heterogeneous by starch gel electrophoresis (van Hell and Schuurs, Figure 1). Furthermore Donini described a preparation of HMG which was homogeneous in the ultracentrifuge, by disc electrophoresis and, when labelled with an iodine marker, by starch gel electrophoresis. It gave no immunological reaction with an antiserum to LH but there were two lines in immunoelectrophoresis using an antiserum to HCG. Certainly a single precipitin line in immunoelectrophoresis does not prove homogeneity but the technique is valuable for demonstrating heterogeneity.

Peckham considered electrophoresis at more than one pH was a useful test of homogeneity. Roos drew attention to the fact that the staining of polyacrylamide gels in disc electrophoresis may fail to detect minor contaminants, especially if they are glycoproteins with high carbohydrate contents.

Radioimmunoassay, in other fields, has been an excellent method

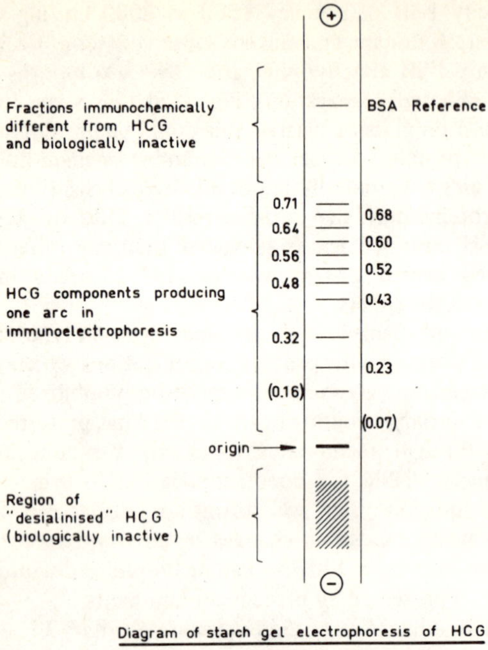

Figure 1. Diagram of starch gel electrophoresis of HCG showing the position of components which are immunologically identical.

for detecting very minor differences in structure: thus porcine and whale insulins can be differentiated although they have the same amino acid sequences and differ only in tertiary structure (Hunter). Even more striking, steroid-protein conjugates can be differentiated by an alteration of only one hydrogen atom in the steroid structure (Stevens).

Analysis

In spite of great differences in their biological activities there is good agreement between the amino acid analyses of preparations of HMG (Donini). The varying activities are probably due rather to slight differences in the N-acetyl neuraminic acid contents (Bahl). This residue is known to be essential for FSH (Butt, Gray) and for HCG activities (van Hell, Donini) but not for LH (Papkoff and Li). Hartree however considered that the agreement between different amino acid analyses was not all that good: in fact her LH analyses agreed just as well with the HMG analyses. Differences between laboratories with present methods should agree within 10 per cent, and in any one laboratory within 5 per cent.

Amino acids in LH are similar to those in HCG except for the higher histidine, glutamic acid, glycine and isoleucine and the lower serine and proline estimates in LH compared with HCG (Hartree). Canfield commented that the estimate of proline is a good guide in following the purification of HCG. A preparation of HCG from the urine of patients with trophoblastic tumours (Dr K. Bagshawe) was similar to normal HCG except for 9.3 residues of arginine per mole compared with 10.8, and 1.8 residues of methionine compared with 2.8. The decrease in methionine however could be explained by the fall in the biological activity of this preparation from the original 19,000 i.u. to the present 8,000 i.u. per mg. (Hartree). The HCG from male patients with testicular carcinoma is also similar in amino acids to normal HCG (Canfield).

Highly purified preparations of HCG differed in their biological activities probably because of differences in their N-acetyl neuraminic acid contents (van Hell, Schuurs). Their electrophoretic mobilities vary but separate amino acid analyses of fractions obtained by starch gel electrophoresis have not been undertaken. Bangham drew attention to the possibility that certain changes in the structure of HCG may occur at different stages of pregnancy: thus in the second and third trimesters the ratio of immunological to biological activity alters. Commercial preparations however are prepared from the urine of women in the first trimester (Schuurs). Were the variations noted therefore due to genetic polymorphism, to proteolytic degradation, to contaminants or artifacts produced during extraction? (Lunenfeld). Canfield considered that if the separate electrophoretic fractions were shown to have the same primary structure they could be assumed to reflect the same gene: if a few peptide bonds were cleaved this would probably have occurred subsequent to synthesis. Some of the carbohydrate chains are incomplete but these are not genetically determined. In the HCG of patients with tumours who have been treated with inhibitors of protein synthesis such as actinomycin D it would not be surprising if some of the carbohydrate chains were less complete.

Sub-units in LH and HCG

It is more difficult to dissociate human LH than ovine LH into sub-units. Dissociation of ovine LH is complete at pH < 3 or pH > 12 and there is up to 15 per cent dissociation between pH 5 and pH 10 (Jutisz). A pH of less than 3 however is required to dissociate human LH (Hartree). Furthermore, in applying the countercurrent distribution system devised for ovine LH to human LH Reichert found it necessary to incubate the hormone in the lower phase for 45 minutes at 37°C prior to partition, otherwise the dissociation was incomplete.

It was generally agreed that after dissociation of ovine or human LH there was considerable loss of biological activity as measured by the OAAD assay. Quantitative results should be interpreted with caution, however, since the sub-units are essentially different molecules: with their molecular weights of about 15,000 factors such as filtration through the kidney may become important and the relative potencies will depend upon the time after administration that the parameters of the assay are taken (Bangham). Re-association of the sub-units leads to restoration of biological activity. There is even some evidence that the final specific activity may be greater than the initial so that the process is a purification step (Hartree). The biological activity of ovine LH after dissociation at pH 3 was only partially restored, however, by re-association possibly because some damage was incurred at the low pH (Jutisz). After reassociation of Hartree's human LH pH 1.3 sub-units 85 per cent of the biological activity was restored (Butt).

Human LH has also been dissociated by incubation in 8M urea for 24 hours (Hartree). It remained dissociated when dialyzed against buffer at pH <5.5 but re-associated at pH 6.0. Ultracentrifugal studies indicated that the re-associated hormone did not have quite the original molecular weight and preliminary OAAD assays (Dr E. T. Bell) suggested that the activity was probably lower than the original.

The nature of the possible sub-unit structure of HCG is less well understood. The observations of Bahl on the C-terminal sequences suggest the presence of two identical chains: peptide mapping of the electrophoretic fractions of Canfield however suggests that native HCG is composed of two non-identical sub-units. This discrepancy was discussed at length but there is at present no satisfactory explanation*. Carboxymethylation at alkaline pH could modify lysine, histidine and tyrosine and cleave methionine residues. The technique is well established however and has been applied satisfactorily to several other proteins. It must be remembered that HCG of 19,000 i.u. per mg. has been shown to be heterogeneous by techniques such as starch gel electrophoresis (van Hell) so the possibility of confusion due to the presence of impurities must still be borne in mind. There is as yet no information on the nature of the linkage of the sub-units of HCG. Jutisz observed no dissociation of HCG when it was treated with urea or guanidine and Bahl noted that HCG forms a salt with urea and then fails to sediment in the ultracentrifuge. The method of using a low pH to dissociate LH does not appear to have been tried for HCG: certainly some differences in behaviour would be anticipated since such treatment would remove sialic acid residues which are essential for the activity of HCG but not for LH (Bahl).

Structural studies

HCG. Extensive structural studies have been reported on HCG (Bahl, O. (1969) *J. biol. Chem. 244,* 575). The preparation was desialyzed and carboxamido-methylated and submitted to tryptic hydrolysis and stepwise cleavage of the sugars from non-reducing ends by specific glycosidases. Five glycopeptide fractions were obtained. Sialic acid and fucose are at the non-reducing ends of the carbohydrate chains and these are joined to polypeptide chains by an N-acetyl-glucosaminyl-asparagine linkage. Galactosamine residues are probably joined by

* See Footnote page 159.

N-acetyl galactosaminyl serine linkages. The heterogeneity shown in the carbohydrates of various preparations of HCG (van Hell and Schuurs) is more likely to be due to differences in N-acetyl neuraminic acid therefore than to other carbohydrates such as glucosamine which is not in a terminal position.

FSH. In addition to the techniques described by Chaplin *et al.*(p.77) Kennedy reported on the hydrolysis of pituitary FSH in 0.01M-HC1 for 60 minutes at 80°C, All the N-acetyl neuraminic acid and fucose units were liberated under these conditions: 0.1M-HC1 at 80°C liberated all the N-acetyl neuraminic acid by 40 minutes. Gas phase analysis of the hydrolysates obtained from a series of increasing acid strengths showed that no more carbohydrate units were liberated until 1.0M-HC1 at 100°C for 120 minutes was employed when small amounts of galactose and mannose were liberated.

Fractionation of the products of pH 1 hydrolysis for 60 minutes at 80°C, on Sephadex G25 gave two peaks absorbing at 220 nm, one corresponding to unchanged molecular weight and one to a molecular weight of 6,000. The two peaks together contained 20 per cent of the original activity as judged by radioimmunoassay. After hydrolysis at pH 2 for 75 minutes at 80°C, only one peak corresponding to unchanged molecular weight was obtained: this contained all the original radioimmunological activity of the preparation.

Bahl observed that the alkaline degradation studies of Chaplin did not exclude the possibility that hexosamine was involved in linkage of the polypeptide to asparagine or some amino acid other than serine or threonine.

There was discussion on whether structural studies were justified on impure material (van Hell, Lunenfeld). Bahl considered that useful information could be obtained from a preparation which did not have full biological activity provided it could be shown to be homogeneous. Negative information, such as the lack of a particular end group or of a certain type of linkage in an impure preparation would also still be valid for the pure hormone (Gray).

Another approach to the structure of FSH is by the investigation of enzymic digests. Saxena and Rathman discussed their experiments with papain and Gray, similar experiments with trypsin. After limited digestion it appeared that small molecular weight fragments were obtained which possessed both biological and immunological activities. The tryptic fragment however has been unstable and its biological activity fell off rapidly. Oxidation with chloramine-T of FSH from which N-acetyl neuraminic acid has been removed also produced small molecular weight material which is active by radioimmunoassay.

Structure-function

Many modifications to the structure of FSH have been shown to result in alterations in its biological activity (Saxena, Rathman, Butt *et al.*). It is important to note that some groups in the molecule, such as neuraminic acid, are easily removed and this may account for the loss of activity following treatment with some reagents rather than to modification of some other group (Bahl).

It is clear that the biological and immunological activities of FSH may vary independently (Butt). It would be interesting to know what effect changes to individual components of the molecule have on its tertiary structure: X-ray crystallography may eventually help here (Canfield).

A modified form of FSH, or fragment of FSH, which was biologically inactive would be of interest if it acted as a competitive inhibitor of FSH. Such a possibility was investigated by Ryle. Her *in vitro* studies showed that FSH free of neuraminic acid had no such inhibiting effect and, in fact, was similar to native FSH in stimulating the incorporation of labelled thymidine into mouse ovaries. This result led to speculation regarding the function of neuraminic acid in the molecule: is it involved in the transport of the hormone from the injection site to the ovary? Its function is probably more complex than this because LH contains much less of this residue and yet still gets to the ovary (Bahl). It may be that the modified structure is less stable than the native molecule and is therefore more susceptible to some enzyme action.

The *in vitro* assay was used in these experiments only with purified materials: its specificity has not been examined with a wide range of substances (Loraine). Ryle did not believe that it was specific: if older animals are used or longer incubation periods are employed the effect is likely to be given by LH. Schuurs considered that the system would be useful for studying the mode of action of antisera to gonadotrophins. This has not yet been tried but Lunenfeld and Eshkol have reported rather similar work using *in vivo* experiments. When mice were given antiserum to gonadotrophin from birth in order to deprive them of endogenous hormone, administration of FSH increased the number of follicular cells in the ovary. When small amounts of LH were added no further increase in cells was observed but when higher levels were given with the FSH the incorporation of thymidine into ovarian DNA was actually depressed.

Grasslin reported on the effect of chemical modifications to the HCG molecule. A preparation of 15,000 i.u./mg was submitted under mild conditions at physiological pH to acetylation and succinylation. Activity was only found to alter with the latter when it was supposed that the change in charge may have led to a change of tertiary structure. Treatment with tetranitromethane which reacts with the hydroxyl group in the o-position of tyrosine led to only 15 per cent loss.

Immunological studies

Odell reported that information gained in his laboratories, and those of others, during the development of radioimmunoassays for the four human glycoproteins over the past six years indicated remarkable immunological (and presumably therefore structural) similarities among these hormones. These data may be summarized as follows:

1. All anti-human TSH sera thus far studied have shown cross-reaction with HCG and human TSH. Figure 2 demonstrates an example of this type of reaction. The fact that highly purified radioiodinated human TSH may be competitively displaced by unlabelled HCG, but that the dose-response curve is very different from TSH indicates that hormones are competing for the same antibody binding sites.

2. Antisera produced against HCG usually bind radioiodinated human LH and HCG indistinguishably and each may competitively displace the other. Paul and Ross have shown (Paul, W. E. and Ross, G. T. (1964) *Endocrinology, 75*, 352) that some anti-HCG sera have less affinity for human LH than HCG, illustrating that minor immunological differences may exist between HCG and LH.

3. The majority of anti-human FSH sera are very non-specific, showing binding of LH, FSH, TSH and HCG and competitive displacement of highly purified iodinated FSH by each of these four hormones.

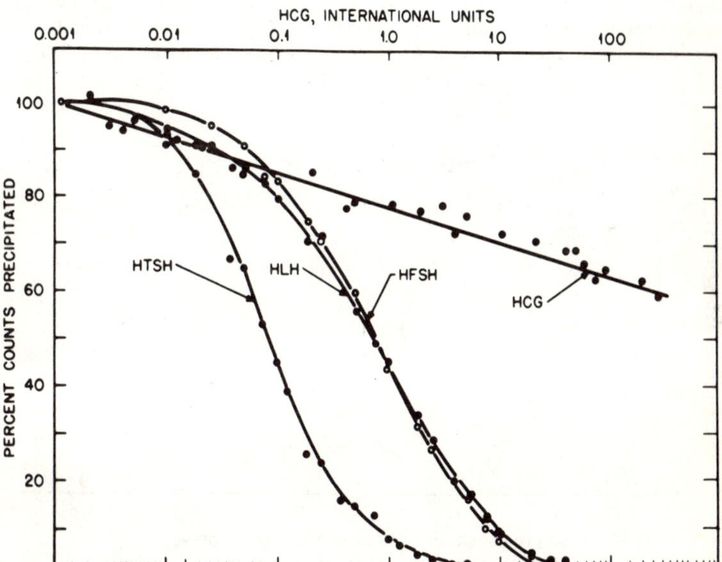

Figure 2. Cross reaction of HCG, HLH and HFSH in the radioimmunoassay for HTSH, using a non specific antiserum. This flat dose response curve for HCG has been observed with all anti HTSH studied. (Odell, W., Vanslager, L. and Bates, R. (1968). In *Radio-isotopes in Medicine, in Vitro Studies,* Ed. Hayes, R. L., Goswitz, F. A. and Pearson Murphy, B. E. U.S. Atomic Energy Commission, Conf. 671111).

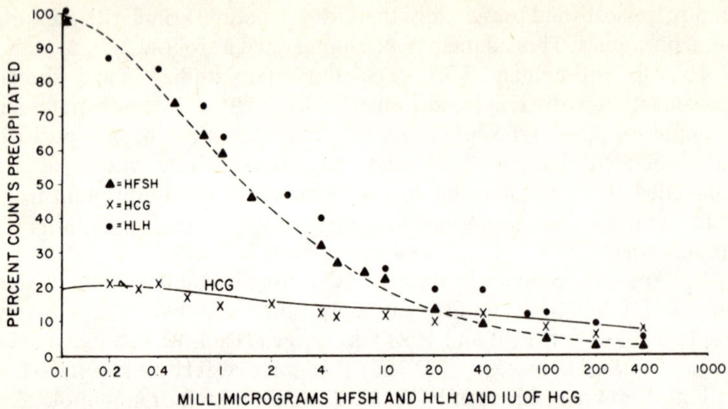

Figure 3. Radioimmunoassay for HFSH using a non-specific antiserum. HLH and HFSH show approximately equal ability (on a weight basis) to competitively displace radioiodinated HFSH from these antibodies. HCG displaces competitively over 80 per cent of the radioiodinated HFSH at doses greater than 0.1 i.u. Between 0.1 and 0.001 i.u. of HCG (not shown on this graph), HCG gives a dose response curve indistinguishable from that produced by HFSH and HLH. Therefore on a weight basis, HFSH, HLH and HCG compete approximately equally for the HFSH binding sites. (Odell, W. D., Abraham, G., Rand, H. R., Swerdloff, R. S. and Fisher, D. A. 1969 *Acta endocr. Copenh.* Suppl. *142,* 54.

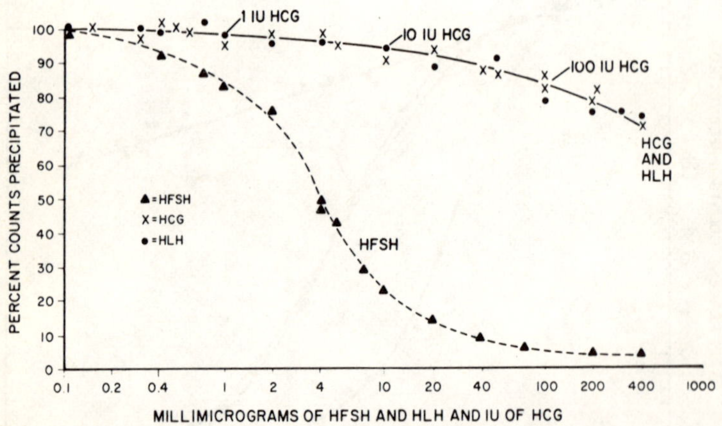

Figure 4. A relatively specific radioimmunoassay for human FSH using an unabsorbed anti HFSH. Note that up to 10 i.u. of HCG and 10 millimicrograms of HLH failed to react significantly in the assay system. Greater amounts gave significant reactions. This antiserum was obtained after three immunizations of a rabbit. The antiserum depicted in Figure 3 was obtained from the same animal after a fourth immunization. This dramatic change in specificity has been previously reported. (Odell, W. D., Reichert, L. E. and Bates, R. W. (1968). In *Protein and Polypeptide Hormones,* Ed. Margoulies, M. Part 1, p. 127, Amsterdam; Excerpta Med.)

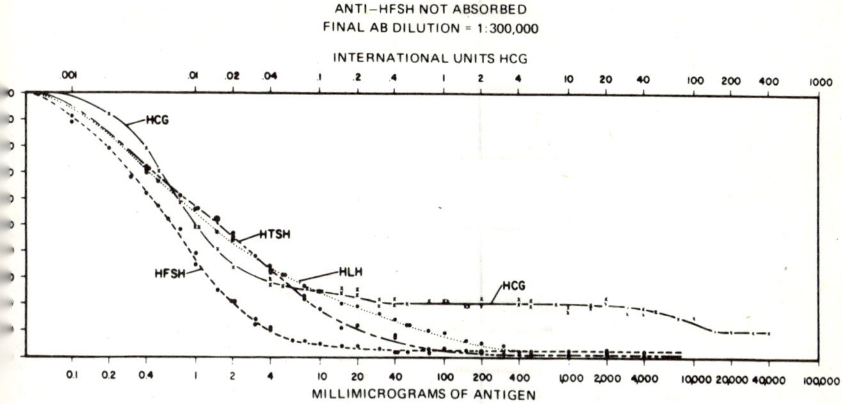

Figure 5. Illustration of an anti FSH serum containing two distinct populations of antibodies, one showing approximately equal reaction between HFSH and HCG, and a second showing much more specific reaction with HFSH only. When used at this antiserum dilution (1:3000,000), this assay is not specific for FSH. (Odell, W. D., Abraham, G., Raud, H. R., Swerdloff, R. S. and Fisher, D. A. *Acta endocr. Copenh.* in press).

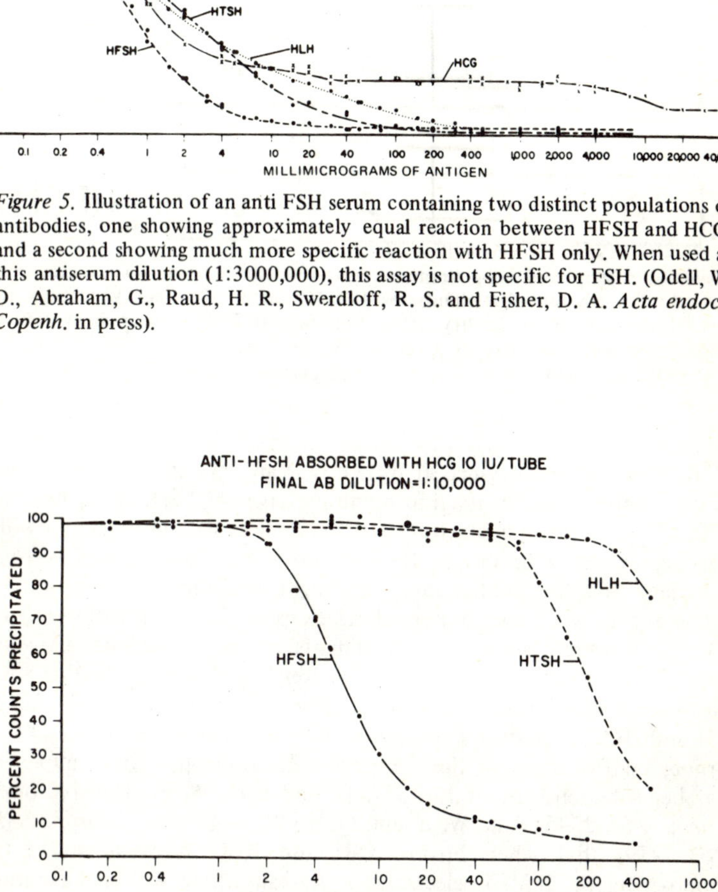

Figure 6. A specific radioimmunoassay for HFSH utilizing a non-specific antiserum against HFSH but now absorbed with HCG. The antiserum dilution was increased to 10,000 and the population of antibodies showing cross reaction with HCG and HLH no longer affects the system. (Odell, W. D., Abraham, G., Raud, H. R., Swerdloff, R. S. and Fisher, D. A. 1969 *Acta endocr. Copenh.* Suppl. *142*, 54.)

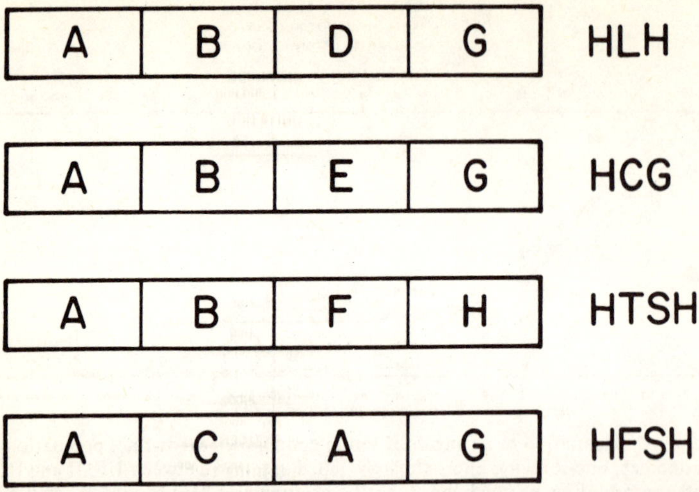

Figure 7. A simplified schematic representation of the immunologic interrelations between human luteinizing hormone (HLH), human chorionic gonadotrophin (HCG), human thyroid stimulating hormone (HTSH), and human follicle stimulating hormone (HFSH). (Odell, W. D., Abraham, G., Raud, H. R., Swerdloff, R. S. and Fisher, D. A. 1969 *Acta endocr. Copenh.* Suppl. *142,* 54.)

4. With varying degrees of difficulty, antisera may be produced which result in highly specific immunoassays for FSH, HCG, LH and TSH. Figures 3 to 6 illustrate four examples of immunoassays with varying degrees of specificity for FSH. From these data, as well as from a variety of other publications, one might construct a very simplified schematic view of immunological relations of these hormones (Figure 7). The so-called 'antigenic determinants' are believed to be relatively small in size, equivalent to molecular weights of perhaps 600 to 800. However it is quite likely that for molecules of the size of TSH, FSH, LH and HCG, a definitive tertiary structure is required to maintain the proper configuration of these antigenic determinants. For example, it has been demonstrated (Odell, W. D. and Paul, W. E. (1965) *J. biol. Chem. 240,* 2045; Paul, W. E. and Odell, W. M. (1964) *Nature, Land. 203,* 979) that when bovine TSH and HCG are irradiated with mono-energetic 1 MEV electrons, a 'hit' anywhere within the entire molecule destroys biological activity, but that immunological activity is much less sensitive; a hit in a smaller portion (one-third to one-half) of the molecule is required to destroy activity and a hit outside this region presumably does not affect immunological activity. However, the estimated size of this portion 'responsible' for immunological activity was still 10 to 20 times greater than estimates of the size of antigenic determinants based on haptene inhibition.

As shown in Figure 7, one may postulate that all four human glycoprotein hormones bear similar immunological sites (A); LH and HCG have several similar sites (A,B,G), but each also contains one specific for that hormone (D,E); FSH has several sites similar to those on the other three hormones (A,G) and at least one unique to FSH (C). All these immunological studies lead to the hypothesis that the four hormones share in part, common primary structures. Perhaps portions of the amino acid sequences are very similar; differences between these hormones may result from minor differences in amino acid sequence or differences in carbohydrate composition also discussed above.

Antisera are usually prepared against crude antigens since they are just as likely to produce specific antisera as more highly purified antigens. Material for labelling however must be as pure as possible and the specificity of radio-immunoassays depends largely on this purity. In this respect it was noted that FSH used for labelling varies considerably in potency: what is the lowest potency to give satisfactory results? Although preparations of 14,000 i.u. per mg have been described (Roos), fractions containing only 2000 i.u. per mg have been used in some radio-immunoassays. Some of these have been demonstrated to contain less than 5 per cent of other proteins (Peckham) so that the low activity may be due to some subtle change in the molecule rather than because of the presence of impurities. An exchange of preparations between laboratories would be useful and this is being organized.

The question of artifacts being produced by the iodination procedure was discussed (Barker). If the biological activity is destroyed, as it is with FSH, this does not matter provided the antigenic activity is retained and there is no change in electrophoretic properties, i.e. iodination damage (Naftolin): the labelled material merely acts as a marker. There are other examples of inactivated antigens being satisfactory in haemagglutination techniques and for raising antisera.

Very little is known about the relationship of urinary LH and HCG. Schuurs has investigated six different antisera to purified HCG by complement fixation and haemagglutination-inhibition techniques to compare the LH potencies of an HMG preparation in terms of HCG. The haemagglutination-inhibition estimates were relatively constant for each antiserum but the complement fixation method showed a variation of almost threefold in the potency estimates. This suggests that the latter technique can detect differences when an HCG standard is used for estimating urinary LH better than the haemagglutination method, and the quality of the antiserum is more important. There is some evidence that the complement fixation technique detects differences in the spatial structure of the molecule better than other methods. Further examples of this are the loss of activity in the complement fixation but not the haemagglutination-inhibition assay of HCG treated with acid (Schuurs) or with chymotrypsin (van Hell).

Radioimmunoassays for FSH applied to neat urine do not appear to be satisfactory and reactions with some small molecular weight substances have been reported (Stevens). This led to a discussion on the nature of FSH as it occurs in urine: does it appear in the form in which it was secreted from the pituitary? Franchimont investigated the inhibition curves of pituitary and urinary FSH using antiserum to pituitary FSH and labelled pituitary FSH and found them to be identical: similar results were obtained by Butt and Lynch. However the inhibition curves were not identical when Franchimont used antiserum to urinary FSH and urinary FSH for labelling. Butt and Lynch found very poor binding of their labelled pituitary FSH with antiserum to urinary FSH. All these results indicate some slight differences between urinary and pituitary FSH.

When bioassay is compared with radioimmunoassay for both FSH and LH in the urine of women during the menstrual cycle the peaks observed are generally much more pronounced as estimated immunologically. The index of discrimination between the methods during the cycle varies considerably even when all variables such as the standards and antisera are taken into account (Stevens). Loraine drew attention to the variable pattern of FSH in the normal cycle: differences between immunoassays done on repeated occasions may have something to do with this. There appears to be a peak in the excretion of FSH at mid-cycle as judged by immunoassay but there is no information on the chemical nature of this material compared to that excretion earlier in the cycle (Bahl).

Radioimmunoassay is useful for monitoring fractionation procedures but it may measure inactive as well as active material so that it can be misleading. Peckham (Figure 6, Peckham, p. 8) showed that in the fractionation of pituitary LH on Sephadex G100 the ratio of the immunological to biological activity varied: the second protein peak showed immunological activity which could not be ascribed to biologically active LH or to cross-reaction with FSH or TSH. It occurred close to the position of growth hormone but was not this hormone and it seems possible that there are other glycoproteins in pituitary tissue which cross-react immunologically.

More attention needs to be paid to statistical procedures for radioimmunoassay. Individual assays may be very precise with λ values of 0.01 to 0.02 but variation between different assays may be great. Odell has found that the coefficient of variation in individual assays was about 2.5 per cent but in repeated assays it may be 15 to 20 per cent and occasionally even 100 per cent. Confidence limits should be calculated for repeated assays and any assays which fall outside these limits should be eliminated (see Rodbard, D., Rayford, P. L., Cooper, J. A. and Ross, G. T. (1968) *J. clin. Endocr. Metab. 28*, 1412).

Standards

A biological standard is a reference material used for the calibration of an activity by comparative assays. Until recently this activity has usually been a biological effect *in vivo* but now a new generation of standards is required for the immunological assay of hormones. There are several important points about standards that should be emphasized (Bangham):

1. It must be homogeneous so that all users have samples that are as similar as possible.

2. It must be stable and in the interests of stability it is not necessarily the purest available material.

3. It must be well characterized by biological, and if necessary, immunoassays,

4. The standard selected for an assay must be as near as possible similar to the substance under test: the validity of comparative assays depends on the assumption that like is compared to like.

Types of standard

1. International: this is available in strictly limited quantities and is expected to last ten years or so. It should be used primarily for the calibration of national or laboratory standards: in research it is employed directly as standard in critical assays. Current existing international (WHO) standards have potency estimates obtained by bioassay and are intended for bioassays: one standard is now available for immunoassay - HGH, which was set up last year.

2. Unofficial 'research standards' set up by national organizations such as the N.I.M.R. These are usually prepared with the same care as international standards and are intended for comparative assays only.

3. N.I.H. preparations: these are prepared in batches from various species of animal and are widely distributed for general physiological and chemical research work, not necessarily for use as biological standards.

Homogeneity of standards

To ensure homogeneity the standard is made up in solution and known quantities are introduced into ampoules and freeze-dried. Some added carrier such as lactose, sucrose or albumin is added so that there is about 5 mg of material in each ampoule. Among other things the carrier helps to prevent adsorption of the active constituent hormone to the glass. The whole batch is subjected to a secondary desiccation over P_2O_5 in a vacuum and the ampoules are filled with nitrogen and glass-sealed. This is essential for stability since rubber stoppers are porous to some gases and to moisture.

Stability

Among the factors likely to cause instability are bacteria, moisture, oxygen, light and heat. The ampoules are of hard glass which has been acid-washed and sterilized but the standards are not guaranteed sterile: if the material is known to stand the treatment it is filtered through a membrane which would be a sterilizing procedure. To ensure complete homogeneity of the material prior to distribution into ampoules the solution is submitted to high speed centrifugation.

It is technically difficult to estimate accurately the moisture content of the small amount of material in such an ampoule: defined in terms of the change in weight after heating at 56°C over P_2O_5 *in vacuo* it is usually about 1 per cent.

To check for stability some ampoules are set aside after sealing and are submitted to elevated temperatures for various times. Their potencies are then compared with standards kept at -20°C. Any change can then be noted and the rate of loss calculated for the temperature of storage (-20°C). International standards deteriorate by less than 1 per cent per year.

Alteration caused by handling may be detected by comparing the original powder with the filling solution kept in liquid nitrogen and the final ampoule. Changes in pH have been reported to occur on freezing buffers containing sodium and potassium mono- and di-phosphates since various salts have different eutectic points: this leads to changes in pH on freezing and this may affect potency (Chilson, O. P., Costello, L. A. and Kaplan, N. O. (1965) *Fedn Proc. Fedn Am. Socs Exp. Biol. 24,* S.55).

The quantity of standard per ampoule is usually in excess of the amount required for one assay. The remainder is often stored at low temperature for future use: it must be remembered that salts or albumin are present and the solution may have a very low eutectic point, e.g., for saline this is -23°C. To minimize losses quick freezing in CO_2 or liquid nitrogen followed by storage at -30°C is safer than slow freezing at -20°C.

Some loss of potency in transit has been reported for some ovine preparations of FSH. With international standards accelerated degradation studies indicate that no significant losses should occur during normal surface or air transit.

Individual standards

HCG. The second international standard appears to be satisfactory for immunoassays although from previous discussions it presumably contains molecules of varying N-acetyl neuraminic acid content. Assays using the standard have been reported to be valid using the prostatic and seminal vesicle weight methods for crude placental extracts but some lack of parallelism has been reported for assays in serum (Loraine). A serum standard may therefore be useful: it would have further application for radioimmunoassays in serum.

A reference preparation of highly purified biologically active HCG would be useful for calibration of the specificity of antisera for the biologically active hormone.

Urinary FSH and LH. The 2nd IRP-HMG was set up for bioassay but is used widely for radioimmunoassay for which it is an excellent standard. It will probably be replaced by a batch of similar but more highly purified material now made routinely; this should not be as toxic as some have found the 2nd IRP-HMG to be, which was an early preparation. There is a need for reference material of highly purified urinary FSH to calibrate the specificity of antisera.

Pituitary gonadotrophin. The M.R.C. LH (ICSH) Standard A has not been subjected to collaborative assays. It has proved to be an excellent standard for radioimmunoassay. It has very low TSH and FSH activity. A second batch of similar material has been prepared, containing just detectable TSH.

A large batch of a relatively simple extract of FSH and LH (prepared from part of a batch known as LER907) is at present being prepared as new standard.

Future prospects

Canfield forecast that probably within 12 to 18 months we should see the complete structure of the LH sub-units – at least of ovine and probably of human – and of the HCG sub-units. One of the reasons for interest in the structure of the gonadotrophins is because of the need to understand their precise mechanism of action and structure is a prelude to this. The preparation of radioactive derivatives is becoming increasingly important in this work since it should be possible to trace these derivatives to their site of action. When one considers the spatial structure of proteins it is clear that iodination may not be the method of choice for this purpose, although it remains the best available for radioimmunoassay. Taking lysozyme as an example the crystal structure shows that the surface of the protein is covered with hydrophilic residues such as arginine, lysine and aspartic acid while the interior is covered with hydrophobic residues: tyrosine is in this hydrophobic environment. Trypsin, at room temperature, does not cleave a single arginine or lysine residue but if the surface is disrupted it may become susceptible to trypsin digestion and following its biological behaviour may be following an artifact. There is only room for one molecule of water inside this crystal structure and therefore a large atom like iodine would disrupt the centre. It would be worth

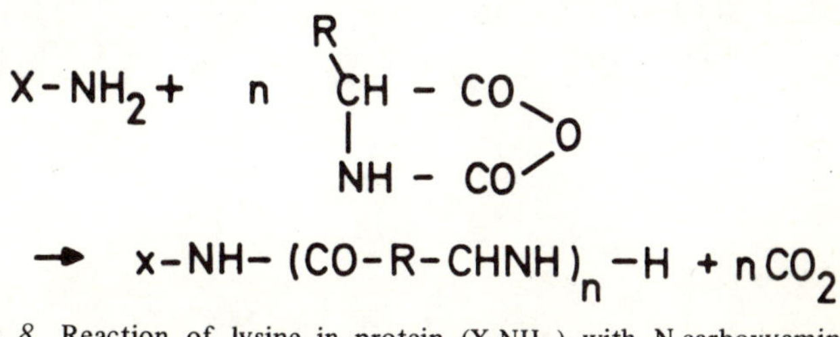

Figure 8. Reaction of lysine in protein (X-NH_2) with N-carboxyamino acid anhydride (e.g. tritiated alanine).

investigating the method used by Fuchs and Sela (Fuchs, S. and Sela, M. (1965). *J. biol. Chem. 240,* 3558) whereby a labelled poly-alanine side-chain is introduced thus preserving the hydrophilic surface of the molecule (Fig. 8). It was possible to introduce 700 residues into rabbit gamma globulin and preserve its capacity to bind antigen. The method has also been applied to HGH (Canfield) and is now being investigated for HCG (Eshkol).

Workshop Meeting on the Development of the Ovary in Infancy

Part I Normal Development

INTRODUCTION

The Workshop Meeting on the Development of the Ovary in Infancy was organized in order to bring together a group of scientists who have been working on one or another aspect of the early development of the ovary. It seemed that the dictum 'Nothing happens in the ovary during infancy' had become obsolete and that evidence was emerging that during the infant and juvenile periods the ovary is not dormant but constantly changing, as it develops from a fairly simple organ at birth to a highly differentiated one at the beginning of maturity. This evidence was coming from studies in morphology, cell kinetics, histochemistry, endocrinology, electron microscopy and organ culture. People working in these different fields met for a three-day session and discussed (1) the normal development of the ovary between birth and maturity, (2) the dependence on and influence of hormones during early development and (3) the hypothalamic-hypophysial-gonadal interactions during infancy.

The papers and summaries of the discussions are contained in this volume. They give evidence that much is already known about the developing ovary but that even more is yet to be investigated. Only a few species have so far come under investigation, but it is already becoming clear that there are multiple ways in which an ovary can develop.

The discussions underlined the constant beginning of growth, throughout infancy, of some oocytes emerging from the large pool of apparently dormant small germ cells, though it was not possible at this point to find or define the signal that stimulates or prevents the start of their growth. Reports on the morphological details which become visible in the electron microscope added greatly to the understanding of developmental processes as well as of the function of the ovary. The mechanism of action of hormones at cellular levels and their influence on DNA, RNA, and protein synthesis in the infant and juvenile ovary were discussed. The sessions clarified and co-ordinated the knowledge accumulating on the importance of the presence of certain hormones for early ovarian development and on the evolving production of hormones in the infant ovary and their influence on the feedback mechanism. One of the purposes of this workshop was to find out what is already known and to define and formulate questions to be explored in order to come closer to an understanding of the dynamics and function of the ovary and its feedback interactions with other organs during infancy.

Hannah Peters
Convenor of the Workshop Meeting
on the Development of the Ovary in Infancy.

Kinetics of Follicle Growth in the Ovary of the Immature Mouse

Torben Pedersen

ABSTRACT

The kinetics of ovarian follicle development in the immature mouse were investigated in autoradiographs prepared after pulse labelling with tritiated thymidine.

The following parameters were determined at the age of 7 days and at weekly intervals up to 35 days:

1. The number of follicles present of the different types shown in Figure 1 (differential count).

2. The time it takes follicles to grow from one stage of development to another. This was calculated from the total number of granulosa cells in these stages and from their doubling times. The doubling time of granulosa cells in a follicle was estimated from (a) the labelling index of the follicle, i.e. the percentage of labelled granulosa cells in the largest cross section of the follicle, and (b) the duration of the DNA-synthesis phase of the individual granulosa cells, as determined from the curve of labelled mitoses, i.e. the percentage of labelled *mitotic* nuclei at different times after pulse labelling. The labelling index of follicles varied not only from one type to another, but also within the same type at different ages. The duration of the S-phase of the individual granulosa cells depended, however, only on the type of follicle to which the cells belonged, not on the age of the animal.

3. The number of follicles which start to develop at these ages.

Follicle development is not constant in the period from birth to maturity. More follicles start to grow in the first two weeks of life than later in the immature period. Moreover, the follicles grow faster early in life than later. These variations are responsible for the variations in the differential counts, which show a maximum number of all types of follicles between the ages of 3 and 4 weeks.

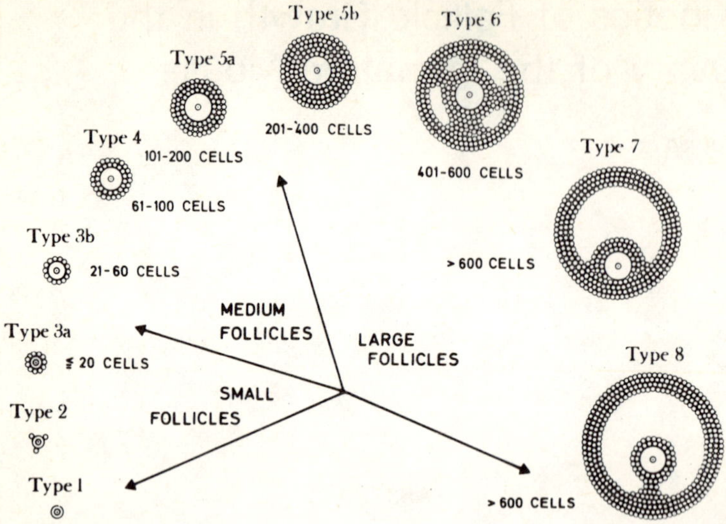

The development, from a follicle with one layer of granulosa cells to one with several layers and antrum formation, takes, in the first half of the immature period, about 10 days, while in the second half it takes about 16 days.

It was calculated that about 850 follicles start their development in the time interval between birth and maturity, and of these more than 600 reach their full development in this period and start either to transform to mature Graafian follicles or to degenerate. A full account of this work has been published in *Acta Endocr., Copenh.* 62 (1969) 117.

GENERAL DISCUSSION

PARTICIPANTS: WEISZ, RYLE, PETERS, MAYER, ESHKOL, LUNENFELD AND ODELL.

1. The precise numbers of cells reported for the follicles of the various types probably do not apply to other colonies or strains of mice.

2. It was pointed out that, as the number of large follicles increased, the growth rate and recruitment of those in the smallest categories declined, suggesting that the larger follicles in some way inhibited growth of the smaller ones.

3. In connection with changes related to age, granulosa cells of the rat vary with respect to their form and enzyme content. Between Days 7 and ca. 18 granulosa cells do not contain Δ-5, 3β-ol dehydrogenase;

at that time the enzyme is confined to the interstitial tissue. By 21 days the granulosa of medium and large follicles contain the enzyme. By 26 days there is a third type of granulosa, characteristic of the adult follicle that is destined to ovulate. As shown by E. Urzua most of this type is pseudostratified and only the pseudostratified portion contains the enzyme. None is seen in the cumulus cells. The granulosa in large follicles is also differentiated with respect to proliferation; in general mitosis stops where steroid-producing enzymes appear. The responses of different types of granulosa cells to gonadotrophins may not be identical. Possibly inhibition of smaller follicles is associated with the differentiation of adult-type steroid-producing follicles.

4. More follicles start to grow per unit time in the early immature period than later. Mechanisms controlling follicle development are apparently not yet co-ordinated.

The Development of the Ferret Ovary

R. Deanesly

ABSTRACT

Serially sectioned ovaries from 24 foetal and 44 early postnatal ferrets have been examined. Ciaccio's method for lipids was applied to nearly half the ovaries so that the development of the glandular interstitial cells could be traced.

Pregnancy in the ferret lasts 42 to 44 days, ovo-implantation occurring at Day 11 or 12. The meiotic phase of the oocytes is postnatal, from 10 to 21 days after birth, but often not complete for another week. When meoisis begins the outer oogonia remain unaffected for some days. By 14 or 15 days post-partum all stages of meiosis can be found and many oocytes are degenerating. Primordial follicles were first seen on Day 16 at the medullary edge of the ovarian cortex.

In the earliest ovaries examined, (24th day of pregnancy), stromal tissue ramifies among the oogonia and this continues as they multiply during the growth of the ovary. The stromal cells, which increase and become more prominent in later stages, are the precursors of the follicular epithelium. There is no evidence of a reserve of identifiable epithelial follicular epithelium. The process of follicle formation can be well studied in 15 to 21 day postnatal ferrets.

Lipid-containing interstitial tissue, consisting of glandular epithelioid cells such as occur in the adult ovary, differentiates from medullary stromal cells from Day 12 post-partum onwards. It is well developed by Day 19, especially adjacent to the hilus, away from the cortical oocytes. The glandular cells proliferate and enlarge and gradually occupy most of the medulla. By Day 28 post-partum, or earlier, the ovary consists mainly of glandular interstitial tissue with an incomplete outer layer of primary follicles. From the reduction of the cortex, it may be estimated that more than half of the original oocytes have degenerated during meiosis. The primary interstitial tissue, similar to that of the adult ovary, develops entirely independently of the follicles or their constituent cells.

A full account of this work will appear in the *Journal of Anatomy* (1970).

GENERAL DISCUSSION

PARTICIPANTS: PETERS, LUNENFELD, WEISZ AND FRANCHI.

1. It was pointed out (Peters) that in five species other than the ferret transformation of the oogonia begins at or proceeds after birth, namely in the rabbit, golden hamster, cat, mink and vole. In all, except the hamster, ovulation is reflex, not spontaneous. It was not known whether these two features were functionally correlated but it was suggested that the abundant interstitial tissue in the ferret might be connected with a functional role in reflex ovulation, in some way comparable to that of old corpora lutea in spontaneous ovulators. In the rabbit. as Hilliard has shown, the interstitial gland responds to LH by secreting 20 α-dihydroxyprogesterone and this steroid is needed as a positive feedback substance to maintain LH release at the level needed to induce ovulation. The interstitial tissue of the adult ferret also produces a progestogen.

2. The possibility of some signal which halts mitosis in oogonia was raised. It was pointed out that in several species the central parts of the cortex were always at a more advanced developmental stage than the outer parts. Possibly something originating in medullary elements inhibits oogonial mitosis. Alternatively, it was suggested that oogonia may be genetically destined to undergo a finite number of mitoses and if they exceed this number they undergo atresia.

Development of the Stroma in the Immature Mouse Ovary

A. Byskov, T. Pedersen and H. Peters

SUMMARY

During the first week of life the undifferentiated stroma cells proliferate and thus increase the stroma mass. Towards the end of the first week differentiation and transformation of some of the stroma cells begins. Some of these cells lie in clusters, often close to blood vessels. At the end of the first week extensions and protrusions of bizarre follicles come to lie within the stroma and might contribute to its differentiated component. However, this is only a temporary phenomenon. The proliferation of stroma cells, as well as their transformation to differentiated cells continues during the second week. However, after that age proliferation of stroma cells does not take place any more and the interfollicular stroma is formed from then on by rests of arrested follicles.

INTRODUCTION AND METHODS

The mouse ovary at birth contains two main cell populations, the oocytes and the stroma cells. The stroma cells fill the centre of the organ and surround the oocyte nests in the cortex. From this pool of undifferentiated stroma cells several highly specialized structures, characteristic of the ovary at the onset of maturity, apparently arise. These include not only the granulosa and theca cells but also the peripheral and central stroma. It is the evolution of the stroma during the infant and juvenile period, which concerns us here.

During the first weeks of life the differentiated stroma might develop in several ways: (1) by proliferation and transformation of the undifferentiated cells present at birth, (2) by the transformation of granulosa and theca cells, (3) granulosa cells may become detached from the follicle and become part of the stroma or (4) by a combination of these processes.

To investigate this problem several approaches have been used. For histological studies ovaries of immature mice were fixed in glutaraldehyde and OsO_4 and embedded in Epon. One micron sections were

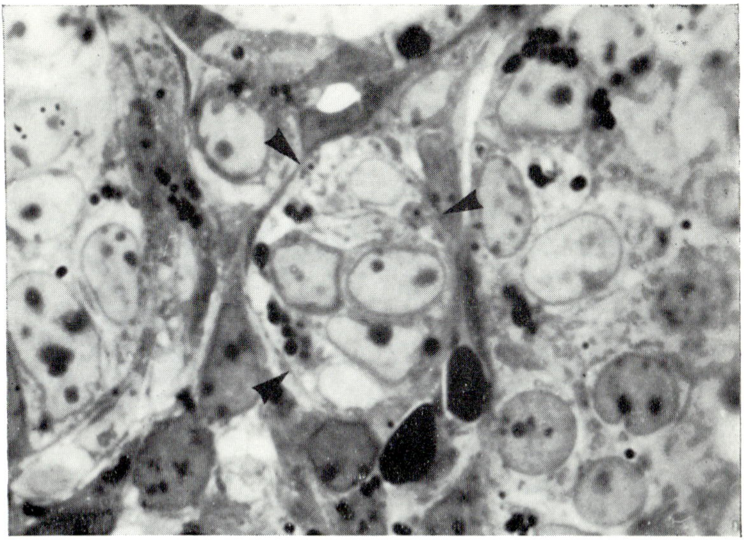

Figure 1. Ovary of a 4-day-old mouse. Undifferentiated stroma cells with few granules surrounded by a membrane and fibroblastic cells: 'stroma clusters'. x 1575 (1 μ Epon section, toluidine blue)

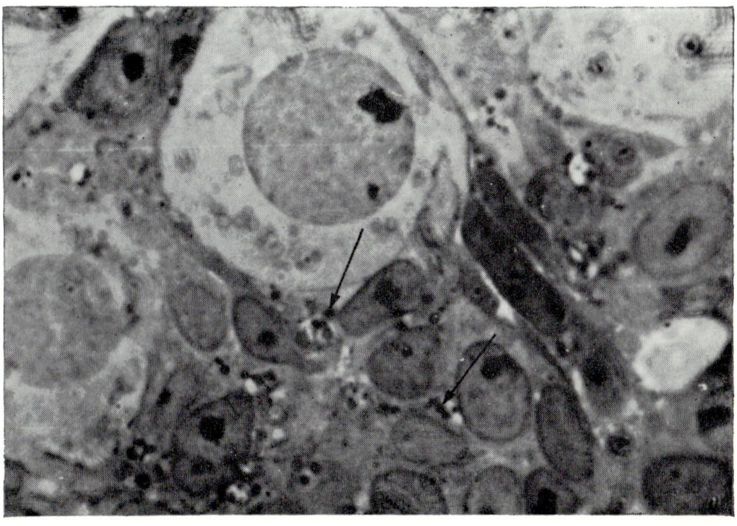

Figure 2. Ovary of a 1-day-old mouse. Pregranulosa and primitive stroma cells containin a few osmophilic granules. x 1575 (1 μ Epon section, toluidine blue)

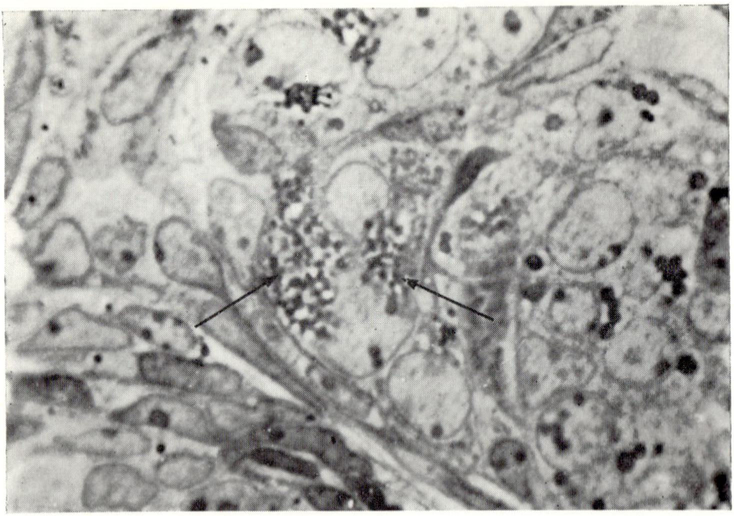

Figure 3. Ovary of an 8-day-old mouse. Many cells in the 'stroma clusters' are filled with large osmophilic granules. x 1575 (1 μ Epon section; toluidine blue)

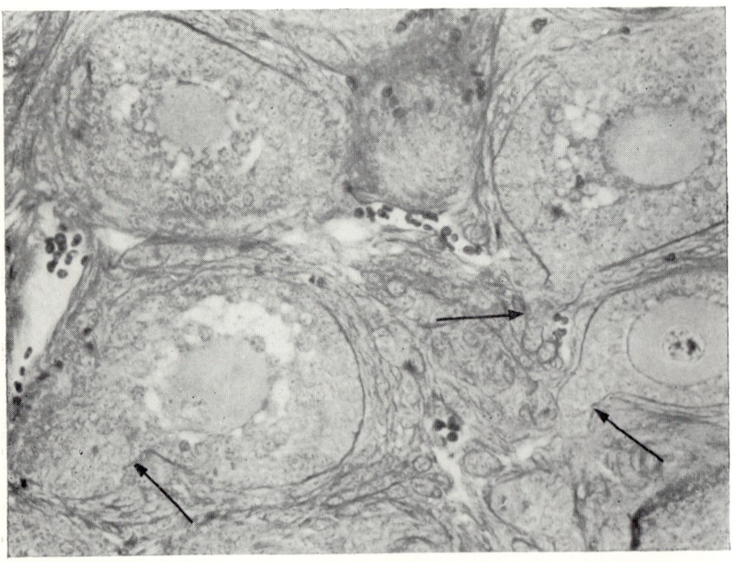

Figure 4. Ovary of a 10-day-old mouse. Many bizarre follicles have distorted granulosa layers, which form extensions into the stroma. x 382 (Heidenhain)

either stained with toluidine blue or left unstained and examined with phase contrast microscopy. The proliferation and movements of cells were determined by autoradiography. Mice of different ages were injected with tritiated thymidine (^{3}HTdR). By preparing autoradiographs (ARG) one hour after pulse-labelling, proliferating cells in the ovary can be identified (Pedersen, 1970). By deferring the preparation of ARG for three or seven days, movements of cell groups in the ovary can be investigated (Peters and Levy, 1966).

RESULTS AND DISCUSSION

The immature period was divided into (1) the first week of life, (2) the second week of life and (3) the remainder of the immature period (i.e. the third to the fifth week). The results will be discussed separately for each of these three age groups.

The first week of life. Most of the stroma cells which form a central core of the ovary at birth have clear cytoplasm, but some contain small osmophilic granules. At the end of the week these cells are often found in groups surrounded by a membrane and fibroblastic cells ('stroma clusters') (Figure 1).

The proliferation of these cells was investigated in autoradiographs. On all days during the first week of life a high percentage of labelled stroma cells was characteristic. This suggests that at this age the cells (those without as well as those with osmophilic granules) proliferate fairly fast. To follow the movement of some of the stroma cells consecutive ARG were prepared. It could be shown that the first cells which attach to the oocytes (pregranulosa cells) arise from the primitive stroma cells already present in the ovary at birth (Peters and Pedersen, 1967). The pregranulosa cells, like some of the primitive stroma cells, contain small osmophilic granules (Figure 2).

The second week of life. During the second week of life the size of the ovary increases considerably, due not only to growth of follicles but also to an increase in the amount of the stroma mass. Cells in many of the stroma clusters become filled with large osmophilic granules; some become vacuolated (Figure 3). The membrane surrounding such differentiating clusters disappears, and they become invaded by blood vessels.

In addition, clusters of cells are seen in the stroma, which, although morphologically not easily distinguished from stroma clusters, differ in their labelling pattern. In ARG prepared during the second week of life, three types of cells can be distinguished in the stroma: (1) cells with or without a few small granules, which become labelled and are found throughout the stroma mass; (2) cells lying in clusters containing large granules or vacuoles which do not incorporate ^{3}HTdR, i.e. do not proliferate; (3) cells lying in clusters (also surrounded by a membrane) which, however, become labelled after pulse-labelling, i.e. which belong to the proliferating cell population of the ovary. Following such clusters of cells through serially sectioned ovaries it became clear that they are part of the distorted follicles characteristically found in ovaries during the second week of life (Dawson and McCabe, 1951; Rennels, 1951; Peters, 1969) (Figure 4), and that these cell groups lying within the stroma mass are in fact proliferating granulosa cells.

It can be concluded that, during the second week, some of the stroma cells still proliferate, thus increasing the stroma mass. Furthermore, an increasing number of cells become differentiated and cease to proliferate. Whether the groups of cells extending from bizarre follicles into the stroma mass, become detached from the follicle and form part of the stroma (Dawson and McCabe, 1951) can as yet not be definitely decided. This is, however, suggested by the fact that the basement membrane surrounding the extensions from follicles is often faulty, interrupted or even absent.

The third, fourth and fifth week of life. During the later part of the immature period the differentiated stroma mass increases markedly. However, little of this increase is caused by proliferation of the stroma cells themselves, as during this period only very few cells in the stroma still incorporate ^{3}HTdR after pulse-labelling. Furthermore the possible contribution to the stroma from protrusions and extensions of bizarre follicles decreases with advancing age. Bizarre follicles are usually not found after the second week of life. The stroma formation in the later part of the juvenile period must therefore have a different history, which can be 'made visible' by consecutive ARG.

In ARG prepared one hour after pulse-labelling only granulosa and theca cells were labelled, suggesting that at this age only cells in follicles are proliferating. Cells in the stroma remain free of the label. ARG prepared three days after ^{3}HTdR-injection showed labelled granulosa and theca cells in all developing follicles but with fewer grains over the nuclei as division of the cells had taken place in the intervening days. Many of the large degenerating follicles had labelled granulosa cells. In ARG prepared one week after the pulse-labelling a marked change in the labelling pattern was noted: the granulosa cells of the medium and large follicles were lightly labelled (due to repeated cell division) but in addition also the strands of stroma between the follicles had become labelled. As this stroma does not contain labelled cells 1 hour after pulse-labelling, these strands must have subsequently arisen from cells which at injection time became labelled, i.e. the granulosa and theca cells. It is therefore concluded that after the second week of life the stroma is formed, as it is in the adult mouse (Peters and Levy, 1966), from rests of follicles, which did not develop beyond a certain stage.

GENERAL DISCUSSION

PARTICIPANTS: BEN-OR, MAYER, DONOVAN, PETERS, LUNENFELD, GREEP, ESHKOL, WEISZ, RYLE, STEGNER, BRANDAU AND FRANCHI.

1. No comparable data are available for the infantile stages of other species. From 14 days onwards, as in adults of other species, mouse stromal cells are derived from cells of atretic follicles.

2. Up to Day 7 those granulosa cells which will eventually pass out of follicular protrusions (and either re-differentiate or degenerate) are not morphologically distinguishable from those cells which will remain part of follicles. The protrusions form between days 7 and 10; about day 10 some of their cells become filled with osmiophilic granules and the basement membrane ruptures. However, enzyme reaction studies on ovaries of 7-day-old mice show no difference between the protrusions and the rest of the granulosa although differentiating interstitial cells are clearly distinguishable. (See Brandau, p.309).

3. Apart from the oocytes, the ovarian cells of the new-born mouse seem very homogenous. Their location appears to determine the direction in which they differentiate, e.g. those near the oocytes become granulosa cells.

4. Regarding the likelihood of cells from follicular protrusions de-differentiating and re-differentiating into stromal cells, it was pointed out that: (a) when transplanted to the eye, granulosa cells (which can be recognised by their characteristic nuclei) will attach to iris tissue and flatten out like stroma cells (Ben-Or); (b) cells of large atretic follicles transform to stromal cells in the adult; (c) granulosa cells regularly transform to corpus luteum cells; (d) during the early stages of oocyte atresia the mitochondria of the surrounding granulosa cells often transform to the type characteristic of steroid-producing cells.

5. The follicular basement membrane seems to be formed by pre-granulosa cells after these have surrounded the oocyte. At birth it is already present around a few follicles. This membrane, seen by light microscopy, is formed partly of bands of collagenous fibrils and must be distinguished from the membrane (glyco-protein complex) seen with the electron microscope which is typical of all epithelia. The term 'basal lamina' has been proposed for the latter.

REFERENCES

Dawson, A.B. & McCabe, M. (1951). *J. Morph.* **88,** 543.
Pedersen, T. (1970). *J. Reprod. Fert.* **21,** 81.
Peters, H. (1969). *Acta Endocr., Ccpenh.* **62,** 98.
Peters, H. and Levy, E. (1966). *J. Reprod. Fert.* **11,** 227.
Peters, H. and Pedersen, T. (1967). *Fert. Steril.* **18,** 309.
Rennels, E.G. (1951). *Am. J. Anat.* **88,** 63.

Electron Microscope Studies of the Interstitial Tissue in the Immature Mouse Ovary

H. -E. Stegner

Electron microscopic studies on the developing immature ovary have provided additional information about the cyto-differentiation of oocytes and somatic cells, including their structural interrelationships. A good knowledge of normal morphogenesis is necessary for current and future experimental work on gonadal function. However, in most of the recent studies major attention is given to the differentiation of the germ cells. Only a few investigations are primarily concerned with the development of the somatic components. Odor and Blandau (1969) have previously given a description of ultra- structural differentiation of the foetal and neonatal mouse ovary up to the third postnatal day. The progressive and important events in interstitial cell development during the first and second week of life require further sequential studies at closely timed intervals. In the present paper a description is given of the ultrastructural differentiation of the mouse ovary from birth to day 12 of postnatal life.

METHODS

Gonads of inbred Swiss Albino mice were taken for examination daily from the eighteenth day of foetal life up to the twelfth day of postnatal life. The timing of pregnancy was based on observation of a vaginal plug. The day of birth is denoted as 'day O'.

The mice were killed by decapitation, the ovaries removed immediately and most were fixed in 4 per cent glutaraldehyde in 0.1M phosphate buffer for 2 hours, washed in phosphate buffer containing 0.2M sucrose, and postfixed in 1 per cent osmium tetroxide. The tissues were dehydrated in a graded series of acetone-water mixtures (dried over $CuSO_4$) and embedded in Vestopal W. In addition some were fixed in a cold mixture containing 2 ml 5 per cent OsO_4, 1 ml Dalton buffer, and 1 ml 3.4 per cent NaC1 for 2 hours, washed in Dalton burffer and embedded in Vestopal W as described above. Thin sections were cut on a Reichert OmU2 ultramicrotome and stained

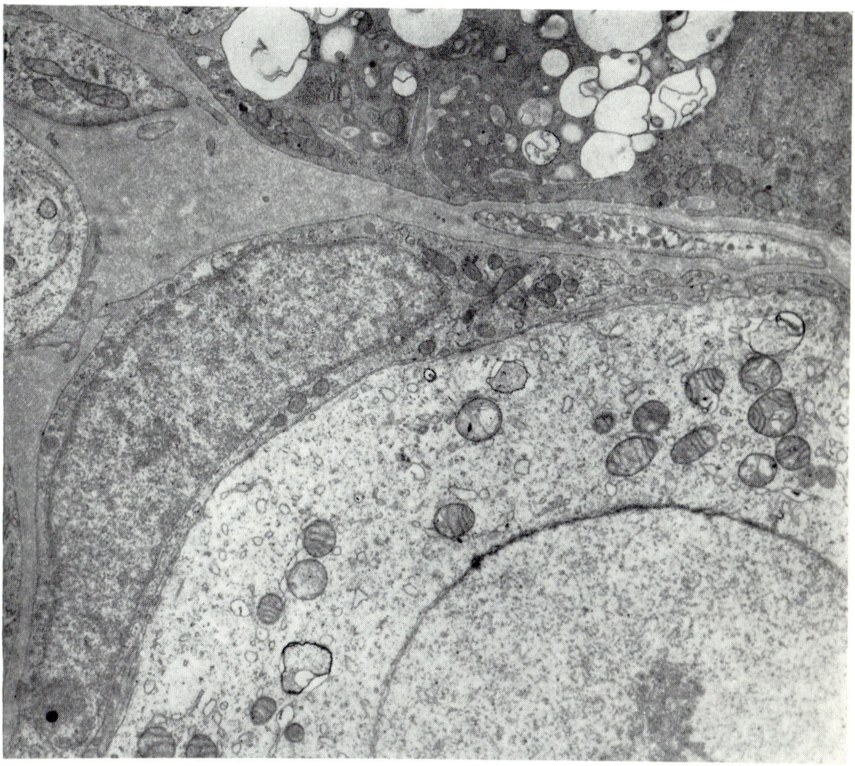

Figure 1. Mouse ovary, 5 days old. Note early granulosa cell closely attached to oocyte's surface. Distinct basement membrane present external to early granulosa cell. × 11025

facing p. 232

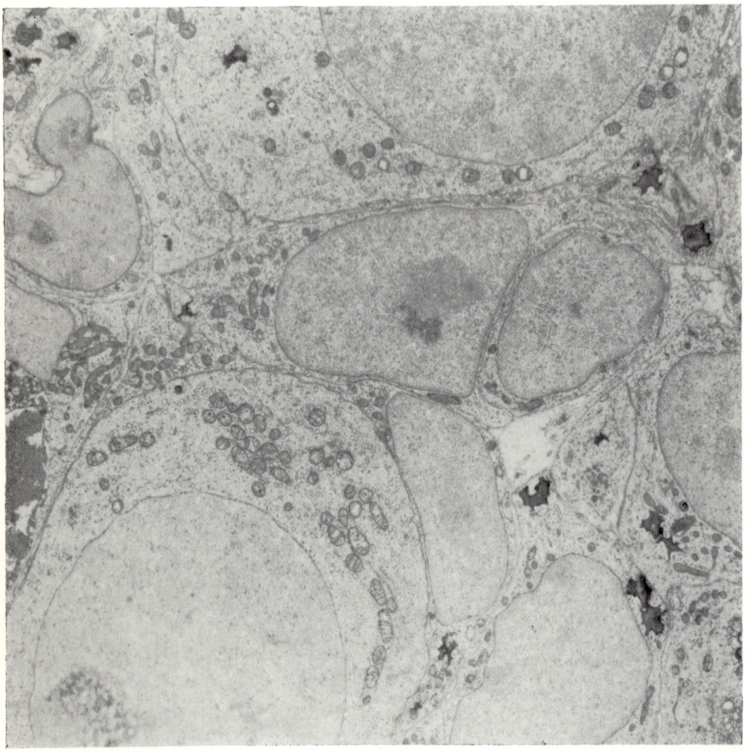

Figure 2. Mouse ovary, 1 day old. Young oocytes separated by pregranulosa cells which are not yet definitively associated with the oocytes. ×6370

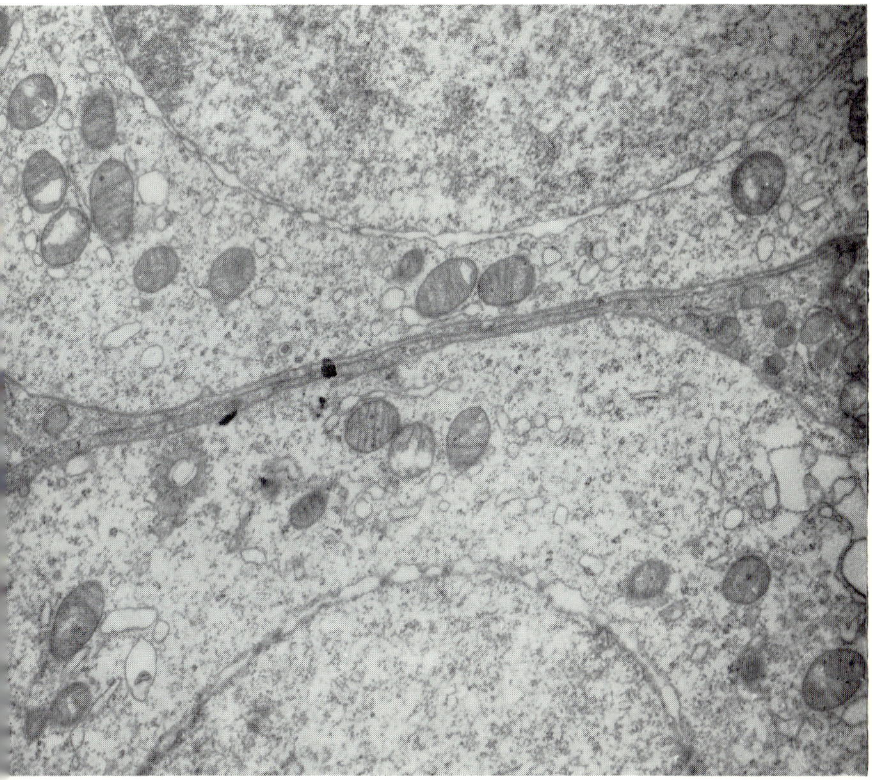

3. Mouse ovary, 1 day old. Separation of oocytes by slender cytoplasmic projections of the pregranulosa cells. ×18,200

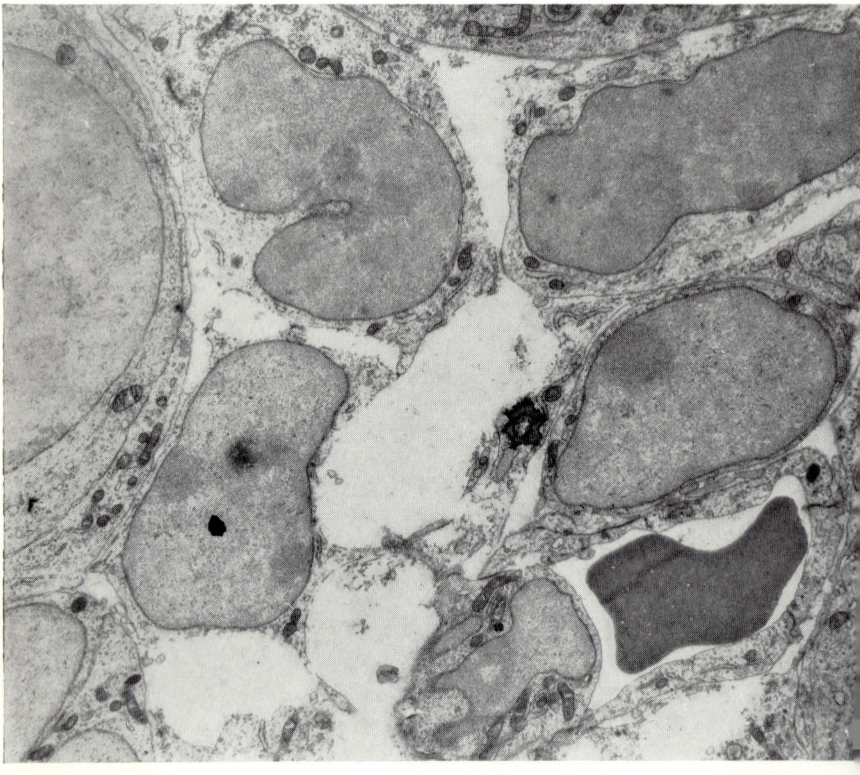

Figure 4. Mouse ovary, eighteenth day of foetal life. Upper third: Young oocyte incompletel enclosed by early granulosa cells. Note undifferentiated stromal cells containing large nucle of irregular shape. Nuclei occupy most of the cytoplasmic matrix. Small capillary, left below ×6860

facing p. 23

with lead citrate (Reynolds). The sections were examined and photographed with a Zeiss EM 9 electron microscope. For the purpose of orientation and survey, sections 0.5 to 1.0 μ in thickness were cut from the same block with a Porter-Blum Ultra-microtome and stained with Toluidine Blue.

RESULTS AND DISCUSSION

There is good evidence from ultramicroscopic studies that the whole population of specialized non germ cells i.e. follicle cells, theca cells and interstitial cells is derived exclusively from the pool of primitive ovarian stromal elements. In ovaries taken on the eighteenth day of foetal life the following somatic components can be clearly defined by structural criteria: (1) early granulosa cells, (2) pregranulosa cells, (3) undifferentiated stromal cells and (4) endothelial cells of capillaries.

Early granulosa cells are intimately associated with the young oocytes. The flat cells, closely adherent to the ovum, are separated by a distinct basement membrane from the entire mass of interstitial tissue (Figure 1). When the e.m. basement membrane has completely enclosed the young follicle, no interstitial elements contribute further to the granulosa epithelium. The early granulosa cells are characterized by a specific assortment of organelles. They are relatively rich in mitochondria, rough membranes and lipid droplets. In contrast, the cytoplasmic pattern of *pregranulosa cells* is similar to that of the undifferentiated stromal cells, i.e. they are poor in internal structures (Figures 2 and 3), but they tend to encircle adjacent oocytes with cytoplasmic prolongations. Although not yet definitively attached to the ovum, their cytoplasmic extensions are actively involved in the process of germ cell separation and follicle formation. The main mass of the stroma consists of *undifferentiated stromal cells.* Some at least of these are potential stem cells for both thecal and interstitial elements which will appear a few days later. In the eighteenth day foetal ovary there are no elements clearly identifiable as precursors of theca or interstitial cells. The nuclei of the undifferentiated stromal cells are irregularly oval in shape, occupying much of the cellular volume, with only a few organelles deposited to either side of the nuclei (Figure 4). One may note small mitochondria and few rough membranes. Ribosomes are highly dispersed. Some endoplasmic vesicles are scattered through the cytoplasm.

During the following days of postnatal development there is further progress in follicle development and differentiation. A single theca layer is first seen in Type 3a follicles (Pedersen and Peters, 1968) external to the basement membrane of the unilaminar granulosa epithelium (Figure 5). At approximately the same time formation of the zona pellucida has begun (Figure 6). At this stage of development granulosa cells of the unilaminar follicles are structurally much more highly differentiated than the surrounding theca cells. Meanwhile the connective tissue has proliferated considerably, dividing the irregular groups of oocytes and follicle cells into branching and cord-like arrangements. Small capillaries formed by closely attached endothelial cells extend to the outer thecal layer of the follicles. Occasionally, in small follicles not completely enclosed by early granulosa cells, they may reach the free surface of the ovum. The most peripheral elements of the thecal layer and groups of undifferentiated stromal cells in the vicinity of the capillaries are the first to become transformed into specialized *interstitial cells* which are considered to play an important role in steroid synthesis. The development of interstitial cells indicates that an appropriate gonadotrophic stimulus must have occurred since they have all the features of cells actively involved in steroid biosynthesis i.e. a well developed endoplasmic reticulum, big round mitochondria of tubular type and lipid inclusions. From histological observations a dual origin of interstitial cells in rats and mice has been proposed (Dawson and McCabe, 1951; Rennels, 1951). A primary type is thought to be developed from granulosa cell cords and a secondary type from the theca interna of atretic follicles. Histochemical findings of Baillie *et al.* (1966) with mouse ovary indicate differentiation of stromal mesenchymal cells into the 'primary type' of interstitial tissue. This can be conclusively confirmed by e.m. observations in the immature ovary. On the fifth day of postnatal life some cells of the theca and the unorganized stromal tissue undergo special transformation. A transitional stage of interstitial cell development recognizable in 5-day-old specimens is seen in Figure 7. Both ribosomes and ergastoplasmic lamellae have accumulated. Aggregates of lipid droplets and enlarged, spherical mitochondria appear at the poles of the cell nuclei. During the following two days of development the cytodifferentiation of the interstitial cells is completed (Figure 8). The structural organization of the cells is now similar to that of other cells actively involved in steroid metabolism such as interstitial cells of the testes or steroid cells of the adrenal cortex.

The cellular site of the different functions in steroid biosynthesis is not fully understood but certain patterns indicate function and are commonly found in steroid secreting cells. In the correlation between the ultrastructure and synthetic functions some cellular components were found to be of special interest; lipid droplets, endoplasmic reticulum and mitochondria. Within the steroid cells the lipid droplets certainly do not represent secretion material. Stimulation does not result in release of droplets from the cell but rather a decrease in size and change in reactivity (Barker, 1951; Rennels, 1951; Claesson, 1954; De Groodt-Lasseel, 1963; Nagay *et al.*, 1967). It is thus suggested that those droplets may be converted into steroid metabolites when the proper stimulus occurs. Regarding their chemical composition no conclusions can be drawn from ultrastructural findings. After osmium tetroxide or glutaraldehyde fixation at least some lipids, especially phospholipids, are retained in the cells (Adams, 1960; Carsten, 1965). Unsaturated lipids i.e. unsaturated triglycerides, phospholipids, cholesterol esters, and fatty acids are intensely blackened with osmium tetroxide, probably because fatty acid double bonds are involved in the reaction with the fixative. Presumably one of the components of the droplets is cholesterol.

The location of the droplets within the cell and their relationship to the other organelles may give some further evidence on their fate during the life span of the steroid producing cell. There are two other types of organelles intimately correlated with lipids: the mitochondria and the endoplasmic reticulum. Lipid droplets are frequently to be found in juxtaposition with mitochondria. It is reasonable to suggest that the function of mitochondria facilitates a release of storage material within the lipid droplets which is passed on to the membrane system of the cell. The enzyme activity necessary for splitting off the cholesterol side chain presumably resides in the mitochondria while most of the other steroid converting enzymes are probably localized in the endoplasmic reticulum (Bjersing, 1967). The structural relationship between the lipid inclusions and the system of smooth membranes favours the possibility that an interaction between these two cell components results in steroid biosynthesis. Frequently the smooth membranes are plastered against the droplets, forming a whorled structure around the lipid inclusion.

There is no direct information from electron microscopic studies regarding the release of steroids from the insterstitial cells. No osmiophilic masses of presumably lipid material could be seen disposed in a manner that suggested they were being extruded from the cell. The membranes of the endoplasmic reticulum could be the means of transport of lipid material within the cell, i.e. the association of lipid droplets and smooth endoplasmic reticulum is probably indicative of both steroid synthesis and intracellular transport. Some authors describe transport of lipids inside the canalicular system of endoplasmic reticulum (Mercer, 1962; Onoe and Ohno, 1965). Guraya *et al.* (1967) demonstrated by cytochemical methods the release of lipid droplets from the rabbit's interstitial cells. The release of preformed lipid droplets was closely accompanied by a considerable loss of cytoplasm and its organelles. The secretory products have been found in the form of discrete bodies between gland cells, in regions adjacent to blood vessels, and in the lumen of the latter. However, these observations have not been confirmed by e.m. investigations.

To sum up, the following steps of interstitial cell development in the immature mouse ovary have been revealed by sequential e.m. studies.

1. Primary interstitial cells of the immature ovary are derived from the pool of ovarian stromal elements.

2. Both the outer layer of theca cells and stromal cell complexes adjacent to the capillaries are preferential sites of interstitial cell development.

3. Transitional stages of developing interstitial cells can first be recognized in the ovaries of 5-day-old mice.

4. Completely differentiated interstitial cells are commonly found in the ovaries of 7-day-old animals. In fully developed interstitial cells the structural equipment is similar to that of other cells actively involved in steroid biosynthesis (that is interstitial cells of the testes, steroid cells of the adrenal gland).

The identification of primary interstitial cells as a special group of somatic cells in the mouse ovary before attainment of sexual maturity is based on the following observations:

1. Structural similarity to other defined types of steroid producing cells.

2. Topographical identity with cells containing sudanophilic birefringent, acetone-soluble, and Schultz positive granules generally considered to act as precursors in steroid formation.

3. Coincidental appearance of morphologically defined interstitial cells and cells exhibiting 3-β-hydroxysteroid dehydrogenase activity within the interstitial stroma of the immature ovary.

4. Marked acceleration of the development of interstitial cells in the immature mouse ovary following exogenous administration of gonadotrophins (see Beltermann and Stegner, 1967).

The fate of the primary interstitial cells and the behaviour of thecal interstitial cells in the adult animal and during gestation are topics of further e.m. studies.

ACKNOWLEDGEMENT

Supported by a grant from the Deutsche Forschungsgemeinschaft.

GENERAL DISCUSSION

PARTICIPANTS: BEN-OR, DONOVAN, FRANCHI, PRESL, PETERS, LLOYD, DEANESLY, WEISZ and GONDOS.

1. Lipid droplets are characteristically seen in granulosa cells of early follicles. They are not present in pregranulosa cells nor in Stage 5 follicles. Similar droplets are seen in the rabbit but are rare in the rat. Lipid inclusions do not necessarily imply steroid synthesis – their

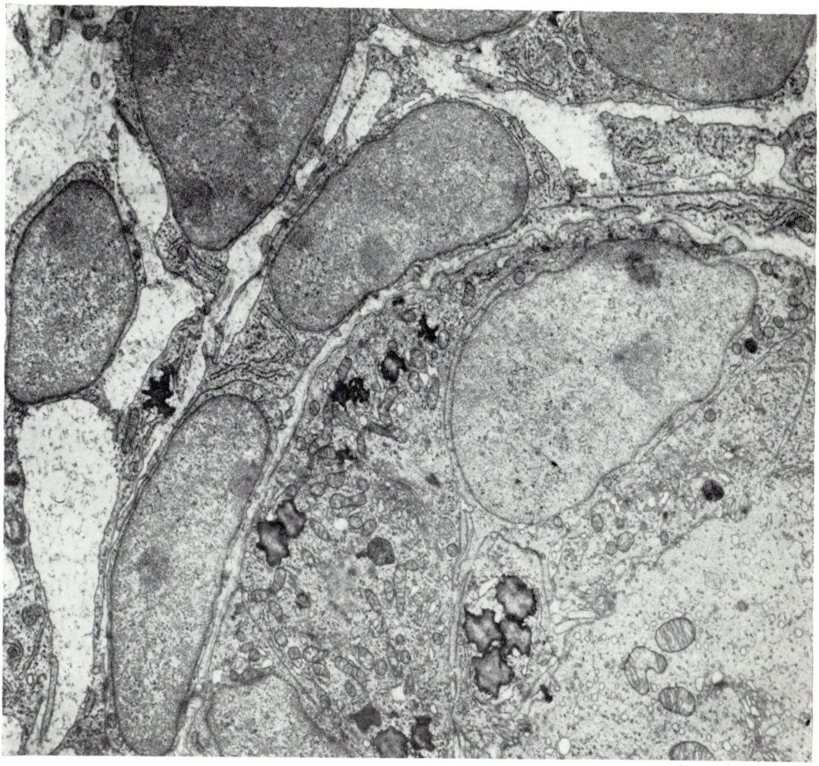

Figure 5. Mouse ovary, type 3a follicle. Note highly developed internal structures of granulosa cells. Single theca layer present external to basement membrane of unilaminar granulosa epithelium. ×7000

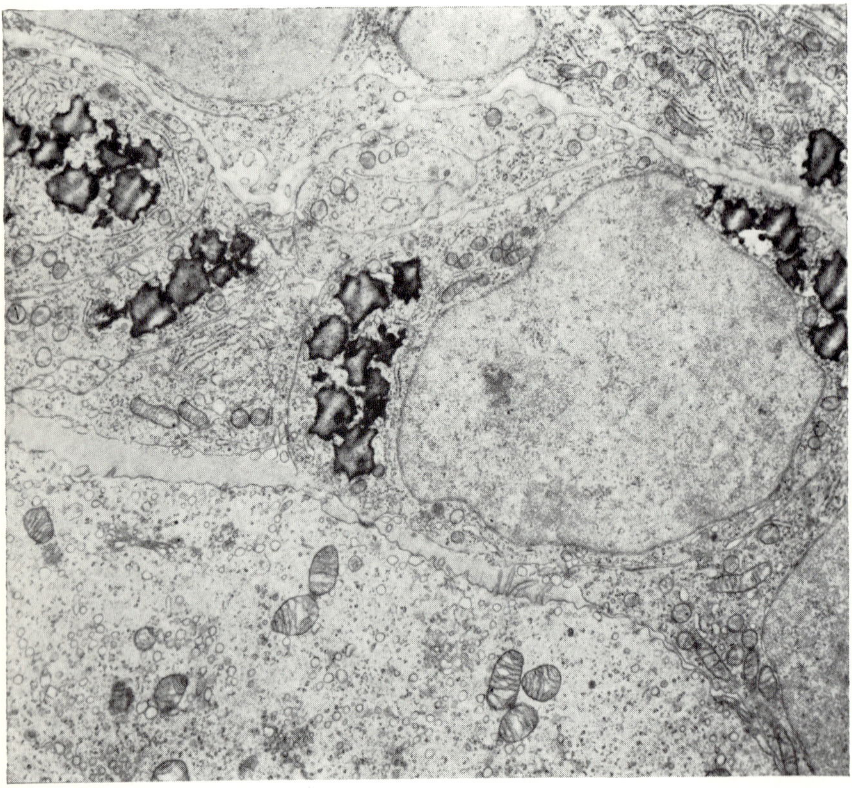

Figure 6. Mouse ovary, type 3a follicle. Space between oocyte and granulosa cells filled by homogeneous fine granular zona substance. ×8,820

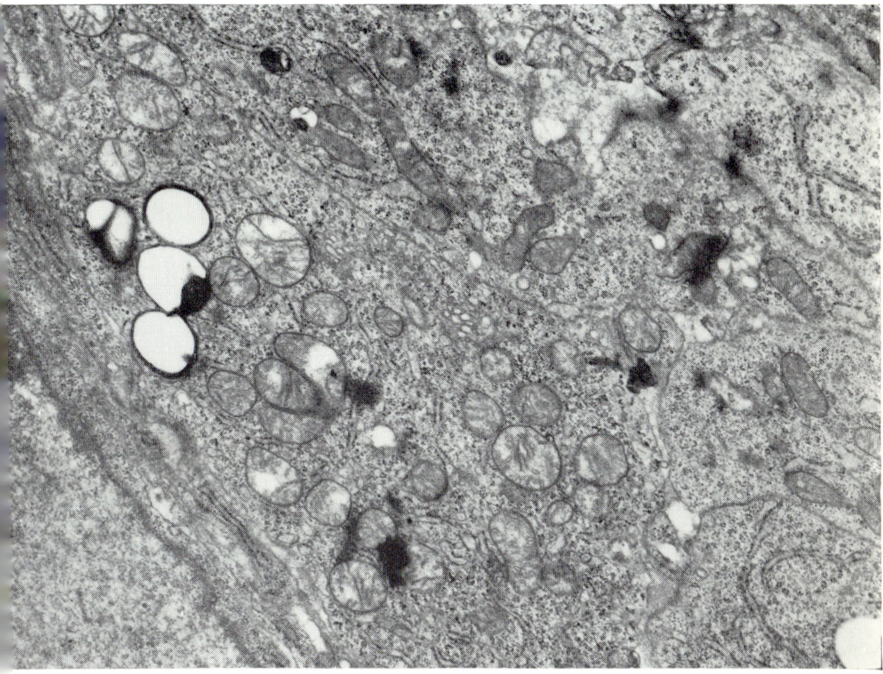

7. Mouse ovary, 5 days old. Transitional stage of interstitial cell development. Note enlargement of mitochondria and accumulation of lipids. ×12,080

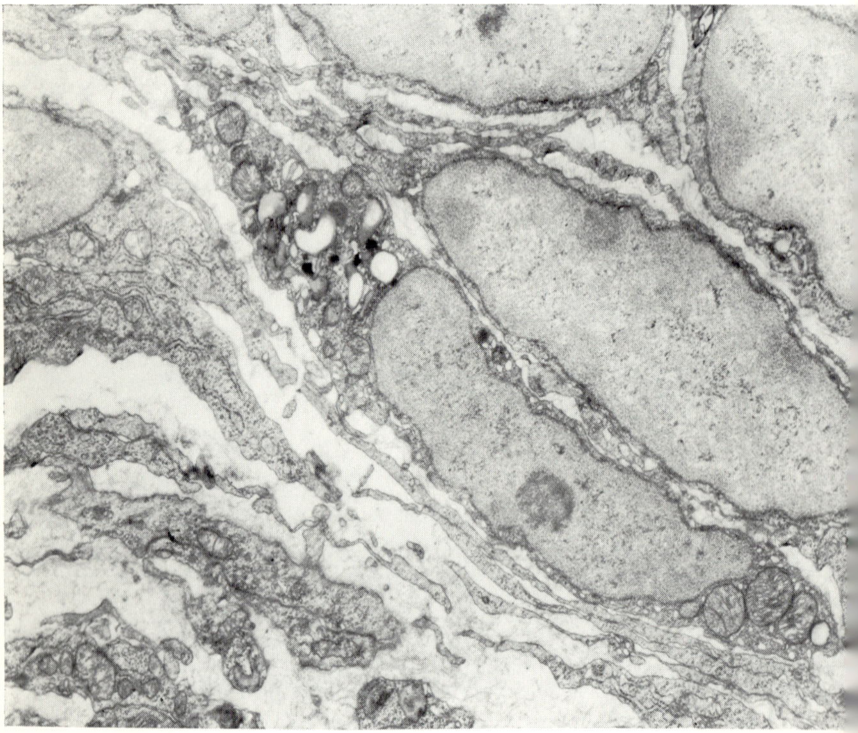

Figure 8. Mouse ovary, 8 days old. Completely developed interstitial cell characterized b round tubular mitochondria, (ii) lipid aggregates, (iii) highly developed endoplasmic reticul

acing p

presence coincides with formation of the zona pellucida and they may be necessary for this process.

2. Homogenates of 5-day-old mouse ovary can convert pregnenolone to progesterone, indicating that Δ5-3β-ol steroid dehydrogenase (3β OHD) and isomerase are present. These enzymes have been demonstrated histochemically in stroma cells of 1-day ovaries but, in other studies, were not found at 4 days. Presence of an enzyme does not necessarily imply activity *in vivo,* e.g. the 'machinery' for earlier synthetic steps may be lacking or precursors may not reach the site of the enzyme. 3βOHD activity is present in the 14-day foetal testis and the 18-day foetal ovary; its presence in these tissues and in the new-born mouse could be due to stimulation by maternal gonadotrophin. It is found in rat theca and interstitial tissue at 8 days, at which time oestrogen can first be detected in the blood by the Ittrich reaction and the first antra appear in the follicles. (Presl).

3. The significance of birth in relation to ovarian development was discussed. In addition to gonadotrophins, the mother supplies steroids and a constant environment. However, in some species, critical stages of ovarian development may precede parturition, which might then be less traumatic.

4. Mouse and rabbit interstitial cells show considerable ultrastructural differences. As in other cells the mitochondria show species-specific characteristics. However, in both species they show the tubular or vesicular arrangement of cristae typical of all steroid-producing cells. Rabbit interstitial cells probably produce more progesterone and less oestrogen than the rat's; the morphological differences may be related to this. Adrenal cortical cells resemble testicular Leydig cells more closely than either resembles ovarian interstitial cells.

REFERENCES

Adams, C.W.M. (1960). *J. Histochem. Cytochem.* **8**, 262.

Baillie, A.H., Ferguson, M.M. & McHart, I. (1966). *Developments in Steroid Histochemistry.* New York: Academic Press.

Barker, W.L. (1951). *Endocrinology,* **48**, 772.

Beltermann, R. & Stegner, H.-E. (1967). *Acta endocr. Copenh.* **57**, 279.

Bjersing, L. (1967). *Z. Zellforsch. mikrosk. Anat.* **82**, 173.

Carsten, P.M. (1965). *Arch. Gynaek.* **200**, 552.

Claesson, L. (1954). *Acta physiol. scand.* Suppl. **113**, 23.

Dawson, A.B. & McCabe, M. (1951). *J. Morph.* **88**, 543.

De Groodt-Lasseel, M. (1963). *Functie en Form van de Interstitiele Cellen van het Ovarium.* Bruxelles: Presses Academiques Européennes.

Guraya, S.S. (1967). *Z. Zellforsch. mikrosk. Anat.* **83**, 187.

Mercer, E.H. (1962). In *The Evolution of Intracellular Phospholipid Membrane Systems in the Interpretation of Ultrastructure.* Ed. Harris, R.J.C. New York: Academic Press.

Nagay, K., Lindlar, F. & Holpman, H.I. (1967). *Z. Zellforsch. mikrosk. Anat.* **79**, 550.

Odor, D.L. & Blandau, R.J. (1969). *Am. J. Anat.* **124,** 163.
Onoe, T. & Ohno, K. (1965). Role of endoplasmic reticulum in fat absorption. In *Intracellular Membranous Structure.* Ed. Seno, S. & Cordry, E.V. Ohayama: Japanese Society for Cell Biology.
Pedersen, T. & Peters, H. (1968). *J. Reprod. Fert.* **17,** 555.
Rennels, E.G. (1951). *Am. J. Anat.* **88,** 63.

Germ Cell Relationships in the Developing Rabbit Ovary

Bernard Gondos

SUMMARY

The regulation of germ cell maturation has been considered in terms of the relationships of germ cells with other germ cells and with granulosa cells. The synchronous and orderly progression from oogonia to oocytes during mammalian oogenesis is explained by the presence of intercellular bridges joining the germ cells into syncytial groups. Groups of germ cells are thereby able to complete oogenesis in an integrated manner. Further coordination of germ cell maturation is considered to be a function of the granulosa cells. Electron microscopic studies have demonstrated a close membrane relationship between granulosa and germ cells in the pre-follicular ovary. Numerous lipid deposits and profiles of endoplasmic reticulum are present in the granulosa cell cytoplasm. The role of the granulosa cells in the regulation of germ cell differentiation deserves further study.

INTRODUCTION

Germ cell maturation during mammalian oogenesis is synchronized and orderly. The changes occurring in the ovarian cortex – mitosis, synapsis, follicle formation – are completed in a relatively short period of time. Each stage is followed by the next in regular succession with minimal overlapping, the cells in the lower part of the cortex generally being at a slightly more advanced stage of development than those above. The transformation of the ovarian cortex from a dense aggregation of germ cells and granulosa cells into a collection of individual follicles separated from one another by bands of stromal connective tissue is a striking demonstration of synchronous differentiation.

In most mammals oogenesis begins in the foetus. An exception is the rabbit, in which differentiation of oogonia into oocytes begins on the first day after birth (Teplitz and Ohno, 1963). The rabbit is therefore a convenient model for studying oogenesis, in that ovaries of animals from the same litter can be studied on successive days. Such observations have provided valuable information on the changes affecting the germ cells as they enter and progress through meiotic prophase prior to follicle formation.

As seen with the electron microscope, the developing germ cell has a low cytoplasm : nucleus ratio. Mitochondria are randomly scattered through the cytoplasm. Endoplasmic reticulum is generally lacking and ribosomes are widely dispersed. As reflected morphologically, the capacity for synthesis and secretion appears limited. The germ cell is therefore dependent on its surroundings for nutrition and support.

It is not surprising that cytoplasmic activity is limited during oogenesis, since this is a time when nuclear activity is maximal. DNA synthesis in germ cells of the rabbit ovary takes place only after birth, reaches a peak during the first four days, then falls off rapidly and is no longer present at 14 days (Peters *et al.*, 1965). The transformation of oogonia into oocytes is also completed by the end of the second week after birth.

Precise regulation of the extensive changes affecting the germ cells during oogenesis is obviously critical. Integration of the complex cellular activities taking place requires considerable coordination. Our investigations indicate that at least two types of coordination are involved. (1) The first of these is an internal synchronization of the germ cells themselves. (2) The second type of coordination is external, depending on close association of the germ cells with adjacent granulosa cells. Both types of relationships apply specifically to the period of oogenesis.

GERM CELL – GERM CELL RELATIONSHIP

Synchronization of germ cell differentiation is a general feature of mammalian oogenesis. Correspondence between age and stage of differentiation of a significant percentage of germ cells is consistently seen. The germ cells, in fact, mature in groups, and each group consists of elements which are at an identical stage of differentiation.

Electron microscopic studies have indicated that this synchronization is related to the presence of intercellular bridges connecting germ cells and resulting in a syncytial organization of the germ cells in the ovarian cortex (Zamboni and Gondos, 1968). Intercellular bridges have been described in the developing ovaries of the rat (Franchi and Mandl, 1962), golden hamster (Weakley, 1967), rabbit (Zamboni and Gondos, 1968), mouse (Ruby *et al.*, 1969) and man (Stegner and Wartenberg, 1963). In all instances, this finding has been restricted to the period preceding follicle formation.

A diagrammatic representation of a typical intercellular bridge is shown in Figure 1. The connected cells are identical in structure and stage of development. The boundaries of the bridge are continuous with

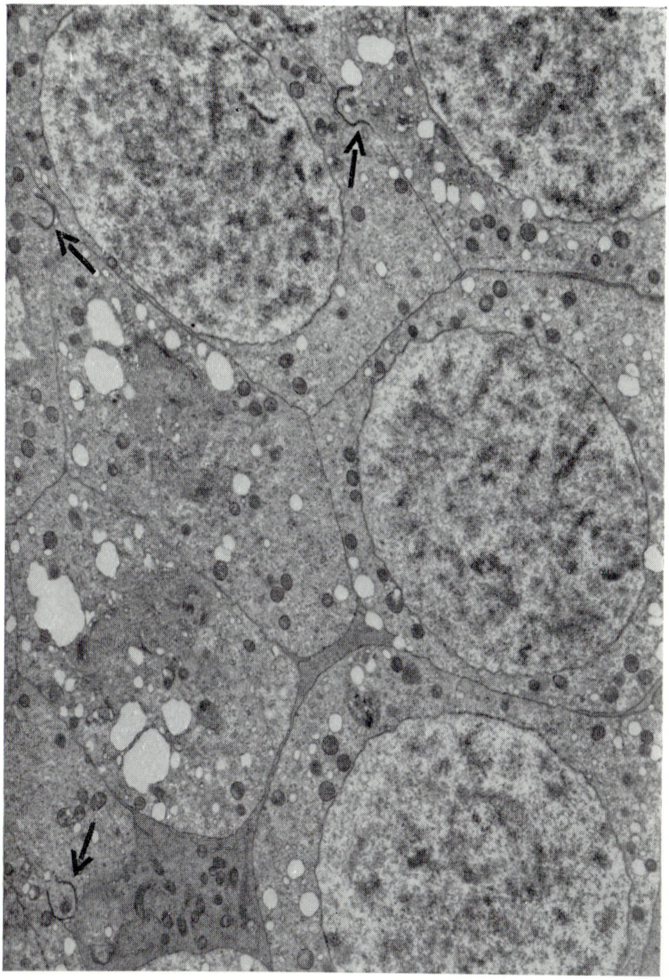

Figure 4. Group of meiotic oocytes, all at the same stage of development with three intercellular bridges (arrows). ×4500

facing p. 240

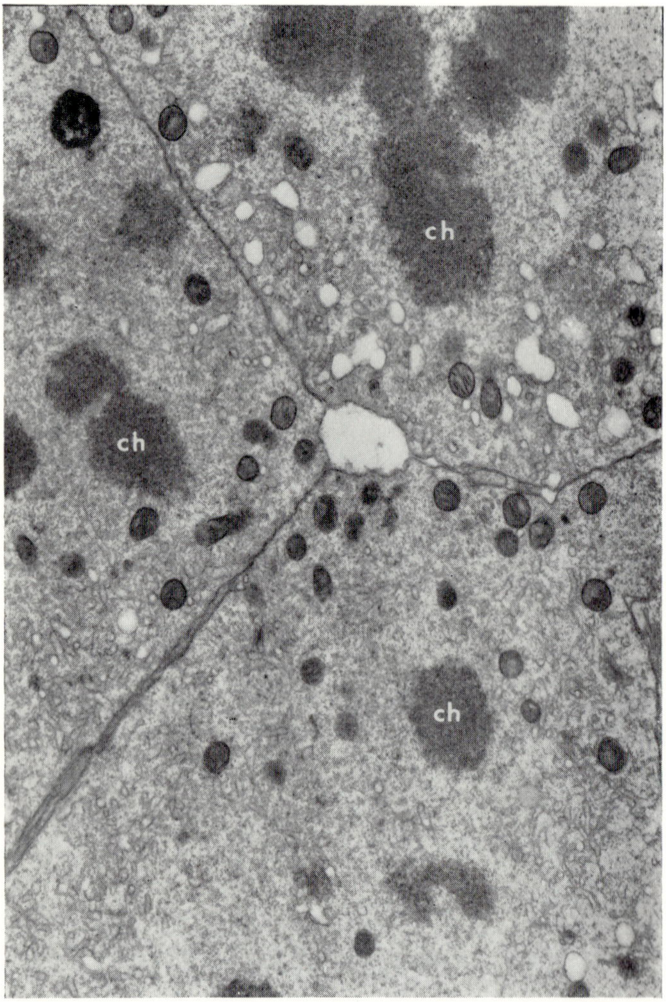

Figure 5. Three adjacent oogonia in mitotic metaphase. Chromosomes, ch ×11,250

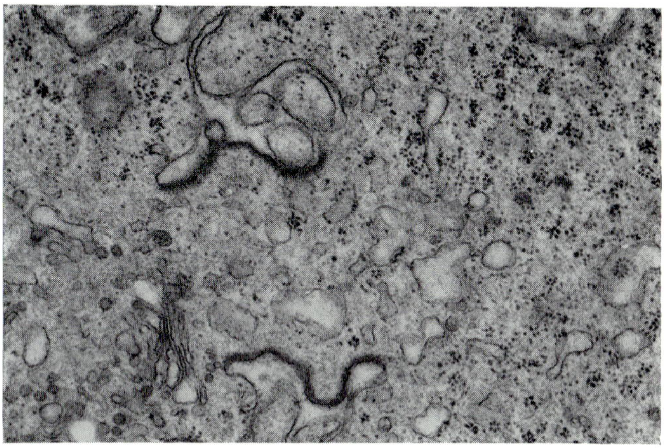

Figure 6. Intercellular bridge sectioned through its long axis. Note similarity of organelles in bridge and connected cells. ×20,500

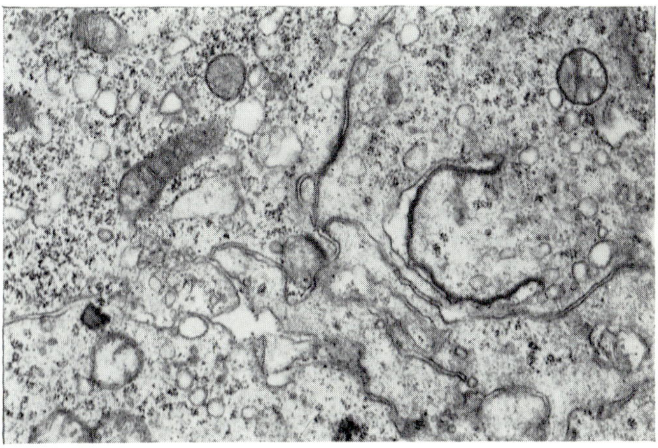

Figure 7. Intercellular bridge sectioned obliquely, appearing closed at one end ×17,000

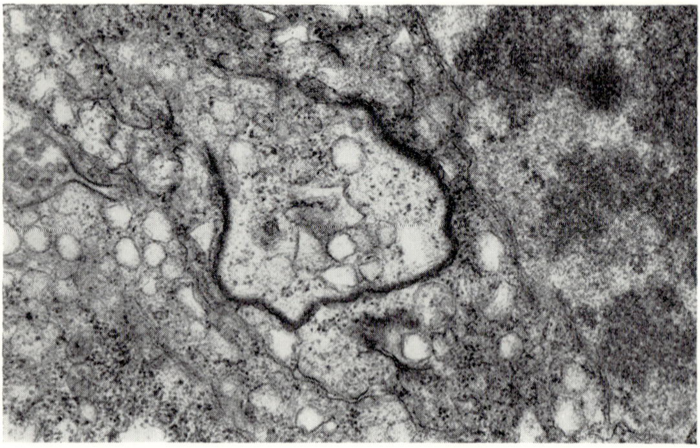

Figure 8. Intercellular bridge sectioned at an angle almost perpendicular to its long axis. ×22,000

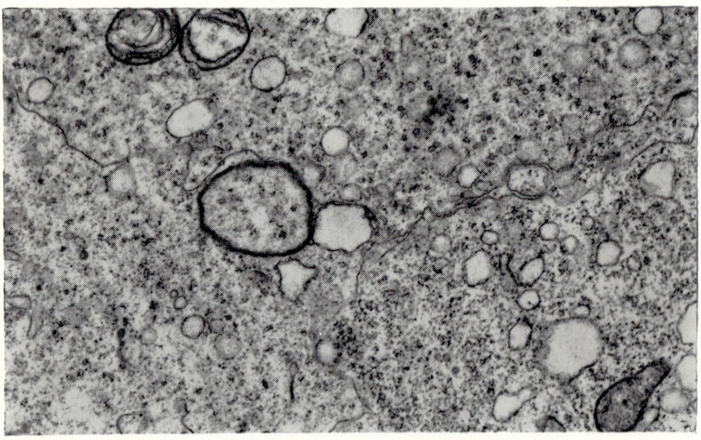

Figure 9. Perpendicular section of bridge indicating cylindrical configuration. ×17,000

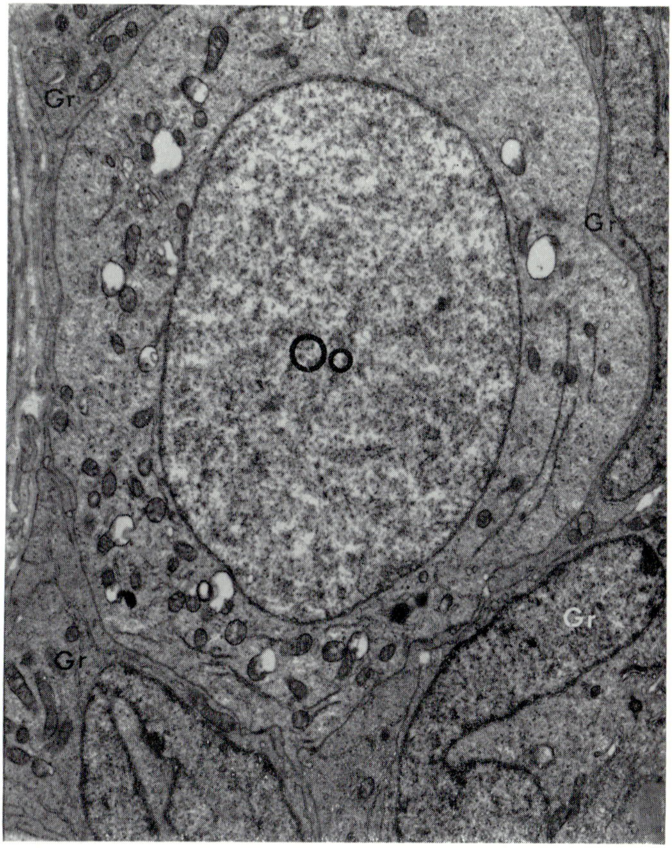

Figure 10. Germ cell (0o) surrounded by four granulosa cells (Gr). Elongated strip of granulosa cell cytoplasm intervenes between germ cell and basement membrane at left. ×5850

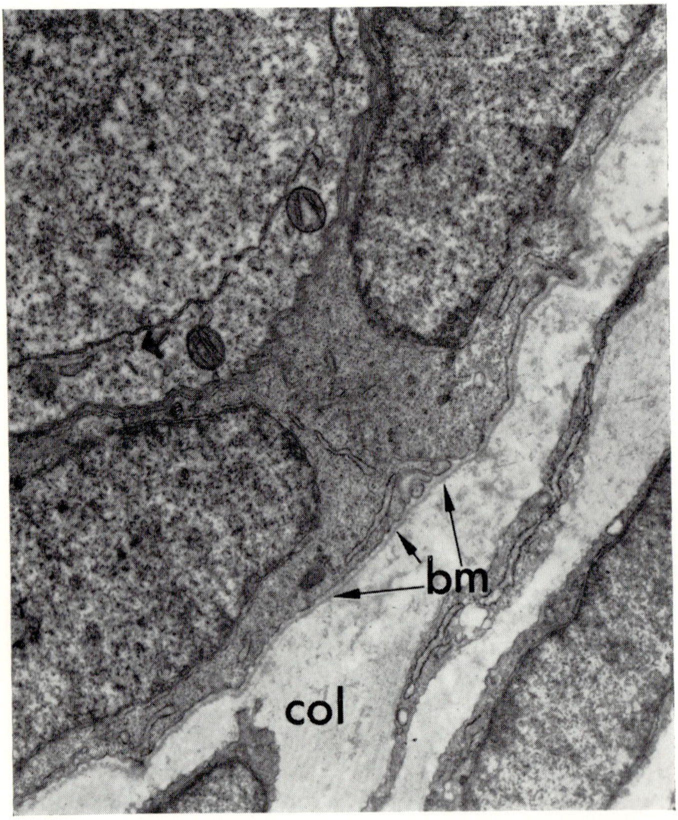

Figure 11. Basement membrane (bm) separates groups of germ cells and granulosa cells from surrounding collagen (col) and stromal cells. ×14,400

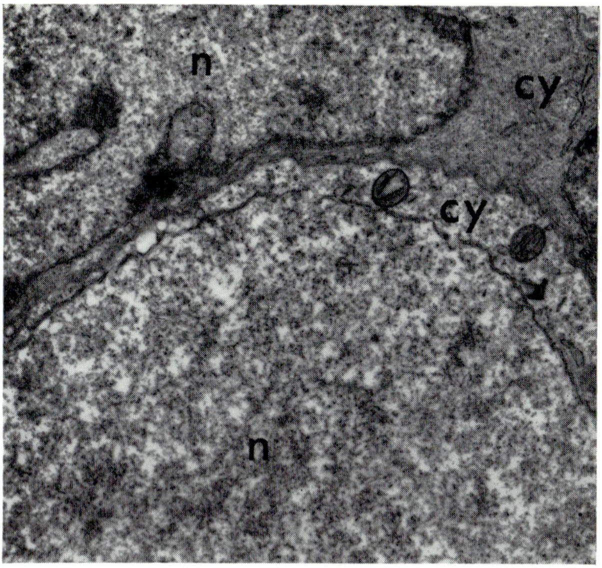

Figure 12. Plasma membranes of granulosa cell (above) and germ cell are closely interwoven. Very little cytoplasm (cy) separates the nuclei (n) of the two cells. ×13,600

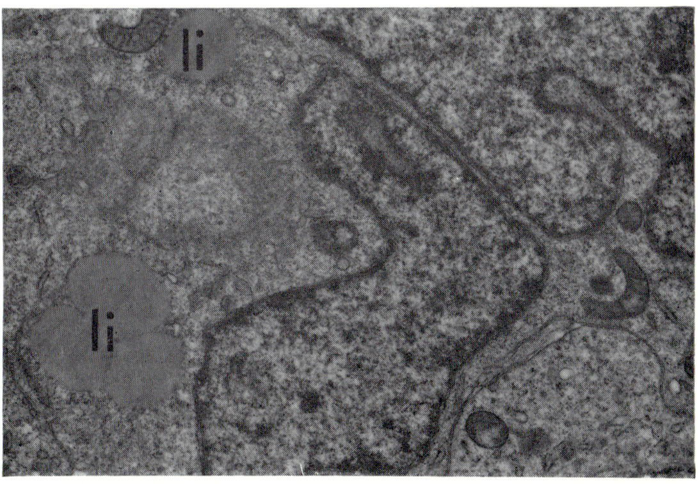

Figure 13. Lipid deposits (li) in cytoplasm of granulosa cell. Note close association of lipid and mitochondrion at upper left. ×16,000

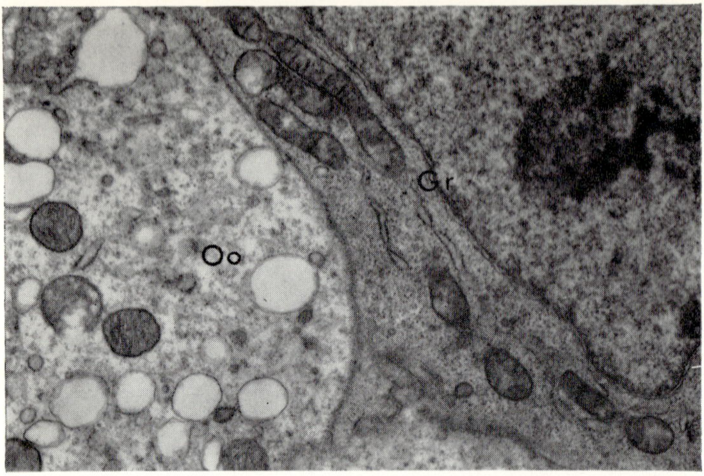

Figure 14. Numerous vacuoles are present in cytoplasm of germ cell (0o) adjacent to granulosa cell (Gr). ×24,000

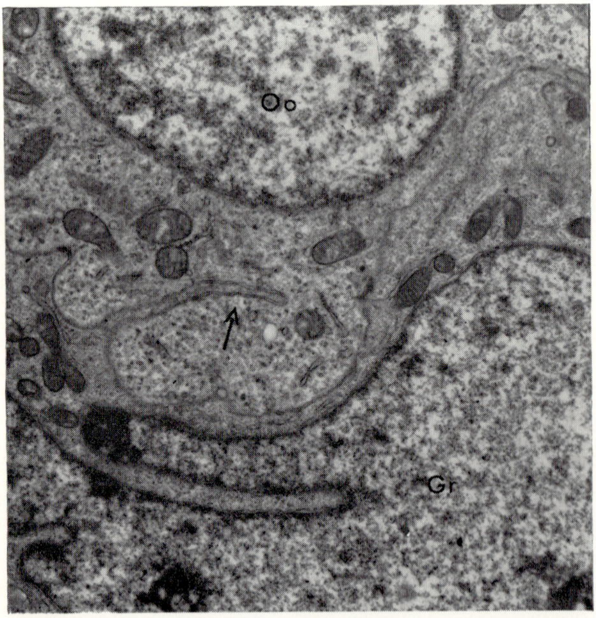

Figure 15. Long narrow cytoplasmic projection (arrow) of granulosa cell (Gr) is shown producing deep indentation of adjacent germ cell (0o). ×12 750

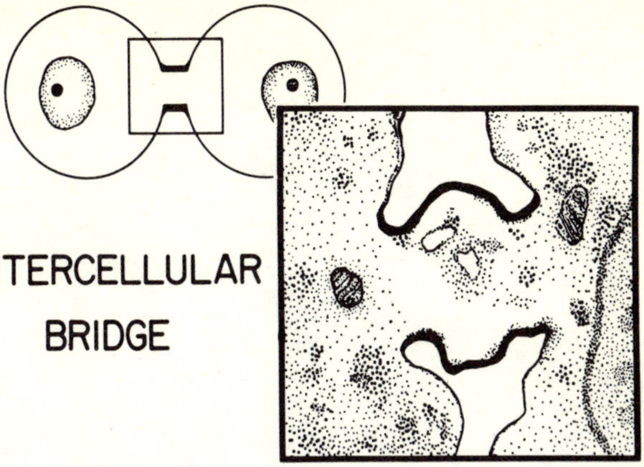

Figure 1. Diagram of intercellular bridge indicating cytoplasmic continuity between connected cells.

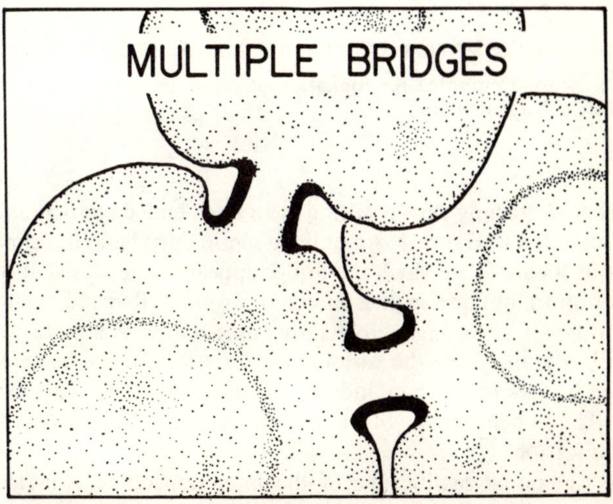

Figure 2. Diagram showing three adjacent cells connected by intercellular bridges.

the plasma membranes of the adjacent cells, but are somewhat thickened and more dense. Cytoplasm in the bridge is similar in all respects to that of the conjoined cells.

Connections between germ cells are easily recognized in electron micrographs, even at low magnification (Figure 4). The appearance of the intercellular bridges, best studied at higher magnification, depends on the plane of section. The bridge shown in Figure 6 is sectioned

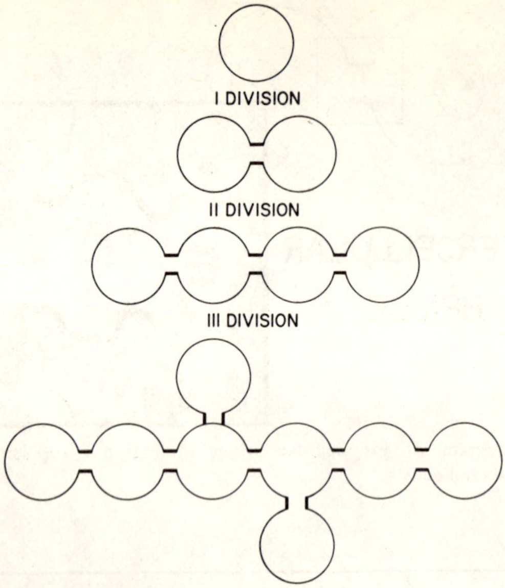

FORMATION OF INTERCELLULAR BRIDGES

Figure 3. Scheme proposed for formation of germ cell syncytium by repeated incomplete mitotic divisions.

parallel with its long axis, resulting in an open-ended configuration, like that illustrated in Figure 1. An oblique section produces the appearance shown in Figure 7, in which the bridge appears to be closed at one end. An even more oblique angle is shown in Figure 8. Perpendicular section of a bridge at right angles to its long axis results in a picture such as Figure 9. By analyzing the various configurations, it can be determined that the bridges must be cylindrical, with a diameter of approximately 0.3μ. The open ends of the cylinder allow free intercellular communication.

The functional significance of the interconnections is indicated by the presence of various types of organelles within them. There is ample evidence that ribosomes, vesicles, strands of endoplasmic reticulum and even mitochondria are passed from one cell to the other. Such organelle transfer appears to be the mechanism whereby synchronous differentiation of adjacent germ cells is accomplished.

The evidence for coordination of cellular activities of pairs of adjacent cells is clear. But how does this explain the more extensive synchronization involving large groups of cells? That a multicellular type of coordination exists is well demonstrated by observations of groups of germ cells at identical stages of division within the cortex

(Figures 4 and 5). The explanation becomes evident after further analysis. Figure 2 shows diagrammatically the connection of a germ cell to two adjacent cells, thus extending the principle of cellular interconnections to more than merely pairs of cells. Verification of this analysis has already been suggested by the multiple bridges shown in Figure 4. Such a finding is not at all uncommon. In fact, multiple connections are easily found. It is obvious that multiple interconnections must be extremely common, since only a limited field of observation is provided by the ultrathin sections used for electron microscopy.

These observations have led to the conclusion that a network of intercellular connections results in the organization of germ cells into multiple syncytial groups (Gondos and Zamboni, 1969). Such a network is felt to result from a series of mitotic divisions in which cell separation is incomplete. That is, mid-bodies of dividing oogonia persist, producing permanent connections of the daughter cells. Incomplete and simultaneous mitotic divisions of two connected oogonia lead to the formation of a syncytium of four oogonia. Incomplete cell separation, repeating itself at each mitotic division, results in the formation of larger syncytial groups of germ cells (Figure 3). It is thus that synchronous differentiation is accomplished.

GRANULOSA CELL–GERM CELL RELATIONSHIP

During oogenesis, the germ cells are in close contact with adjacent granulosa cells. These two cell types are arranged in irregular cords and nests within the ovarian cortex. In the rabbit ovary between birth and the second week of life, granulosa cells line up around the germ cells, separating them from one another and from the limiting basement membrane which encloses each large cellular aggregate (Figures 10 and 11).

With the electron microscope it can be seen that the association of granulosa cells and germ cells is extremely close (Figure 12). Plasma membranes of adjacent cells are closely applied, leaving an intercellular distance of less than 300°A This tight relationship is also maintained while the germ cells are dividing. In addition to their close apposition, the membranes of adjacent cells are thrown into multiple interdigitating folds by the presence of short, narrow cytoplasmic projections. In some cases, elongated projections of the granulosa cells cause marked indentation of the membranes of adjacent germ cells, extending almost to the point of nuclear contact (Figure 15). These deep extensions generally arise in regions of the granulosa cell cytoplasm which are filled with mitochondria.

The cytoplasm of the granulosa cells during this period of differentiation is much more complex than that of the germ cells

(Gondos, 1969a). Characteristic elongated mitochondria, with multiple transverse cristae, are numerous. Ribosomes are densely distributed, both free and attached to prominent strands of endoplasmic reticulum. Many of the cells contain lipid droplets, sometimes in coalescent aggregates (Figure 13). Golgi complexes are well developed. The granulosa cells appear fully capable of synthetic and secretory activity.

The function of the granulosa cells in the developing ovary has received little attention. It is generally thought that granulosa cells provide nourishment for the oocyte during oogenesis (Raven, 1961). In the mature follicle, they support oocyte maturation (Donahue and Stern, 1968) and, as granulosa lutein cells, are involved in steroid production (Ryan and Smith, 1965). Electron microscopic studies in the rabbit have indicated that granulosa cells are responsible for the elaboration of the zona pellucida (Merker, 1961). In considering the granulosa-germ cell relationship, it is important to recognize that after follicle formation has occurred the zona pellucida provides a means for transfer of nutrients to the oocyte. Anderson and Beams (1960) considered that nutrients reach the oocyte by diffusion and pinocytosis. In the pre-follicular stage there is no intercellular medium. Nevertheless, it seems likely that pinocytotic transfer occurs at this stage as well. Varying-sized vacuoles are commonly seen in the oocyte cytoplasm immediately adjacent to the granulosa cell membrane (Figure 14). Such an arrangement suggests that the granulosa cells may synthesize materials required by the oocyte and transfer them directly across the narrow intercellular space.

Although the mechanisms are unknown, there is at least morphologic evidence that the granulosa cells are involved in supporting germ cell maturation during oogenesis (Gondos, 1969b). Investigation of endocrine aspects of ovarian development may be particularly useful in approaching this problem. According to Gillman (1948), the foetal pituitary and adrenal glands have critical roles in gonadal development. How these roles are mediated has not been clarified. Recently, Merker and Diaz-Encinas (1969) showed that stimulation of juvenile rabbits with gonadotrophins resulted in luteinization of theca cells. In studies on the mature rabbit ovary, Blanchette(1966) compared the ultrastructural changes which occur in the granulosa cells of the normal follicle prior to ovulation and under the influence of exogenous gonadotrophins. The profound effect of gonadotrophins on differentiation of the granulosa cells was evident.

Further studies are necessary to clarify the role of granulosa cells in relation to germ cell differentiation. Of special interest is the possibility of some type of endocrine interaction. Enzymes involved in steroid metabolism have been found in the granulosa cells of the human ovary as early as the fourth month of foetal life (Goldman *et al.*, 1966). In the light of current knowledge on the complex endocrine function of granulosa lutein cells, it is tempting to consider that granulosa cells in the developing ovary may have a hormonal influence on germ cell maturation.

GENERAL DISCUSSION

PARTICIPANTS: FRANCHI, DEANESLY, BEN-OR, PEDERSEN, LLOYD AND WEISZ.

1. It was stressed that the 'arrested division' of oocytes does not inhibit their other activities, e.g. they enter meiotic prophase. When oocytes degenerate they appear to do so in synchronised clusters. This degeneration and the accompanying phagocytosis reaches a peak about Day 8 to 11, coinciding with the peak of meiosis. Bridges between oocytes occur in the human, monkey and rat as late as the pachytene stage. Light microscope observations suggest that they may occur in the mouse also. In these species degeneration occurs simultaneously in groups of oocytes and may affect somatic cells within the same basal lamina. It is possible (a) that the bridges are disrupted within a 'cell nest' when the pre-granulosa cells grow and extend processes between the oocytes; and (b) that the basal lamina surrounding the 'cell nest' then invaginates and eventually divides up the individual follicles.

2. In the male, bridges occur at all stages from spermatogonia to spermatids. They have been described in seminomas and are to be expected in other germ cell tumours but not in somatic cell tumours. The connections between somatic cells indicated by measurements of electrical charge are probably much finer than the bridges described here.

3. In a number of species, as pre-granulosa cells become transformed into young granulosa cells, electron-dense areas of contact appear between them and the oocyte. They are particularly numerous where cytoplasmic projections occur. It was suggested that they may act as anchors or help prevent the escape of intercellular fluid.

4. The need for an adequate nomenclature defining the cellular components of the immature ovary, and acceptable to both light and electron microscopists, became evident during a prolonged discussion centred on the term 'pre-granulosa cell'. This is applied to those cells which first surround the oogonia and produce the basal lamina that separates each 'cell nest' in the sex cords from the rest of the ovary. Although they resemble stromal cells under the light microscope, their fine structure is like that of the granulosa cell in several respects and is different from that of the stromal cell. If the oocytes degenerate the encircling cells may become phagocytic or fibrous stroma cells (Deanesly). Biochemical studies might help to determine their specific nature. Granulosa cells may be defined as those attached to a single oocyte and separated by an intact basal lamina from the rest of the ovary.

REFERENCES

Anderson, E. & Beams, A.W. (1960). *J. Ultrastruct. Res.* **3**, 432.

Blanchette, E.J. (1966). *J. Cell Biol.* **31**, 501.

Donahue, R.P. & Stern, S. (1968). *J. Reprod. Fert.* **17**, 395.

Franchi, L.L. & Mandl, A.M. (1962). *Proc. R. Soc. Med.* **157**, 99.

Gillman, J. (1948). *Carnegie Contrib. Embryol.* **32**, 81.

Goldman, A.S., Yakovac, W.C. & Bongiovanni, A.M. (1966). *J. clin. Endocr. Metab.* **26**, 14.

Gondos, B. (1969a). *Anat. Rec.* **165**, 67.

Gondos, B. (1969b). *J. Embryol. exp. Morph.* **23**, 419.

Gondos, B. & Zamboni, L. (1969). *Fert. Steril.* **20**, 176.

Merker, H.J. (1961). *Z. Zellforsch. mikrosk. Anat.* **54**, 677.

Merker, H.J. & Diaz-Encinas, J. (1969). *Z. Zellforsch. mikrosk. Anat.* **94**, 605.

Peters, H., Levy, E. & Crone, M. (1965). *J. exp. Zool.* **158**, 169.

Raven, C.P. (1961). *Oogenesis. The Storage of Developmental Information* London: Pergamon Press.

Ruby, J.R., Dyer, R.F. & Skalko, R.G. (1969). *J. Morph.* **127**, 307.

Ryan, K.J. & Smith, O.W. (1965). *Recent Prog. Horm. Res.* **21**, 367.

Stegner, H.E. & Wartenberg, H. (1963). *Arch. Gynaek.* **199**, 151.

Teplitz, R. & Ohno, S. (1963). *Exp. cell Res.* **31**, 183.

Weakley, B.S. (1967). *J. Anat.* **101**, 435.

Zamboni, L. & Gondos, B. (1968). *J. Cell Biol.* **36**, 276.

Part II

Gonadotrophins and Development

Ovarian Development in Infant Mice. Dependence on Gonadotropic Hormones.

Aliza Eshkol, Bruno Lunenfeld and Hannah Peters

INTRODUCTION

Between birth and maturity the mouse ovary undergoes a series of developmental changes. At birth the mouse ovary is an organ which consists mainly of small oocytes and stroma cells. At maturity follicles in all stages of development are present in the organ. In the infant and juvenile ages changes take place which involve oocyte growth, multiplication of granulosa cells, formation of the basement membrane, formation and differentiation of theca cells and accumulation of follicular fluid (Peters, 1969). Concommitant with follicular development differentiation of some of the stroma cells into interstitial cells occurs (Stegner, 1970), and the vascular system develops.

It has been shown that in the juvenile period (after the second week of life), gonadotrophins stimulate processes involved in follicle development and secretory activity (Ben-Or, 1963; Hertz, 1963). However, whether gonadotrophins stimulate any of these processes in early infancy is still an unsolved question. No evidence of follicular stimulation in mice during the first 12 days of life was found after injection of pregnant mare serum gonadotrophins by Pfeiffer and Hooker (1942), and the ovary of rats was reported refractory up to the age of 10 days even with large doses of pituitary gonadotrophic extracts by Hertz (1963). Price and Ortiz (1944) noted an 'increased mitosis in the theca and granulosa leading to slight follicular enlargement' in rats when gonadotrophins were injected between the ages of 4 and 10 days, and Eayrs (1951) found that follicular development was advanced in rats killed at 9 days of age after having been treated with equine gonadotrophins from the day after birth.

The aim of the present study was to investigate the pattern of ovarian development in infancy in the absence of gonadotrophins. This was done by studying the effects of gonadotrophic deprivation from birth. The specific roles of follicle stimulating hormone (FSH) and FSH plus luteinizing hormone (LH) were investigated in animals deprived of endogenous gonadotrophins.

MATERIALS AND METHODS

For the deprivation of endogenous circulating gonadotrophins an antiserum capable of neutralizing mouse gonadotrophins was used. This antiserum (aRG) was prepared by immunizing rabbits with rat pituitary gonadotrophins. (The potency of the aRG was similar to that described by Lunenfeld *et al.*, 1967.) The antiserum was found to cross-react with mouse but not with human gonadotrophins (Lunenfeld *et al.*, 1967). Daily subcutaneous injections of 0.2 ml antiserum were given to 12 newborn Swiss albino mice for 7 or 14 days.

To study the specific effect of FSH on the developing ovary, the animals were deprived of endogenous gonadotrophins by daily aRG injections, and received simultaneously biologically 'pure FSH'. For the preparation of FSH, human menopausal gonadotrophins (HMG) containing FSH and LH was used (Donini *et al.*, 1966). The biologically 'pure FSH' was prepared by adding to HMG an antiserum in order to bind the LH activity (Donini *et al.*, 1966; Eshkol and Lunenfeld, 1967). Daily subcutaneous injections of 0.5 i.u. in 0.1 ml saline were given 30 minutes after the aRG injection. An interval of 30 minutes between the two injections was used in order to ensure the presence of the aRG in the circulation at the time of the hormone injection. To study the combined effect of FSH plus LH, HMG in a dose containing 0.5 i.u. FSH plus 0.5 i.u. LH was given daily about 30 minutes after the aRG injection. Two or three mice were used in each experimental group.

The animals were sacrificed when they were 7 or 14 days old. The ovaries were removed and fixed in Bouin's solution, serially sectioned at 5 μ and stained either with haematoxylin-eosin or with Heidenhain's azan stain.

In order to evaluate to what extent follicles had developed under normal or experimental conditions, follicles in the different stages of development were counted. Follicle development was defined: (1) by the diameter of the oocyte and (2) by the number of granulosa cells surrounding the oocyte, counted in that section of the follicle which included the nucleolus of the oocyte. (This section often represents the largest diameter of the follicle). The follicles were divided into those which had 10 to 20, 21 to 40, 41 to 60 and 61 to 100 cells surrounding the oocyte. This subdivision was necessary to recognize specific disturbances in development. Differential counts were done in every fifth section of the ovary and the total number of follicles determined by using the following formula (Mandl and Zuckerman, 1951).

$$\text{Number of follicles} = \text{number counted} \times 5 \times \frac{\text{thickness of section}}{\text{thickness of section} + \text{size of nucleolus}}$$

In each experiment ovaries of control littermates were used for comparison.

RESULTS

Ovaries of 7-day-old mice.

Control animals. The ovary of the newborn mouse is a solid organ containing mainly small oocytes and stroma cells. In the Swiss albino mouse most of the oocytes already have one or two cells attached to their surface at birth and about 100 small oocytes can be recognized with a complete ring of 10 to 20 granulosa cells. Occasionally follicles with more than 20 granulosa cells were seen.

In the 7-day-old mouse follicular development has progressed (Table I): one-third of the growing follicles had 20 to 40 cells on the largest diameter, one quarter of them had 41 to 60 cells and some of them had 61 to 100 cells surrounding the oocyte. The granulosa cells in the growing follicles are arranged in an orderly way and have much cytoplasm. Spaces (probably early foci of follicle fluid formation) occur between some of the cells (Figure 1A). A basement membrane and early development of a theca layer is seen in the growing follicles.

Animals deprived of gonadotrophins. Three mice were injected daily with 0.2 ml aRG. The first injection was given on the morning after their birth (during the night before), and the last one 6 days later. They were killed at the age of 7 days, 24 hours after the last injection.

The ovaries of the treated animals were considerably smaller than the controls. (From the ovary of a treated animal 140 serial sections of 5 μ thickness were obtained compared to 210 sections of the same

TABLE I

Number of follicles in ovaries of 7-day-old Mice

	Type of follicle				
	Small follicle	Medium follicle			Total
No. of granulosa cells on largest cross section of follicle	10–20	21–40	41–60	61–100	
a) Control	36	55	40	25	156
b) aRG	84	96	5	0	185
c) aRG + FSH	82	130	54	11	277
d) aRG + HMG	54	112	54	18	238

thickness from the control ovaries). Histological examination of the ovaries of the gonadotrophin-deprived animals showed that their development was different from the normal. There were many large oocytes, similar in size and number to those of the control, but instead of being surrounded by two or three layers of orderly granulosa cells, their envelope consisted usually of a single layer of crowded cells which seemed poor in cytoplasm. Detailed analysis of the structure of follicles showed that: (1) the granulosa cells were altered: they seemed smaller, probably due to less cytoplasm which caused the nuclei to appear crowded together; (2) even the largest follicles at that age had no spaces between the granulosa cells; (3) the thecal layer in some follicles was altered: beyond the granulosa cells, where in the normal mouse thecal cells reside, an empty space limited by an azan positive thin membrane was seen (Figure 1 B).

In order to determine at what stage follicular development had been altered, differential counts of follicles were done (Table I). Although the total number of growing follicles was not altered, follicular development in the aRG-injected mice had not progressed as far as in the controls (Table I, b). No follicle had more than 60 cells and only about 3 per cent of all follicles had more than 40 cells.

Animals injected with aRG and FSH. Two mice were injected from birth daily with 0.2 ml aRG and O.5. i.u. FSH. They were sacrificed at the age of 7 days, 24 hours after the last injection.

The histologic examination of the ovaries under low power magnification showed that they already looked different from the ovaries of aRG-injected animals and more like the controls. Closer investigation showed the presence of follicles with several layers of regularly arranged granulosa cells. The nuclei were not crowded as after aRG injection but were spaced similarly to those in ovaries of control animals. Many mitoses among the granulosa cells of the growing follicles were noted (Figure 1 C).

Differential counts of follicles revealed that the absolute number of follicles with more than 20 granulosa cells was increased compared to the ovaries of the aRG-treated animals. However, as in the aRG- treated animals, these represented one half of the growing follicles. The number of follicles with 41 to 60 cells surrounding the oocyte was considerably larger in the FSH-injected animals than in the gonadotrophin-deprived animals. The FSH-treated animals showed a considerable number of follicles with more than 60 cells which were totally absent in the ovaries of aRG- treated mice.

Animals injected with aRG and HMG. After daily injections of aRG and HMG, the appearance of the ovaries in the 7-day-old animals seemed similar to the ovaries of the aRG and FSH-treated mice. Also the differential counts of the follicles in these two groups were similar (Table I; Figure 1 D).

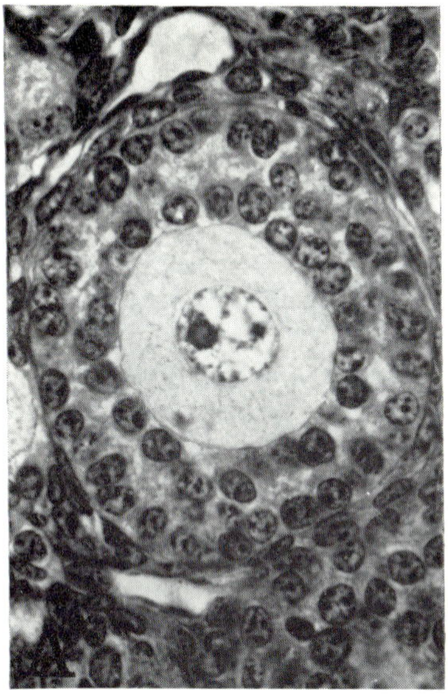

A. Control. Granulosa cells have much cytoplasm; there are often spaces between the cells

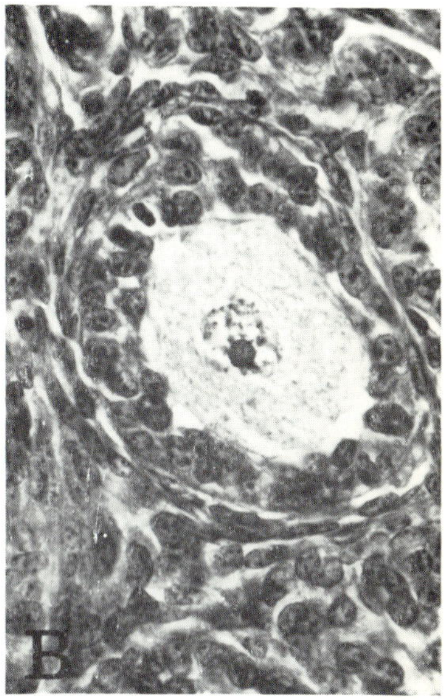

B. aRG treated. The size of the oocyte is not altered. Granulosa cells crowded, with little cytoplasm

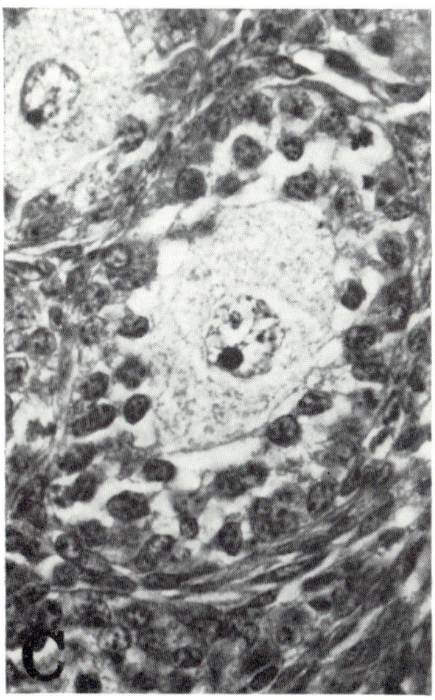

C. aRG + FSH treated. Granulosa cells with ample cytoplasm

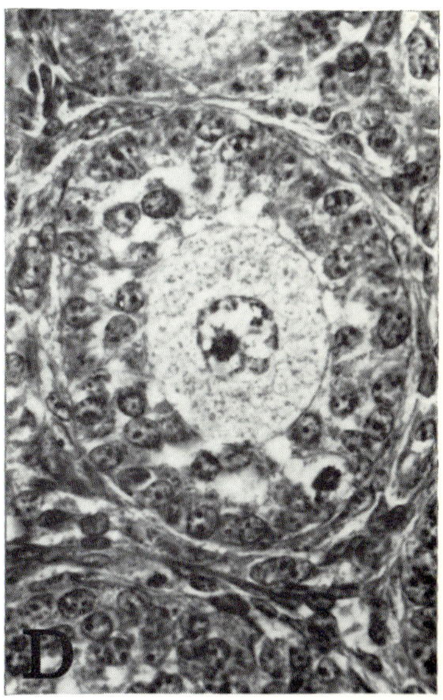

D. aRG + HMG treated

Figure 1. Ovaries of 7-day old mice (Haematoxylin-eosin)

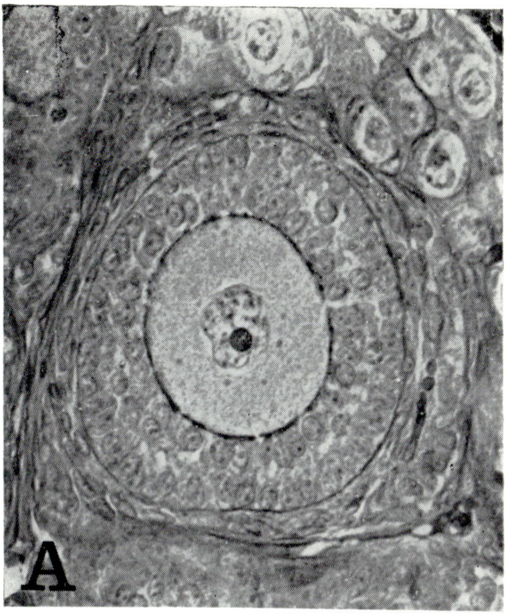

A. Control. Well developed granulosa layer. Theca layer with many cells

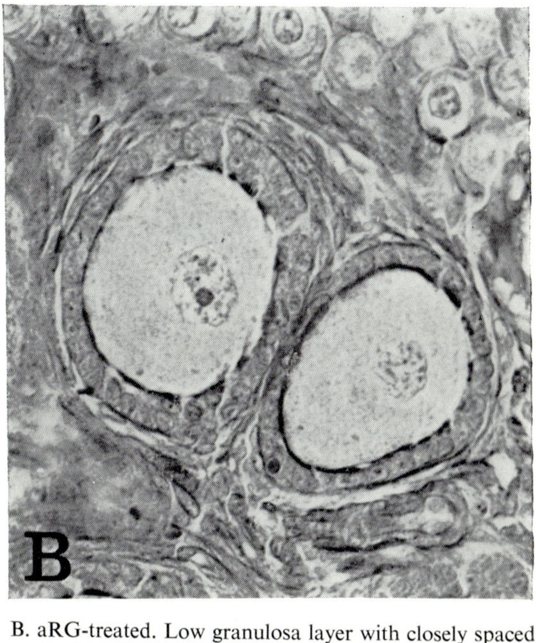

B. aRG-treated. Low granulosa layer with closely spaced nuclei. Poorly defined basement membrane. 'Empty' theca space

Figure 2. Ovaries of 14-day-old mice(Azanstain)

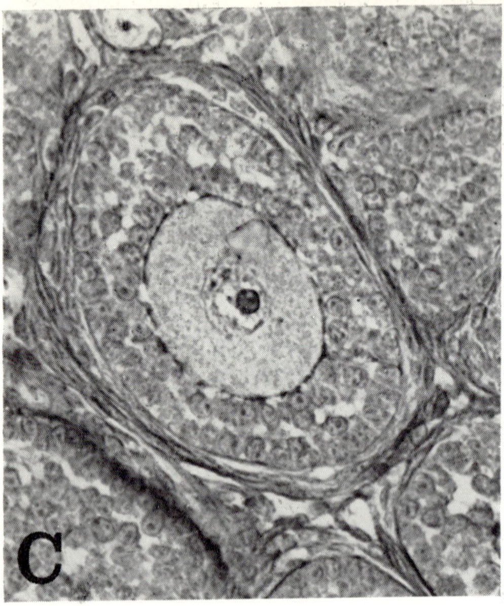

C. aRG + FSH. Granulosa layer restored. Theca layer thin. Few blood vessels

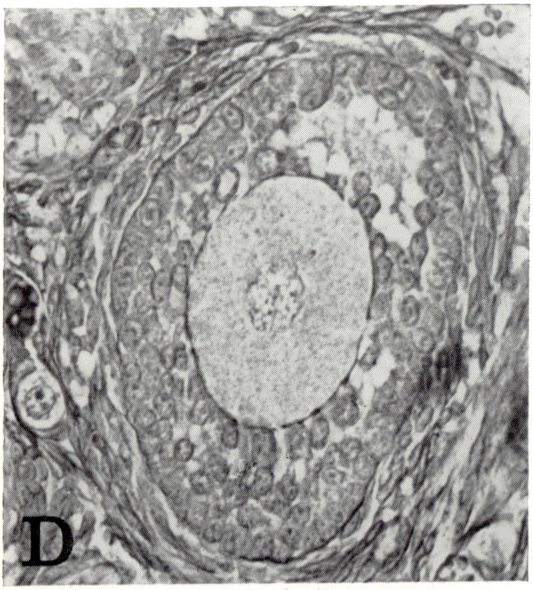

D. aRG + HMG. Beginning antrum formation. Wide theca layer. Many blood vessels

Figure 2. Ovaries of 14-day-old mice(Azan stain)

facing p. 253

Ovaries of 14-day old mice.

Control animals. Between the ages of 7 and 14 days, the ovaries develop further with respect to follicle size, theca development and vascularization. In the growing follicles the granulosa cells are arranged in an orderly way. The cells have ample cytoplasm and the nuclei of the cells are at some distance from each other. Some of the spaces between individual cells are larger than in the 7-day-old mouse, probably due to an increased fluid production. The basement membrane is well developed. The thecal layer consists of several rows of elongated cells among which fine fibrils of connective tissue are visible. Red blood cells are often seen in the thecal layer (Figure 2 A).

The stroma consists mainly of small cells with dark nuclei. Areas of cells with pale nuclei of widely varying shapes and sizes ('cloudy areas') are present. Capillaries are recognized mainly in the centre of the organ but vessels with irregular dilatations, as seen in the later juvenile period, were not noted in the ovaries of 14-day-old animals.

Animals injected with aRG. The ovaries of the animals deprived of gonadotrophins for the first two weeks of life are considerably smaller than ovaries of normal animals. The smaller size of the organ is apparently the result of an inhibition of follicle growth. There are considerably fewer follicles with more than 60 granulosa cells than in the controls (Table II b).

Histological examination shows an abnormal appearance of the granulosa cells and the theca layer, as well as the stroma. The appearance of the granulosa cells in the follicles seen in the ovaries of aRG-injected animals differs markedly from normal cells. The nuclei of the cells are densely packed together with little cytoplasm and no spaces between them. This forms a shell around the oocyte which is narrower than the envelope of a normal follicle containing the same number of granulosa cells (Figure 2 B).

The basement membrane was poorly defined and in some follicles partially or even totally absent. In the theca layer only a few poorly developed cells were seen which were irregularly arranged with empty spaces between them. The periphery of the thecal layer was delineated by fibrous material simulating a membrane. Between the granulosa cells and this membrane-like structure a space was left containing only a few thin (probably theca) cells. Red blood cells were not seen in these spaces. The small oocytes appeared closely packed with only a few cells between them. The vascular system and the stroma appeared poorly developed.

Animals injected with aRG and FSH. The addition of FSH to the daily injection of aRG caused a considerable change in the development of the ovary. The overall microscopic appearance of the ovary was similar to that of the control animals. The number of follicles with more than 60 cells was greatly increased compared to the aRG-treated animals and approached normal values. The granulosa cells

TABLE II

Number of Medium Follicles in Ovaries of 14-day-old Mice

No. of granulosa cells on largest cross section of follicle	21–60	61–100	(x)	101–200	(x)	Total
a) Control	97	97	(0)	25	(0)	219
b) aRG	63	31	(0)	2	(0)	96
c) aRG + FSH	40	50	(0)	22	(0)	112
d) aRG + HMG	66	76	(8)	36	(25)	178

x Number of follicles with antra

appeared normal and were arranged in concentric layers around the oocyte. They had ample cytoplasm and spaces were noted between them. The basement membrane seemed intact but the theca layer was uneven in thickness and thinner than in the control (Figure 2 C). The thickened membrane-like structure limiting the thecal space in the aRG-injected animals was not seen in the aRG and FSH-injected animals. The cortex appeared similar to that of control ovaries with well spaced small oocytes and many cells between them. However, the capillary system still seemed poorly developed and 'cloudy areas' in the stroma were not noted.

Animals injected with aRG and HMG. Following administration of HMG together with aRG for the first 14 days of life, development of the ovary progressed further than after addition of FSH. Antrum formation was noted which was not seen in the ovaries of 14-day-old animals of either the control or other experimental groups. The thecal layer was wider and seemed to contain more cells. Red blood cells were seen within the theca layer (Figure 2 D). A large number of groups of irregularly shaped pale cells – described as 'cloudy areas' in the control animals – were seen between the follicles.

The centre of the organ, which in the normal 14-day-old mouse is solid, showed in the aRG and HMG-injected mice the beginning of a 'lace-like appearance'. The capillary system was well developed with vessels of irregular shape and dilatation, a development seen in the normal animal only later, during the third week of life (Peters, 1969).

DISCUSSION

The results show that, when endogenous gonadotrophic action is blocked, the development of the ovary in the first two weeks of life is not normal. The size of the organ is markedly reduced. However, the growth of oocytes seems undisturbed. Proliferation of granulosa cells is not entirely inhibited but is severely altered. The fact that granulosa cell proliferation does not seem to keep in step with apparently normal oocyte growth suggests that these two processes are not governed by the same stimuli. Oocyte growth does not seem to be dependent on either gonadotrophins or normal granulosa cell proliferation. However, whether a growing oocyte residing in an abnormal follicle envelope is functionally competent cannot be decided by these experiments.

The disturbance of follicular development in the absence of endogenous gonadotrophins is expressed by a reduction in the number of granulosa cells in a given follicle as well as by the change in the appearance of the individual cells and their juxtaposition. Whether granulosa cell proliferation is altered because fewer granulosa cells actively proliferate and/or because the generation time of the individual proliferating cell is lengthened has not been investigated. The disturbance in the development of the individual granulosa cell is characterized by an apparently abnormal development of the cytoplasm, their tendency to crowd together, and the lack of spaces between cells. The crowding of the nuclei seems to suggest that the cytoplasm is diminished. This might be due to impairment of metabolic processes dependent on gonadotrophic stimulation. The results indicate that, already in infancy, endogenous stimulation is necessary for the normal development of follicles.

That a few follicles with more than 40 to 50 granulosa cells developed in ovaries of aRG treated mice may be due to: (1) follicular cell proliferation not being entirely dependent on gonadotrophic stimulation; (2) the administration of insufficient anti-serum; (3) the presence of small amounts of free hormone due to the dissociation of the hormone-anti-hormone complex.

Whether the observed retardation of ovarian development was due either to FSH or to LH deprivation could not be decided without simultaneous substitution therapy since the aRG contains antibodies to both hormones. Substitution therapy with HMG (not neutralized by aRG) showed that gonadotrophins were responsible for the organization and spacing of granulosa cells, the integrity of the basement membrane, the development or maintenance of the thecal cells and the development of the vascular system. When FSH (devoid of LH activity) was substituted in the gonadotrophin-deprived animals, ovarian development reached a stage similar to that of the controls. The follicular cells in the growing follicles had a normal appearance. The crowding and lack of organization observed in the ovaries of the gonadotrophin-deprived animals was not seen in the animals which received FSH stimulation. The basement membrane was intact but the thecal layer was uneven in thickness. The number of thecal cells was less than in the normal animals and they were thinner and more elongated in appearance. The interfollicular tissue was deficient in cells and blood vessels.

When HMG was administered for 14 days, simultaneously with the aRG, follicular growth was not significantly different from that of the animals which received FSH. However, many of the growing follicles had antra of different sizes. Antrum development was not observed in the control littermates. The basement membrane was complete and the thecal layer had become larger. In contrast to the small number of thin elongated thecal cells seen in the FSH-treated animals, the thecal cells of the HMG-treated mice resembled those of the control animals but were round in shape. As noted in the controls, red blood cells were seen in the theca layer and the interfollicular tissue was rich in stromal cells and blood vessels.

The sequence, time and amount of endogenous gonadotrophic stimulation, which are governed by complex feedback mechanisms, obviously could not be reproduced in this experiment. However, the observations described seem to imply a number of effects which the gonadotrophins have on the infant ovary:

1. (a) FSH seems primarily responsible for follicular cell proliferation and organization;
   (b) FSH enables normal development of the basement membrane to occur.
2. The addition of LH to FSH promotes:
   (a) antrum formation;
   (b) the enrichment of the thecal layer;
   (c) an increase in the amount of interfollicular tissue;
   (d) the enhancement of vascular development.

The effects observed in animals which received FSH and LH, as compared with those which received only FSH, cannot be attributed to the direct action of LH alone. The combined effects of both hormones on their respective receptors and the structural and metabolic interrelationship of the various ovarian components must be taken into consideration.

Whether the differentiation of stroma cells is controlled by hormonal stimuli is still being investigated.

ACKNOWLEDGEMENTS

The human menopausal gonadotrophins used in the preparation of FSH was supplied by Instituto Farmacologico Serono, Rome, under the name of Pergonal. This study was supported in part by Ford Foundation Grant No. 67-470.

GENERAL DISCUSSION

PARTICIPANTS: BEN-OR, LUNENFELD, PETERS, RYLE, GREEP, ODELL AND LLOYD.

1. The data suggested: (a) that antiserum (aRG) blocked development beyond a certain point, and (b) that FSH both stimulated the further development of some smaller follicles into larger types and also initiated growth of a new set of small follicles. Alternative interpretations are possible. Antiserum alone increased the number of small follicles as though it had removed some inhibition on their development. The number of follicles starting to grow may depend on the size of the pre-existing pool of very small ones. If antiserum prevents the normal degeneration of follicles in this pool, added FSH would have a particularly marked effect. HMG may have increased atresia in small follicles (Table I), so reducing the pool and diminishing the rate of

recruitment to larger types. Although FSH more than restored the total numbers it decreased the proportion in the 61 to 100 cell stage. Replicate data would help to assess more accurately the suggested effects.

2. In follicles of 14 day ovaries which were exposed to antiserum and FSH, the zona pellucida was often unusually small, and sometimes disrupted. This also occurs in control ovaries, in follicles which are beginning to degenerate, but is much less common.

3. The dose of HMG used was probably excessively high; antra were present, (a) in more follicles than normal and (b) in follicles of types 4 and 5a.

4. There were no treatment effects on oocyte diameter nor on body weight up to 14 days. By 21 days, however, body weights in animals receiving aRG were less than in the controls which received normal rabbit serum.

5. Investigations by other workers with hypophysectomized animals indicated no ovarian dependence on FSH up to about 10 days, or early antrum formation (Ben-Or). The present results are not in agreement with this, though the effects of deprivation at 7 days are relatively slight, and some quite large follicles are present. The few follicles with 41 to 60 cells could, however, result from incomplete binding of FSH by the injected antiserum and were probably not due to anti-rabbit antibody formation since no anti-aRG antibodies were found in males after injection of aRG for 70 days. Mouse ovaries grown *in vitro* with no hormone for 4 days still show mitotic figures, suggesting a slow basal growth rate which is accelerated by FSH stimulation.

REFERENCES

Ben-Or, S. (1963). *J. Embryol. exp. Morph.* **2,** 1.

Donini, P., Puzzuoli, D., D'Alessio, I., Lunenfeld, B., Eshkol, A. & Parlow, A.F. (1966). *Acta Endocr. Copenh.* **52,** 186.

Eayrs, J.T. (1951). *J. Endocr.* **7,** 271.

Eshkol, A. & Lunenfeld, B. (1967). *Acta Endocr. Copenh.* **54,** 91.

Hertz, R. (1963). In *The Ovary,* p. 120. Ed. Grady, H.G. & Smith, D.E. Baltimore: Williams and Wilkins.

Lunenfeld, B., Eshkol, A., Baldratti, G. & Suchowsky, G.K. (1967). *Acta Endocr. Copenh.* **54,** 311.

Mandl, A. & Zuckerman, A. (1951). *J. Endocr.* **7,** 190.

Peters, H. (1969). *Acta Endocr. Copenh.* **62,** 98.

Pfeiffer, C.A. & Hooker, C.W. (1942). *Anat. Rec.* **84,** 311.

Price, D. & Ortiz, E. (1944). *Endocrinology,* **34,** 215.

Stegner, H. (1970). This volume, page 232

Ultrastructural Ovarian Changes after Deprivation of Endogenous Gonadotrophins and the Effect of FSH and HMG in the Deprived Immature Animals

H. -E. Stegner, B. Lunenfeld and A. Eshkol

In previous papers it has been demonstrated that an antiserum (AS) to rat pituitary gonadotrophins is capable of neutralizing circulating endogenous mouse gonadotrophins (Lunenfeld *et al.* 1967). Deprivation of gonadotrophins even from birth is possible by continuous AS administration. From the results of these experiments some conclusions could be drawn on the effects of gonadotrophins on the infantile ovary. Both from histological findings and from data on total follicle counts in the ovaries of control and treated animals, it became apparent that differentiation of the follicles, even in the early postnatal phase of development, depends on gonadotrophic stimulation.

Moreover, since antiserum to rat pituitary gonadotrophins does not cross react with human gonadotrophins, the specific effects of human FSH and LH preparations administered to deprived animals could be investigated. The histological changes observed in gonadotrophin deprived animals and in those receiving AS plus human menopausal gonadotrophin (HMG) or human follicle stimulating hormone (FSH) preparation have already been described (Eshkol *et al.* 1970).

Although the observations made in this histological investigation demonstrated specific effects of FSH and LH preparations on the immature ovary, a deeper insight into reactions at the cellular level which depend on gonadotrophins may be expected from experiments employing both light and electron microscopy. The first can provide evidence on the rate of follicular growth or atresia while the latter supplies a detailed analysis of cytodifferentiation and cell degeneration.

MATERIAL AND METHODS

Newborn Swiss Albino mice from an inbred strain were used for e.m. studies. Full descriptions of the injected materials are given elsewhere (Lunenfeld *et al.* 1967; Eshkol and Lunenfeld, 1967; Eshkol *et al.* 1970). The animals used in the experiments were grouped as follows:

Group A : 5 days old	Group B : 8 days old
1. Controls	1. Controls
2. AS	2. AS
3. AS+HMG	3. AS+FSH
	4. AS+HMG

In each series injections were given daily from birth to the day of sacrifice. The animals were killed by decapitation, the ovaries removed immediately and fixed in 4 per cent glutaraldehyde in 0.1 M phosphate buffer for 2 hours, washed in phosphate buffer containing 0.2 M sucrose, and post-fixed in 1 per cent osmium tetroxide. The tissues were dehydrated in a graded series of acetone-water mixtures (dried over $CuSO_4$) and embedded in Vestopal W. Thin sections were cut on a Reichert OmU2 ultramicrotome and stained with lead citrate (Reynolds). The sections were examined and photographed with a Zeiss EM9 electron microscope. For the purpose of orientation and survey 0.5 to 1.0 μ sections were cut from the same block with a Porter-Blum ultramicrotome and stained with Toluidine Blue.

RESULTS

Animals 5 days old, AS treated. (Figures 1, 2 and 3). There seems to be no difference in the development stage of the treated animals as compared to controls. Most of the oocytes are separated by pregranulosa cells lying attached to the oocyte's surface membrane. The pregranulosa cells are still deficient in specialized internal structures. The spindle shaped nucleus occupies most of the cytoplasm. The ultra-structure of the germ cells and the somatic components seems to be as in the controls. The oocytes are well preserved, rounded in shape, their nuclei containing large reticular nucleoli. The most obvious difference in the AS treated animals is an unusually large amount of an amorphous fine granular substance filling the dilated intercellular spaces of the stroma. As a similar substance is found in the capillaries it may be concluded that it represents a transudate.

Animals 5 days old, AS + HMG treated. Following combined treatment with AS and HMG, 5-day-old specimens exhibit marked responses to the stimulating hormone. Both the light and the electron microscope show marked differences between treated and untreated animals. In the ovaries of the treated mice follicular development has clearly progressed. Numerous follicles have reached stage 4 of

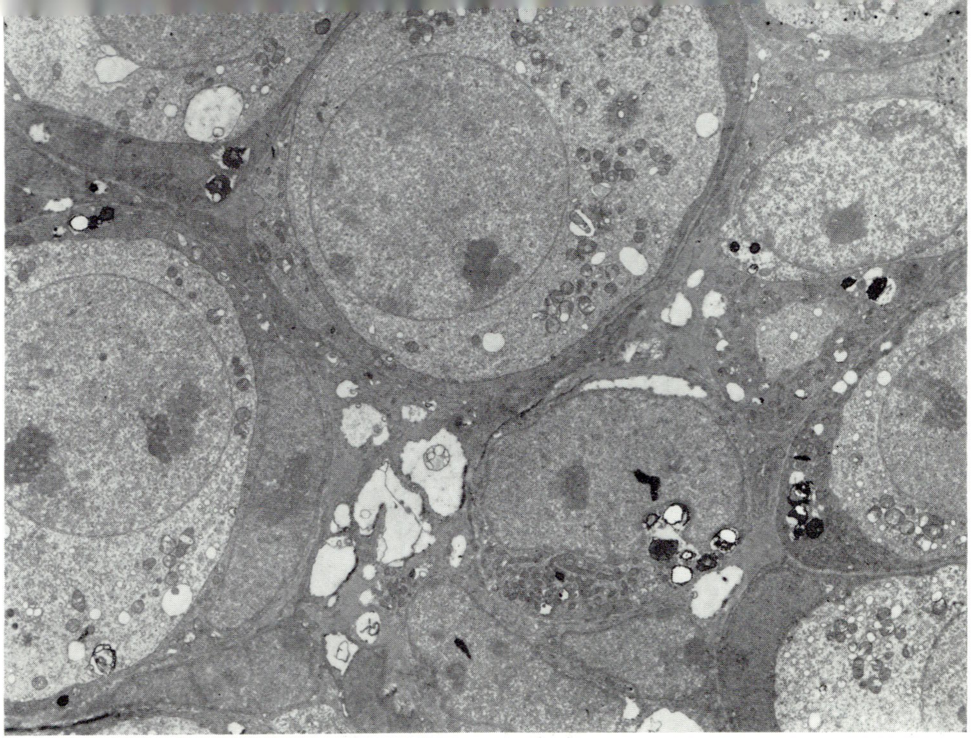

Figure 1. Mouse ovary, 5 days old, control. Oocytes are separated by pregranulosa cells lying attached to the oocytes' surface membranes. $\times 3750$

*facing p.*260

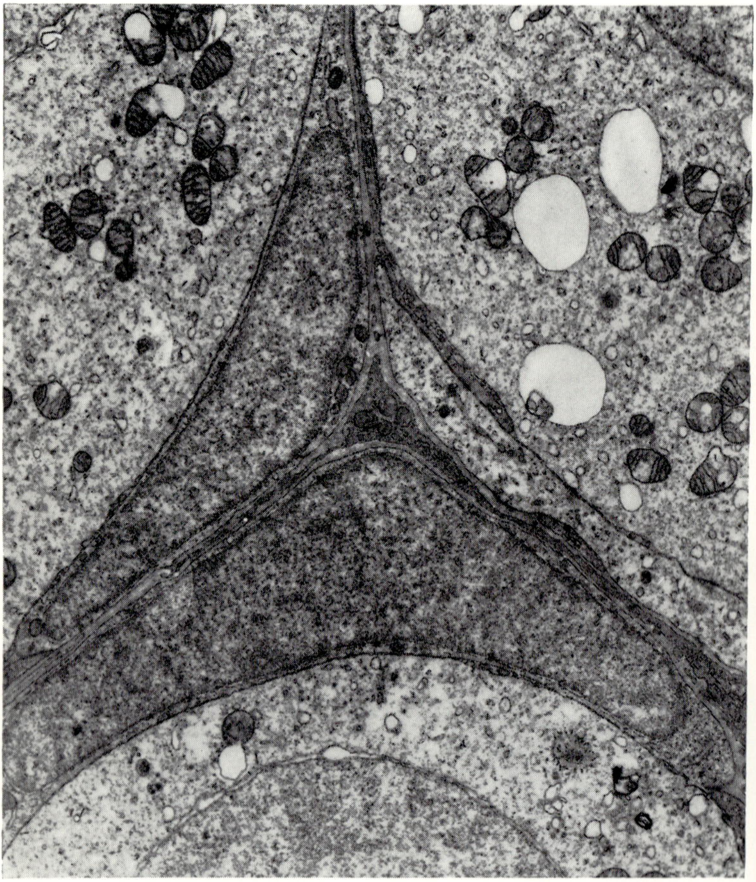

Figure 2. Mouse ovary, 5 days old, control. Note pregranulosa cells intimately associated with the oocyte's surface. Cytoplasmic processes of pregranulosa cells extend between adjoining germ cells, promoting separation of the oocytes. ×9750

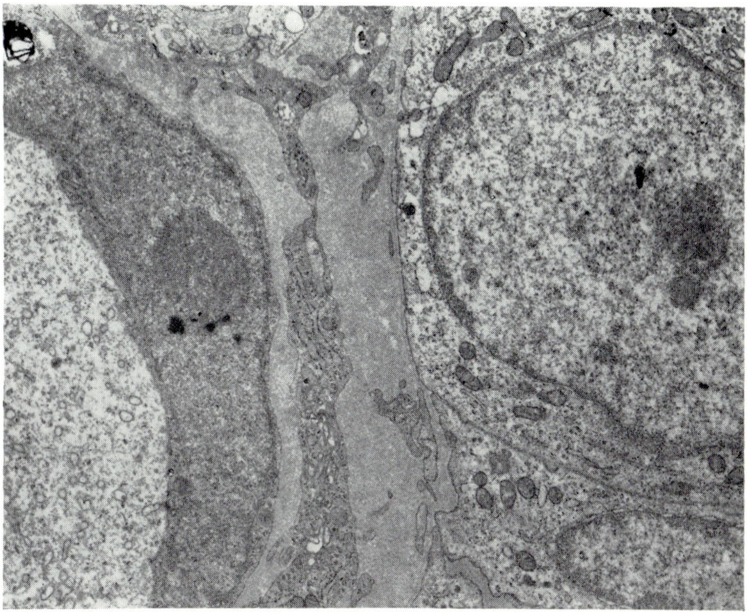

Figure 3. Mouse ovary, 5 days old, AS-treated. Structural development as in control. Note distinct basement membrane underlying granulosa cells of a unilaminar follicle (above). Fine granular material fills the intercellular spaces of the stroma. ×9590

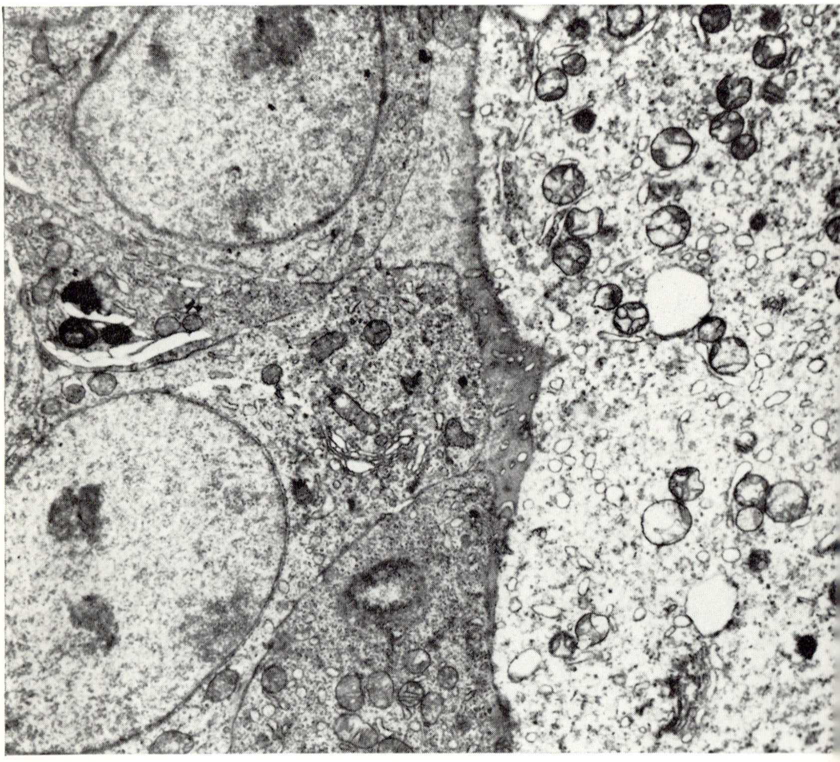

Figure 4. Mouse ovary, 5 days old, AS + HMG-treated. Section from a unilaminar fo
Granulosa cells are cuboidal or cylindrical, their ultrastructural organization indicates
secretion. Note large Golgi apparatus. Formation of zona pellucida has begun. ×975

facing

development (Pedersen and Peters, 1968). The granulosa cells are cylindrical or polyhedral in shape and contain abundant mitochondria, rough membranes, and a well developed Golgi apparatus (Figure 4). Thus the cells are equipped with organelles indicative of active synthesis of secretory material. The intercellular spaces are filled with fine granular material which is condensed at the oocyte's surface membrane for elaborating the zona pellucida. Interstitial cells, characterized by highly developed endoplasmic reticulum and lipid accumulations are present in the outer theca layer and in groups between stromal cells.

Animals 8 days old, AS treated. While in 5-day-old mice injected with AS the ultrastructural pattern of the ovary is similar to that of controls, marked differences are apparent in specimens which received AS continuously up to the eighth day. In these animals follicular development is significantly retarded. Numerous follicles are found in different stages of atresia but there is no evidence from e.m. studies on the rate of oocyte atresia. Most of the follicles are in stage 3b of development. With regard to structural relationships of the somatic components there is good evidence of severe disturbance. It seems that the signal which induces normal cell association was lacking or did not act in all places. The granulosa cells are stuck together and their cytoplasmic matrix appears to be condensed (Figure 5). This impression is probably strengthened by the high density of ribosomes. The surface membranes of the closely attached granulosa cells are indistinct. Although the experimental material was fixed in the same conditions as the controls the preservation of cellular structures is less satisfactory. Nevertheless the cellular structure of the granulosa cells is quite well developed and does not show degenerative changes. In addition to increased ribosome density both rough membranes and mitochondria are abundant, but the Golgi lamellae are poorly developed. Thus the organelles present in the granulosa cells indicate that they are equipped with all the structures required for secretion, but that this function has not yet started; the cells seemed to be 'retaining cells'. This suggestion finds further support in the lack of secretion material normally found within the intercellular spaces of the granulosa epithelium. Even in bilaminar follicles the granulosa cells are still tightly apposed to the surface membrane of the ovum. In only a few places are the membranes separated to enclose irregularly shaped spaces (lakes) filled by zona-like granular material. The arrangement of the follicular envelope is irregular. The thecal elements are crowded and closely attached to the basement membrane of the granulosa epithelium. The ultrastructural organization of the theca cells is poorly developed. 'Dark' and 'light' cells are to be distinguished, both containing only a few organelles. No typical interstitial cells can be recognized amongst the population of stromal and thecal elements.

Animals 8 days old, AS + HMG treated. Continuous HMG

application up to day 8 is highly effective in stimulating the different structural components of the infantile ovary. The ovaries are in an even more advanced stage of development than controls of the same age. Numerous follicles have reached stage 5a of differentiation.

The cell membranes and the different cell layers are clearly defined and regularly arranged (Figure 6). Oocyte differentiation is co-ordinated with that of the related follicle apparatus. Disintegration of Balbian's yolk nucleus and dispersion of Golgi lamellae into the periphery of the ovum is accompanied by the onset of formation of the zona pellucida. Numerous microvilli of the oocyte's surface membrane indicate endocytotic (pinocytotic) activity. The structural equipment of the granulosa cell is that of an actively secreting cell. The apical portion is occupied by a large and compact Golgi apparatus. Lipid droplets are abundant and stored predominantly in the basal portion of the cells. The ergastoplasmic lamellae are distended, most of them being transformed into vesicles filled with a fine granular substance considered to be a specific secretion material which may contribute to the formation of the zona pellucida. The theca cells external to the follicular envelope are spindle shaped. Their cytoplasm contains extended and occasionally ramifying ergastoplasmic lamellae, moderate numbers of mitochondria, lipid droplets and dense bodies. Typical interstitial cells characterized by the triad of lipid accumulation, tubular mitochondria and highly developed agranular reticulum are found predominantly in the outer theca layer or in groups in the connective tissue near the capillaries (Figure 7).

Animals 8 days old, AS + FSH treated. Combined treatment with AS + FSH did not appear to stimulate the ovaries of 8 day old mice. This disagrees with the light microscope observations. The ultrastructure of the ovaries is almost the same at that observed in specimens which received AS alone. Both follicle development and cytodifferentiation of the interstitial elements show marked retardation. Only a few follicles have reached the bilaminar stage. The granulosa cells are crowded. All the cytoplasmic components are condensed, their membranous structures being indistinct (Figure 8). No antra are visible and interstitial cells are absent.

DISCUSSION

Morphological findings in the AS treated mice. The fact that follicle development is not retarded in specimens receiving AS continuously up to day 5 suggests that there may be little gonadotrophic secretion during this period of postnatal life; AS application would be ineffective if there were no substrate to be neutralized. It seems unlikely that the lack of significant alterations in the ovarian ultrastructure following five days of AS treatment is due to insensibility of the immature gonad. The target cells of the immature ovary are capable of response to

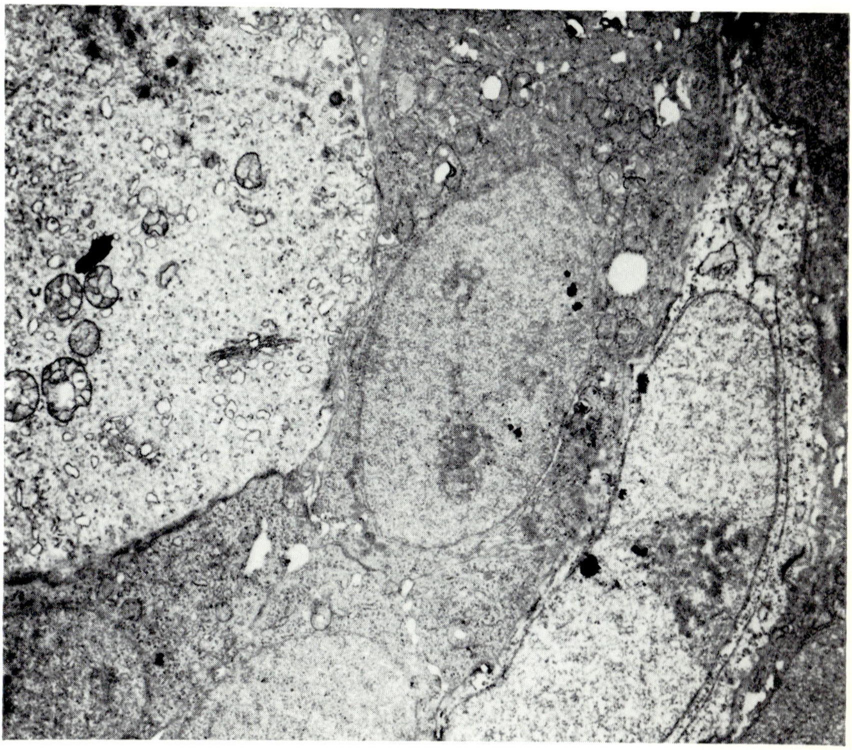

Figure 5. Mouse ovary, 8 days old, AS-treated. Section from a unilaminar follicle. Granulosa cells are stuck together, cell membranes are indistinct, cytoplasm seems to be condensed. Delayed formation of zona pellucida. Note poorly developed internal structures of theca cell underlying granulosa cell layer. ×8250

facing p. 262

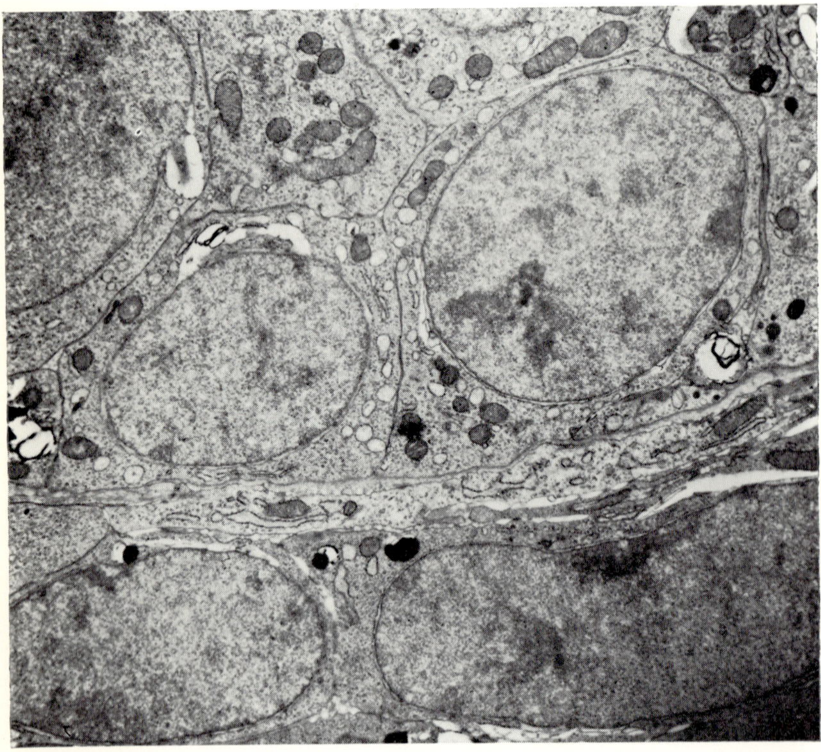

Figure 6. Mouse ovary, 8 days old, AS + HMG-treated. Basal portion of a type 5a follicle. Cell layers are regularly arranged, cell membranes are clearly defined. Note extended rough membranes filled with fine granular material. Secretory function of the granulosa cells is evidently stimulated. ×7000

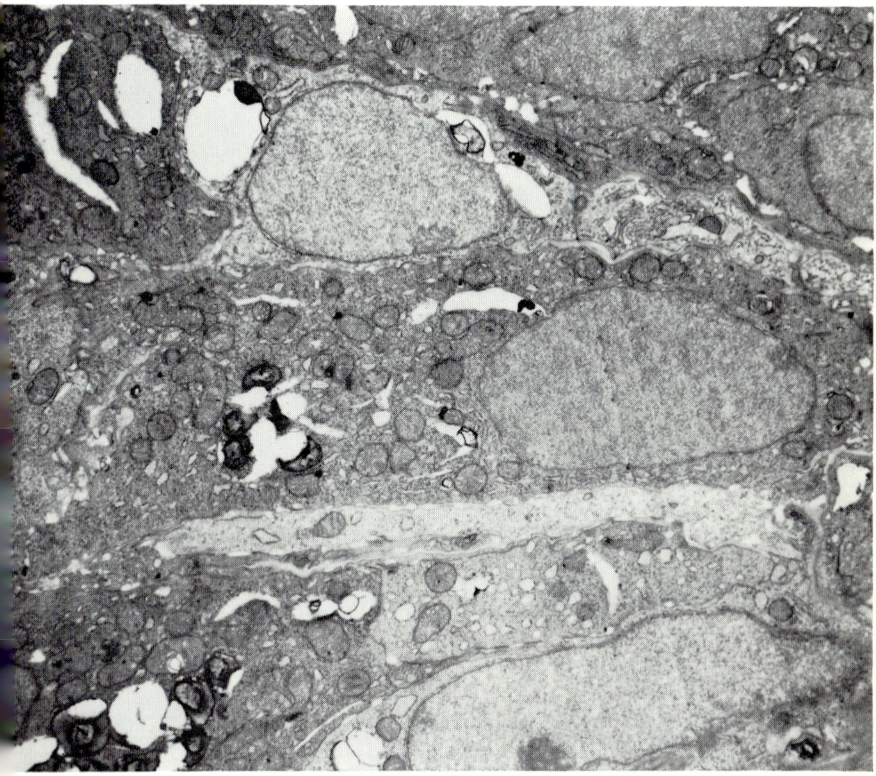

igure 7. Mouse ovary, 8 days old, AS + HMG-treated. Interstitial cells. Note marked evelopment of endoplasmic reticulum, accumulation of both lipids and mitochondria. ×7000

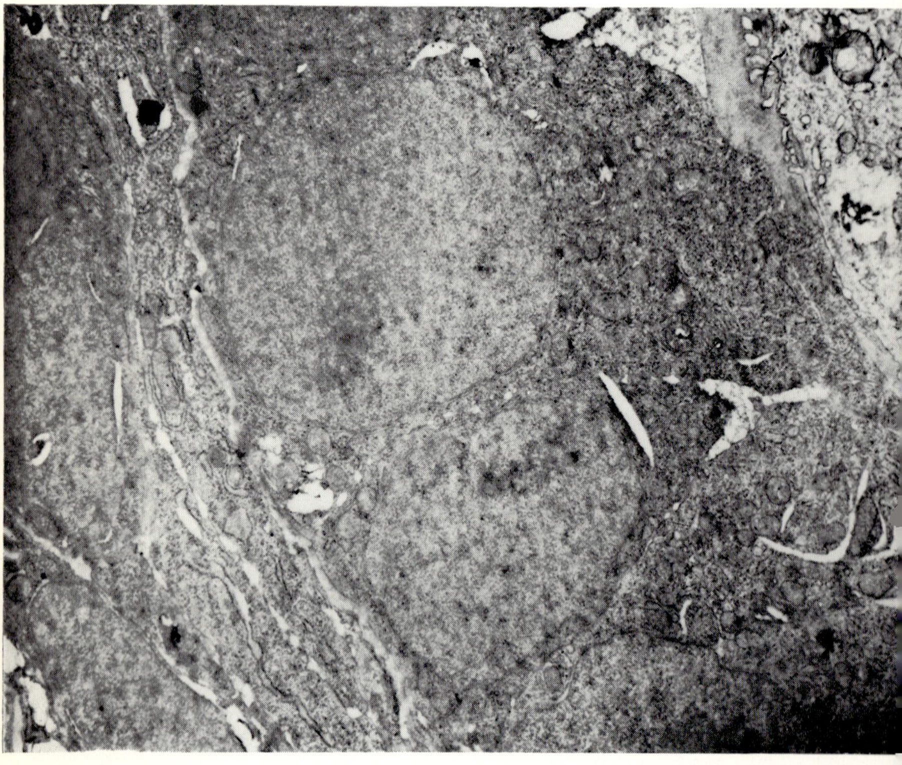

Figure 8. Mouse ovary, 8 days old, AS + FSH-treated. Section from a unilaminar follicle. Structur pearance is similar to that of AS-treated animals. Granulosa cells are crowded, their cytoplasm condensed. Formation of zona pellucida is delayed. Follicular envelope is ill defined. × 8250

facing

gonadotrophins even at a very early age (Beltermann and Stegner, 1968) though sensitivity depends on age. Ben-Or (1963) noted that 6 to 12 day ovaries were less sensitive to FSH than were 15-day and older specimens. In contrast to the effect in 5-day-old animals marked deviations of gonadal development occur when antiserum treatment is continued up to day 8. Both follicle formation and cytodifferentiation of the interstitial cells are severely affected. The most obvious findings in deprived animals 8 days old can be summarized as follows:

1. Inhibition of secretory activity of the follicle cells (giving rise to the formation of 'retaining' follicle cells).
2. Delayed (abnormal) formation of the zona pellucida.
3. Abnormalities of the follicular envelope.
4. Disturbance of normal cell spacing and association of the different somatic components (cell disorder).
5. Failure both of theca cell differentiation and of interstitial cell development.
6. Delay in vascular development.

From these results it is concluded that normally the release of adequate amounts of gonadotrophin is initiated between days 5 and 8 of postnatal life. Thus AS treatment is effective only when continued up to this critical period of development. This presumption is supported by several experimental findings i.e. appearance of completely differentiated interstitial cells at the end of the first week, presence of 3-β-hydroxysteroid-dehydrogenase activity in immature mice as early as 5 days old (Presl *et al.* 1965; Baillie *et al.* 1966; Kuppers, 1967; Schlegel *et al.* 1967) and detection of oestrogens in the blood by day 10 of postnatal life (Presl *et al.* 1967).

Ultrastructural effects of human FSH and HMG on immature mice deprived of endogenous gonadotrophins. The retardation of ovarian development due to AS cannot be ascribed to either FSH or LH deprivation, since AS contains antibodies to both hormones. What then are the specific effects of either FSH or HMG on the immature ovary in deprived animals? Combined treatment with AS + HMG led to even more advanced gonadal differentiation by 5 days than in controls of the same age. Several follicles had reached stage 4 of development. The granulosa cells were spaced in an orderly fashion and their membranes were clearly defined. Their ultrastructural organization indicated active secretion. Fine granular secretion material was present within the intercellular spaces of the granulosa epithelium. Many of the growing follicles had antra of different sizes. Interstitial cells, characterized by an extensive development of the endoplasmic reticulum and accumulation of lipids, were present in the outer theca layer and in groups between stroma cells. In contrast to light microscope observations described in the previous paper, FSH administered simultaneously with AS neither promoted follicular growth nor stimulated differentiation of the interstitial tissue. However, the experimental conditions were somewhat different and the lack of gonadal stimulation may be due to a dose of FSH too low to compensate for the AS effect. Since the electron microscope observations suggest that FSH, at least in the dosage used here, was unable to compensate for the lack of endogenous gonadotrophin no conclusions can be drawn regarding specific effects of highly purified FSH on the immature gonad. Nevertheless the combined effects of FSH and LH seem to be necessary for physiological development of the mouse ovary, although only after day 5 of postnatal life, since deprivation of endogenous gonadotrophin by antiserum leads to significant deviation in gonadal development only when continued up to the 8th day.

ACKNOWLEDGEMENT

Part of this investigation was supported by Ford Foundation Grant No. 67-470 to B. Lunenfeld.

GENERAL DISCUSSION

PARTICIPANTS: GONDOS, RYLE, LUNENFELD, BEN-OR, FRANCHI, PEDERSEN, LOUWERENS, WEISZ AND LLOYD.

1. Mice aged 5 and 8 days were used instead of 7-day-old ones as in the previous paper because, at 5 days, electron microscope studies show important developmental changes in the interstitial tissue. This suggests a change in endogenous gonadotrophin levels because even at 1 or 2 days after birth the mouse ovary will respond to (human) gonadotrophin containing LH activity by precocious development of interstitial tissue.

2. The areas selected for e.m. study were near the medulla and, in the 8-day-old antiserum-treated mice, included abnormal follicles with large oocytes but relatively few granulosa cells. These 'crowded' granulosa cells had closely aligned cell membranes, little intercellular space, no organelle pattern indicating active secretion and, in most areas, no zona pellucida formation, the zona being absent from pre-laminar follicles. Nevertheless, the oocytes grow to normal size, although dispersion of the Golgi apparatus may have been delayed. In hypophysectomized animals oocytes of normal size also occur but it is not certain whether their growth takes place after or before hypophysectomy.

3. It was stressed that the antiserum could only eliminate circulating endogenous gonadotrophin. Exposure to gonadotrophins *in utero* may well have influenced earlier development, so even antiserum treatment from birth cannot be regarded simply as preventing normal stimulation and development; it may cause a definite interruption. Such 'blocking'

at a certain stage of a programmed developmental process might make subsequent normal responses to gonadotrophic stimulation impossible.

4. If, as these results suggest, LH is necessary for the production of the zona pellucida, it is conceivable that it is required for all synthetic-secretory activities of the ovary and that FSH is concerned only with the induction of mitosis.

REFERENCES

Baillie, A.H., Ferguson, M.M. & McHart, J. (1966). *Developments in Steroid Histochemistry.* New York: Academic Press.

Beltermann, R. & Stegner, H.-E. (1968). *Acta endocr. Copenh.* **57**, 279.

Ben-Or, S. (1963). *J. Embryol. exp. Morph.* **11**, 1.

Eshkol, A. & Lunenfeld, B. (1967). *Acta endocr. Copenh.* **54**, 91.

Eshkol, A., Lunenfeld, B. & Peters, H. (1970). This volume, page 249.

Kuppers, S. (1967). *Acta histochem.* **27**, 267.

Lunenfeld, B., Eshkol, A., Baldratti, G. & Suchowsky, G.K. (1967). *Acta endocr. Copenh.* **54**, 311.

Pedersen, T. & Peters, H. (1968). *J. Reprod. Fert.* **17**, 555.

Presl, J., Horsky, J., Herzmann, J., Mikula, I. & Henzl, M. (1967). *J. Endocr.* **38**, 201.

Presl, J., Jirasek, J., Horsky, J. & Henzl, M. (1965). *J. Endocr.* **31**, 293.

Schlegel, R.J., Faries, E., Russo, N.C., Moore, J.R. & Gardner, L.I. (1967). *Endocrinology,* **81**, 565.

Development of the Ovary Under Different Experimental Conditions

S. Ben-Or

Maturation and maintenance of ovarian functions are under the control of pituitary gonadotrophins. Administration of follicle stimulating hormone to the prepubertal animal accelerates the process of ovarian maturation. This response to the hormone is, however, lacking in the infantile animal and, up to a certain age, which varies according to species, sexual maturity cannot be accelerated by exogenous gonadotrophic stimulation (for references see Price and Ortiz, 1944). Various factors may be responsible for this phenomenon: it may originate in the environment provided by the infantile body, from properties intrinsic to the infantile ovarian tissue, or from the interaction between them. Our studies on the development of the immature ovary of the mouse under different experimental conditions were aimed at clarifying the contribution of these various factors to the phenomenon. They include the following:

1. A follow up of the morphological and functional development of the ovary *in situ* and the effect of pregnant mare serum gonadotrophin (PMSG) stimulation on these processes (Ben-Or, 1963).
2. Studies on the development of the infantile ovary as a graft transplanted into adult spayed hosts (Ben-Or, 1965), which will be mentioned only briefly.
3. Transplantation of the juvenile ovary into infantile hosts, which will be reported here in detail.
4. As a consequence of the results obtained with these three procedures, studies were also made on the developmental potential of the isolated primary follicle under the transplantation conditions used in (2) and (3).

Effect of PMSG on the development of the immature ovary in situ. The life span of the mouse from birth to maturity has been divided into three periods according to divisions made by Rennels (1951) for the rat: (1) *The infantile period* from birth to 6 days when the ovary is refractory to PMSG stimulation; (2) *The early juvenile period* from 7 to 15 days, when the reaction to PMSG results in a deviation from the normal developmental pattern, namely, the secretory activity of the ovary is enhanced while follicular growth is impaired, (compare Figures 1 and 2); (3) *the juvenile or prepubertal period* from 16 to about 30 days, when the first ovulation occurs. At that stage PMSG accelerates the maturation of the ovary; enhanced oestrogen secretion is accompanied by follicular growth, Graafian follicles develop and full maturity is reached.

During the infantile and early juvenile periods organogenesis of ovarian tissue takes place. The processes involved in the specific tissue organization are independent of gonadotrophin (Smith 1930; Hertz, 1963). The acquisition of competence to react to PMSG is correlated with the time the ovary becomes dependent on gonadotrophin for its further development, namely at the prepubertal period. During the development of competence, in the early juvenile period, the interstitial tissue acquires competence to react to gonadotrophins some time before the follicular tissue. PMSG stimulation at that stage, interferes therefore with the normal relations between these two tissue components, and causes the deviation from the normal developmental pattern.

Development of the infantile ovary in transplantation conditions in adult spayed hosts. The process of ovarian development under transplantation conditions in the adult spayed host, though generally accelerated, remains essentially very similar to the development *in situ* under PMSG stimulation. The most interesting feature (for the present discussion) observed in grafts from infantile donors is the divergence of functional activity from morphological maturation. Ovarian capacity to produce and secrete hormones is greatly enhanced while follicular growth is retarded. Figure 3 presents one of these grafts; an ovary from a 1-day-old donor after 12 days in the anterior chamber of the eye of an adult spayed female host. The graft was removed for histological analysis when the host's vagina recornified as a result of the graft's hormonal activity.

This divergence from the normal developmental pattern was not observed in grafts taken from donors at the early juvenile period. At that stage, when signs of oestrogen secretion by the graft were evident in the host, the graft contained different stages of well developed follicles including Graafian follicles. Thus the abnormal response to gonadotrophins which is evident *in situ* in the ovary at the early juvenile period takes place earlier when the ovary is exposed to the endogenous gonadotrophic stimulation in the adult environment.

Infant mice as hosts for ovarian homografts. The comparison between the reactivity of young ovaries to PMSG *in situ* and as grafts in adult spayed hosts, shows that the competence of the juvenile follicles to respond to their trophic hormone is enhanced under the transplantation conditions. The earlier maturation of the juvenile ovary in the adult spayed host appeared to be due to more favourable growth conditions rather than to the mere availability of gonadotrophins. The aim of the present study was to clarify whether the young animal system provides favourable conditions for PMSG action. For this purpose ovaries from 18-day-old mice, highly reactive to PMSG, were transplanted into 3-day-old mice.

The local strain of albino mice that served for the earlier studies was used. The 18-day ovary, freed from its capsule, was cut into four to six parts. Using a tuberculin syringe which contained 0.05 ml saline, the ovary was picked up at the tip of a No. 21 needle, and injected subcutaneously into the flank of the 3-day-old host. The grafts were recovered after 7 or 10 days when their total age was 25 or 28 days and the age of the host 10 or 13 days respectively. Long term experiments were not performed since, after this time lapse, the host's own ovaries achieve competence to react to PMSG, thereby complicating the experimental conditions and the interpretation of the results. The grafts, the ovaries of the host and part of its uterus were fixed for histological examination; 32 homologous transplantations were carried out. They were divided into two groups, with and without exogenous PMSG stimulation.

The degree of success with which these subcutaneously transplanted grafts 'take' varies greatly and as a result so does their development. Two adjacent parts of a graft, taken from the same ovary, could appear quite different. Some ovarian fragments adhered to the subcutaneous tissue, others to the muscle underneath while many others were found free under the skin and appeared like 'white grafts' which proved to be degenerating (Lawrence, 1959). In this transplantation site the grafts become encapsulated by the host's connective tissue. It appears that this 'capsule', when uninterrupted, prevents the infiltration of lymphocytes into the graft. The immune response, observed in this short term experiment was in general not intense.

Group I. PMSG (4 or 6 i.u.) was injected intra-peritoneally immediately after the transplantation. This stimulation was repeated three days later. Nine out of 20 grafts were successful. The typical ovarian organization was maintained and well differentiated follicles of all different stages were found. The graft continues to grow in the normal pattern under these transplantation conditions. In five grafts well developed Graafian follicles were found (Figure 4), three of which were functional in secreting oestrogen, as evidenced by the host's uterine response (Figure 5). The reaction of the host's own ovaries to these conditions did not differ from the typical unbalanced response of the early juvenile ovary to PMSG stimulation discussed above and presented in Figure 2.

Group II. In the second experimental group, the transplantation conditions were the same except that the exogenous gonadotrophic stimulation was omitted. Five out of 12 grafts 'took', became vascularized and appeared well organized at the end of 10 days after transplantation. In two such grafts, Graafian follicles were also found but with degenerative features. There were no signs of secretory activity by these grafts in the uterus of their hosts.

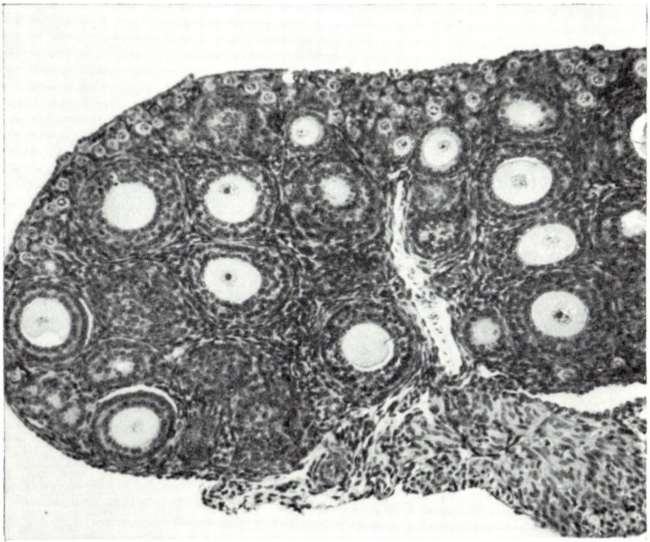

Figure 1. 13-day ovary, control. Haematoxylin eosin. ×136

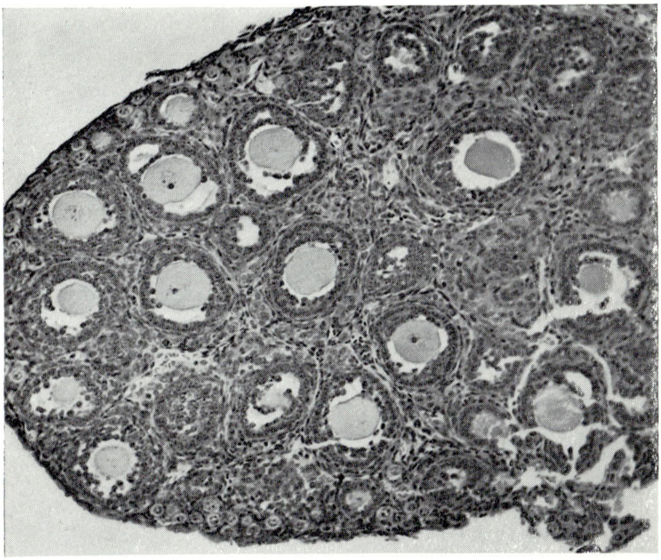

Figure 2. 13-day ovary, three days after PMSG. Note the lutein like appearance of the interstitial cells. Gomori's chrom-alum phloxine stain. ×136

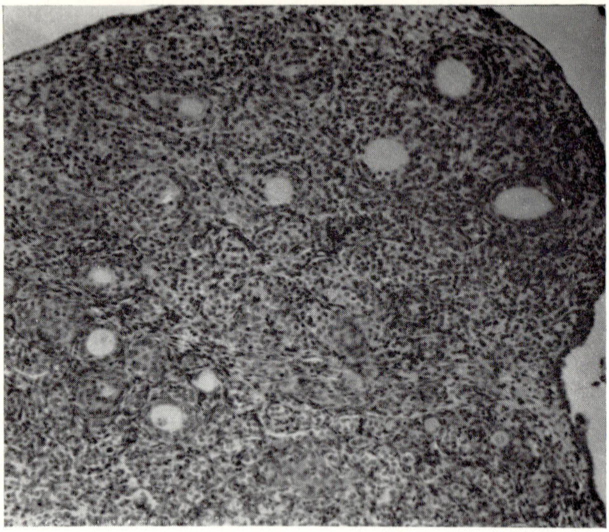

Figure 3. Ovarian homograft from a new-born donor in the anterior chamber of the eye of an adult spayed host, at the end of the latent period. Note the hypertrophied interstitial cells and impaired follicular growth. Haematoxylin eosin. ×136

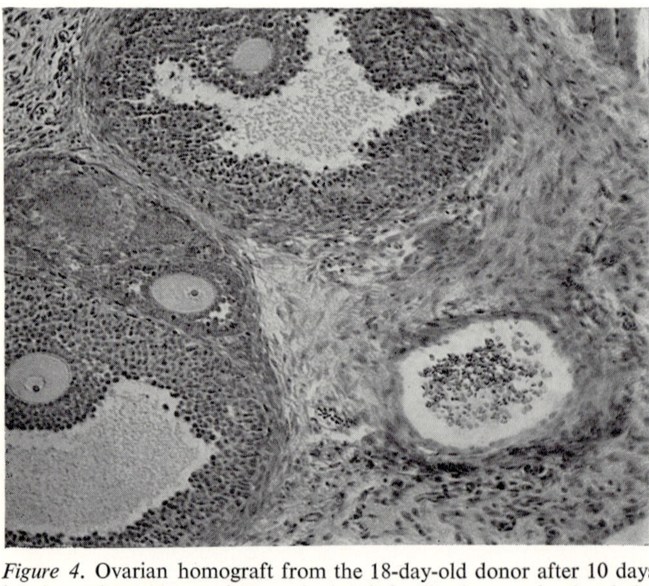

Figure 4. Ovarian homograft from the 18-day-old donor after 10 days in the subcutaneous tissue of the infantile host, stimulated by exogenous PMSG. Haematoxylin eosin. ×119

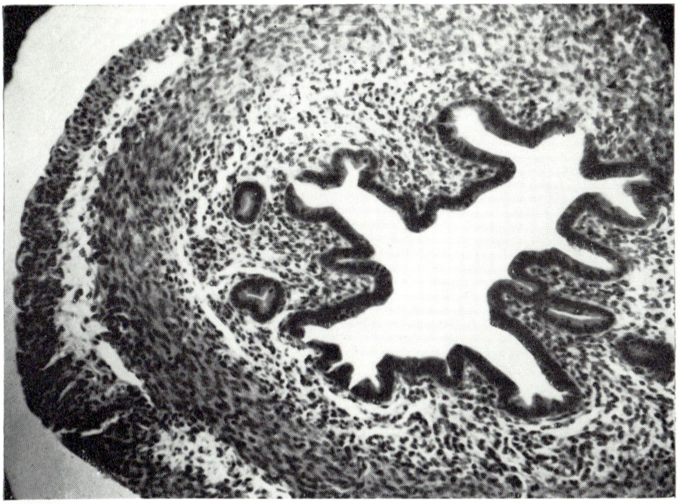

Figure 5. Uterus of the 13-day host bearing the ovarian homograft shown in Figure 4. Haematoxylin eosin. ×136

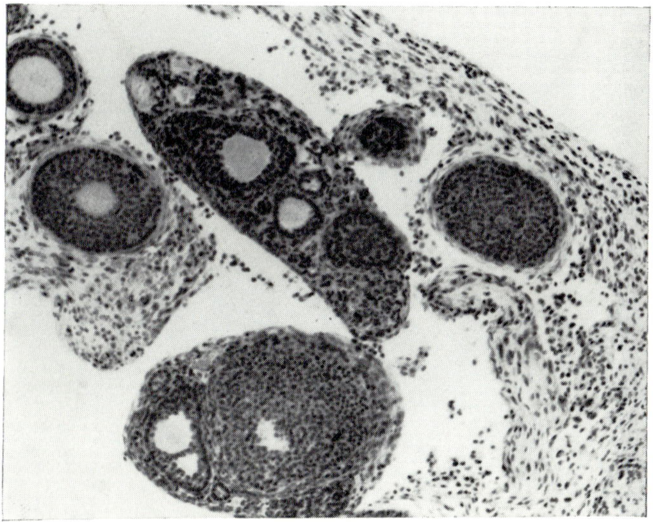

Figure 6. Homografts of isolated follicles from the 18-day-old ovary, in the subcutaneous tissue of the infantile mouse, after exogenous PMSG stimulation. Note the difference between the isolated follicles and those embedded in ovarian tissue. Compare also Figure 4. Haematoxylin eosin. ×113

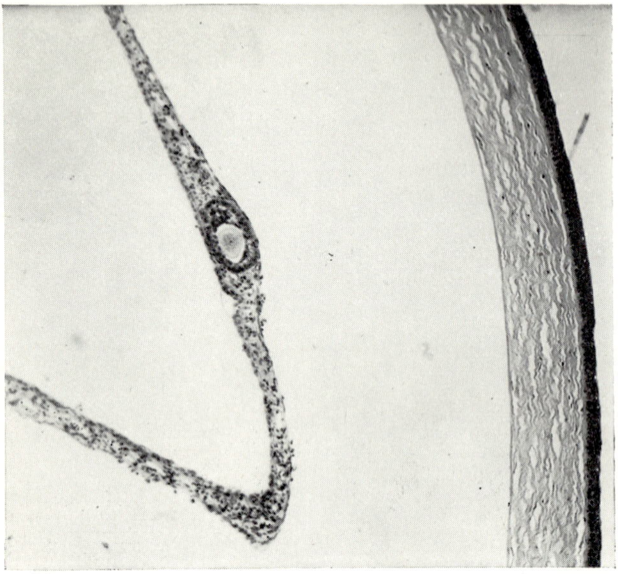

Figure 7

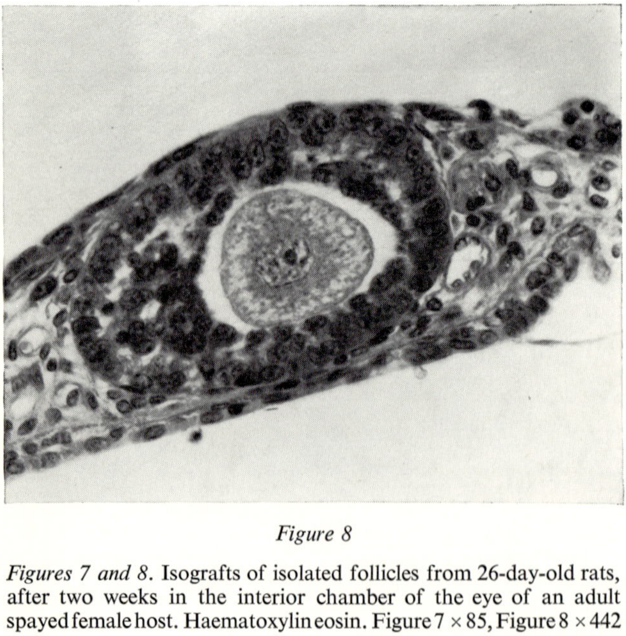

Figure 8

Figures 7 and 8. Isografts of isolated follicles from 26-day-old rats, after two weeks in the interior chamber of the eye of an adult spayed female host. Haematoxylin eosin. Figure 7 × 85, Figure 8 × 442

facing p. 269

The developmental potential of the isolated primary follicles. The abnormal developmental pattern induced in the ovaries by too early gonadotrophic stimulation, *in situ* and under transplantation conditions in adult spayed hosts, suggested that the interaction during development between ovarian mesenchyme and follicular tissue should be investigated. This has, for the time being, been attempted only indirectly by following the development of isolated primary follicles under conditions which support ovarian growth and maturation.

The prepubertal ovary of the mouse (albino, local strain) and the rat, (R III pure strain) were used as the donors of follicles. The follicles were released from the ovary by two different methods:

1. By perfusion of the ovary, while *in situ,* with saline containing 0.01M EDTA. The ovary was then transferred to a watch glass containing saline, where gentle mechanical pressure was applied, (Galon and Ben-Or, in preparation);

2. By gentle proteolytic action of 0.1 per cent pronase, according to Cole *et al.* (1966). The suspension was concentrated by low speed centrifugation and drawn into a syringe.

The follicles that were obtained from the mouse's ovary were transplanted to the subcutaneous tissue of infant mice, followed by an injection of 6 i.u. PMSG. The follicles obtained from the rat were transplanted into the anterior chamber of the eye of adult spayed rats. Some of the follicles that were transplanted into the subcutaneous tissue of the infant host were found adhering to the skin and looking healthy. Different follicular stages remained well organized under these conditions for at least two weeks. But, in contrast to the follicles which remained embedded in ovarian tissue and continued to develop up to the Graafian stage, the development of isolated follicles appeared to be arrested (Figure 6).

Some of the follicles that were transplanted into the anterior chamber of the eye, were found embedded in the iris tissue of their host (Figures 7 and 8). Despite the rich vascularization and the high gonadotrophic content of their host, they did not show any signs of further development.

DISCUSSION

Ovarian function is controlled by various factors which operate within the organism. The stimulatory effect of gonadotrophin is mediated by several factors, some of which have been shown to enhance gonadotrophic action (Hopkins, 1968), others to antagonize it by affecting ovarian responsiveness, (Talbert *et al.* 1957).

Substances that antagonize or inhibit gonadotrophic action have been found in the anterior pituitary (Woods and Simpson, 1961), in the pineal body (Reiss *et al.* 1963; Soffer *et al.* 1965; Das Gupta, 1968) and in the urine (Soffer *et al.* 1965; Landau *et al.* 1965).

The relatively prolonged immature state of the ovary *in situ* could be due to some systemic inhibition while the acceleration of the maturation process, occurring in young ovaries transplanted into an adult environment, could result from release from this inhibitory control. Any effect of such substances must, however, depend on the state of the ovarian tissue. The experiments presented here show that maturation of the 18 day ovary of the mouse can take place in the immature host. The effect of exogenous PMSG stimulation, given to the infantile host, was fully manifested in the competent tissue of the graft, while the reaction in the host's own ovaries was limited and unbalanced. The nature of ovarian response to gonadotrophins depends therefore mainly on the stage of ovarian tissue differentiation. A certain degree of tissue differentiation is a prerequisite for the stimulatory effect of PMSG, *in situ* as well as under transplantation conditions in adult hosts. This state is reached at the end of the early juvenile period when fully organized primary follicles with a well differentiated theca are found.

During the infantile and early juvenile periods organogenesis of ovarian tissue occurs. Inductive processes, similar to those operating in embryonic systems, take place and new biochemical properties are acquired which prepare the ovary to respond to FSH and LH action. A premature interruption by gonadotrophins changes the sequence of the programmed events and causes deviation from the normal developmental pattern.

The studies presented here indicate that the interaction between ovarian interstitial cells and follicular tissue is essential for development. This is suggested by the following observations: the hypertrophy caused by PMSG in the competent 18-day ovary, disappears 24 hours after stimulation (in correlation with oestrogen secretion), and reactivity is transferred to the granulosa cells. The first apparent response in the granulosa cells is intense mitotic activity. Simultaneously, there is a decline in hypertrophy of the interstitial cells. In the stimulated early juvenile ovary, there is no apparent change in the mitotic index of the granulosa cells and hypertrophy of the interstitial cells persists both *in situ* and under transplantation conditions in adult spayed hosts, where signs of oestrogen secretion were evident. It appears as if the reactive granulosa cells counteract and control the reactivity of the interstitial tissue.

Furthermore, it appears that isolated follicles do not respond to gonadotrophins. They do not continue to develop, under the conditions studied, but at the same time do not disintegrate either, as do follicles under premature PMSG stimulation. Thus it appears that the gonadotrophic stimulus is indirectly transferred to the granulosa cells and that ovarian mesenchymal tissue plays an essential role in this process. When competent granulosa cells receive the proper stimulus from the follicular theca they first respond by intense cell division and only then undergo characteristic maturation. DNA synthesis precedes the new stage of differentiation, a phenomenon known from many other developmental systems (Ebert, 1969). The abnormal reaction of the early juvenile ovary to PMSG may result from the fact that the granulosa cells have not yet undergone the required number of mitotic cycles which is a prerequisite for the proper response.

GENERAL DISCUSSION

PARTICIPANTS: LUNENFELD, RYLE, STEGNER, WEISZ, ESHKOL AND FRANCHI.

1. PMSG contains both FSH- and LH-like activities in the ratio of about 1 i.u. FSH to 25 i.u. LH (Louwerens). It has a significantly longer half-life after injection than either pituitary or urinary FSH or LH.

2. The action of PMSG was well demonstrated in this work. Its LH-like component produced a marked effect and the stromal tissue of very young ovaries was highly responsive to it. This may have resulted in the production of some substance inhibiting follicular growth i.e. inhibiting the action of the FSH-like component.

3. Interstitial gland tissue of very young ovaries may be functionally different from that of older ovaries. When stimulated they may perhaps produce a substance which inhibits follicle growth. This could explain why fragments of the competent 18-day ovaries transplanted to 3-day mice respond to PMSG by normal growth while the hosts' ovaries respond abnormally. Possibly in the absence of exogenous stimulation the interstitial gland tissue of the very early juvenile period is non-functional (Weisz).

REFERENCES

Ben-Or, S. (1963). *J. Embryol. exp. Morph.* **11,** 1.

Ben-Or, S. (1965). *J. Embryol. exp. Morph.* **14,** 111.

Cole, R.J., Edwards, R.G. & Paul, J. (1966). *Devl. Biol.* **13,** 385.

Das Gupta, T.K. (1968). *J. Endocr.* **41,** 607.

Ebert, J.D. (1969). In *Current Topics in Developmental Biology* Vol. 3, Chap. 15. Ed. Moscona, A.A. & Monroy, A. New York: Academic Press.

Galon, Y. & Ben-Or, S. In preparation.

Hertz, R. (1963). In *The Ovary.* Ed. Grady, H.G. & Smith, D.E. Baltimore: William and Wilkins.

Hopkins, T.F. (1968). *J. Endocr.* **41,** 345.

Landau, B., Landau, R.J. & Schwartz, A.D. (1965). *Clin. Endocr.* **25,** 339.

Lawrence, H.S. (1959). *Physiol. Rev.* **39,** 811.

Price, D. & Ortiz, E. (1944). *Endocrinology,* **34,** 215.

Reiss, M., Davis, R.H., Sideman, M.B., Bauer, I. & Plichta, E.S. (1963). *J. Endocr.* **27,** 107.

Rennels, E.G. (1951). *Am. J. Anat.* **88,** 63.

Smith, P.E. (1930). *Am. J. Anat.* **45,** 205.

Soffer, L.J., Fogel, M. & Rudavsky, A.Z. (1965). *Acta Endocr. Copenh.* **48,** 561.

Soffer, L.J., Fogel, M. & Rudavsky, A.Z. (1965). *Israel J. Med. Sci.* **1,** 1267.

Talbert, G.B., Di Pillo, F. & Gordis, L. (1957). *Endocrinology,* **61,** 611.

Woods, M.C. & Simpson, M.E. (1961). *Endocrinology,* **68,** 647.

Some Further Observations on the In Vitro Growth of Infantile Mouse Ovaries in Response to Gonadotrophins

Margaret Ryle

SUMMARY

Ten ovaries, one from each of ten 15-day-old mice, were allocated at random to ten different gonadotrophin treatments *in vitro.* The other ovary from each mouse was fixed at dissection. Five cultured ovaries received follicle stimulating hormone (FSH) and five did not. Luteinizing hormone was added to six of the dishes, to four on day 0 and to two on day 3. The cultured ovaries were fixed on day 3 or day 6. FSH appeared to increase the number of follicles with three or more layers of cells present after three days and the number with four or more layers of cells after six days. LH appeared to increase the number of follicles present after six days providing it was in the medium from day 0. It is possible that LH increased follicle numbers by preventing degeneration during the first three days of culture as well as by stimulating growth. Larger follicles developed in the presence of LH than were found in the ovaries fixed at dissection, even when the LH was added on day 3. FSH did not appear to stimulate the development of larger follicles. Antral spaces were found in only two ovaries: those cultured with both FSH and LH.

INTRODUCTION

Morphological studies on the ovaries of 16- to 17-day-old mice, cultured for four days, have been described previously (Ryle, 1969b). They led to the following conclusions:

1. Follicle stimulating hormone (FSH) stimulated the growth of follicles of all sizes but particularly those with less than four layers of granulosa cells.

2. Luteinizing hormone (LH) mainly stimulated the growth of larger follicles though it also had a slight effect on smaller ones.

3. FSH markedly increased the number of follicles with small antral cavities whereas LH had no such effect.

4. The minimum effective dose of FSH was between 0.1 and 1.0 i.u./ml.

The aims of the present experiment were to study the effects of FSH and LH on follicle growth over a more prolonged period of time than previously, using ovaries taken from younger mice, to test for interactions between them, and to study the effect of adding LH after an initial exposure to FSH alone. The results are, however, preliminary and represent only the first of three sets of replicate treatments, so that a full statistical evaluation is not yet possible.

METHODS

Details concerning (1) the breeding and rearing of the mice, (2) the general culture procedure, and (3) the FSH used, have already been described (Ryle, 1970). The concentration of FSH was 0.5 i.u./ml of medium, which is the level at which the uptake of thymidine is nearly maximal. The concentration of LH was also 0.5 i.u./ml. It was the same preparation as that used previously and contained less than 0.0003 i.u. FSH to 1 i.u. LH (Ryle, 1969b). The medium contained 10 per cent instead of 5 per cent serum from a hypophysectomized sheep, to provide more optimal conditions for the tissue.

The ovaries of ten 15-day-old mice from a single foster litter were allocated at random to the ten treatments. As each mouse was killed one ovary was put in a culture dish and the other was fixed immediately and used as a control. Five of the ten culture dishes contained FSH; five did not. Two dishes in each of these sub-groups had no LH, one of the ovaries being fixed after three days' culture and one after six days' culture. One dish of each sub-group received LH after three days' culture and the ovary was fixed after six days. Two dishes of each sub-group contained LH from the start, one ovary being fixed after three days and one after six days. The schedules are shown in all Figures and Tables.

All ovaries, cultured and control, were fixed for two hours in Bouin's aqueous fixative, embedded in paraffin, serially sectioned at 10μ, mounted, and stained with Weigert's haematoxylin and eosin. The slides corresponding to each ovary were given a code number and examined 'blind', As in the previous experiment (Ryle, 1969b), each follicle having four or more layers of cells was noted, the number of its cell layers was recorded, and its mean diameter in the plane of the oocyte nucleolus was measured. The mean diameters of, (1) all four-layered follicles, and (2) all five-layered follicles in each ovary were calculated. In addition, follicles with only three layers were counted, but not measured, and approximate counts of one-layered and two-layered follicles were made. These categories correspond roughly to those of Pedersen and Peters (1968) as follows: one-layered, Type 3b;

two-layered, Type 4; three-layered, Type 5a; four-layered and five-layered, Type 5b; six-layered and over, Type 6. Antral spaces were noted where these occured in the larger follicles.

RESULTS

1. Diameters of four- and five-layered follicles. The average of the mean diameters of the four-layered follicles in the ten control ovaries was $74 \pm 1.2\mu$. That for the five-layered follicles was $83 \pm 1.4\mu$. The mean diameter of the four-layered follicles in every cultured ovary is shown in Table I but that of the five-layered follicles is shown only for ovaries cultured for six days, as three of those fixed after three days had none of this category. The results indicate no general change in cell size due to growth *in vitro.* They do, however, suggest that the mean cell height in four-layered follicles was slightly reduced in the presence of FSH. Any effect on the histograms of Figures 1 and 2 due to this difference is likely to be very slight.

2. Distribution of follicles of different diameters. Figures 1 and 2 show the numbers of follicles in a series of size classes based on the measurements of follicle diameters in each ovary. The diameters of those in Class (i) ranged from approximately 65 to 71μ, those in Class (ii) from 72 to 78μ, etc. Since follicles with less than four layers were not measured they are not included here. Figure 1 represents those

TABLE I

Mean diameters of 4-layered and 5-layered follicles in cultured ovaries

		4-layered follicles		5-layered follicles
Duration of culture		3 days	6 days	6 days
	−LH	74	74	79
−FSH	LH from Day 3		74	79
	LH from Day 0	72	74	82
	−LH	70	74	79
+FSH	LH from Day 3		70	82
	LH from Day 0	70	70	82

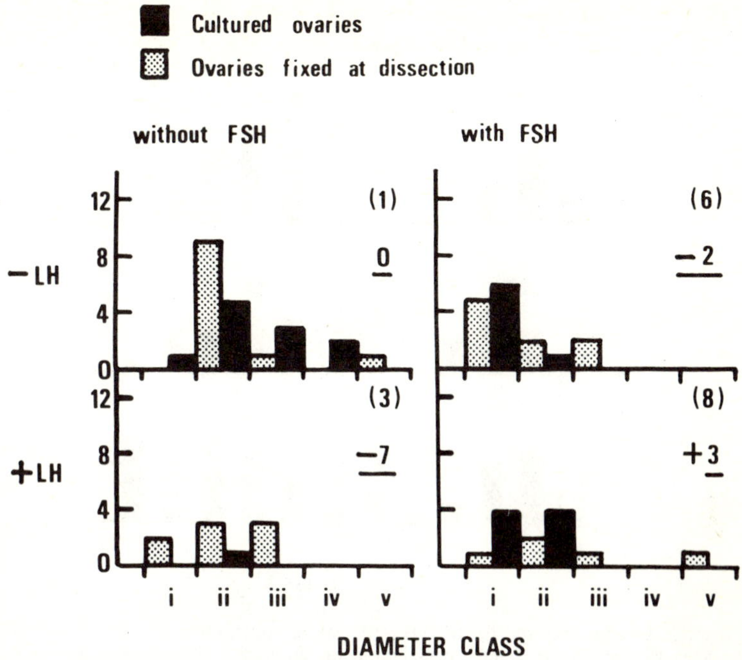

Figure 1. Ovaries cultured for 3 days with paired control ovaries. Distribution of follicles in size classes based on diameter measurements. Treatment numbers in brackets. Differences in numbers of follicles underlined.

ovaries which were fixed after three days' culture; Figure 2 represents those fixed after six days' culture. Each ovary is compared with its corresponding control and the difference in number of follicles is shown for each pair.

Only two of the follicles in the control ovaries exceeded Class (iv) in diameter. Figure 1 shows that little change had occurred in the size distribution of the follicles present after three days' culture. However, Figure 2 shows that after six days the range of follicle sizes had been extended beyond that of the controls in all ovaries exposed to LH, including those exposed to it for only the last three days. FSH seemed to make little difference to the range of follicle diameters.

3. Numbers of follicles with four or more layers of granulosa cells. In two of the four ovaries fixed after three days' culture fewer follicles were present than in their corresponding controls, but after six days all cultured ovaries, including that which received no hormone, had more follicles than their controls; these differences are shown in Figures 1 and 2. The total number of follicles with four or more layers in each cultured ovary is shown in Table II. After six days each ovary receiving FSH had more follicles than the corresponding one without it but LH

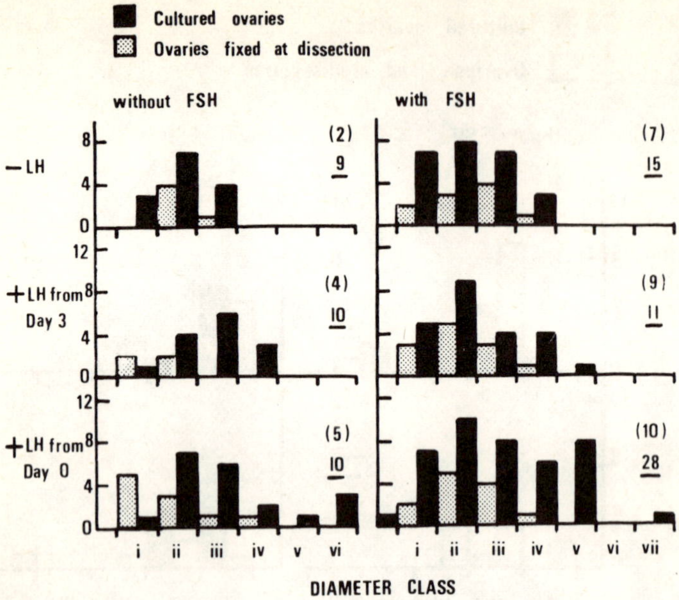

Figure 2. Ovaries cultured for 6 days with paired control ovaries. Distribution of follicles in size classes based on diameter measurements. Treatment numbers in brackets. Differences in numbers of follicles underlined.

TABLE II

Numbers of follicles with 4 or more layers of granulosa cells in cultured ovaries

		Duration of culture	
		3 days	6 days
	– LH	11	14
– FSH	LH from Day 3		14
	LH from Day 0	1	20
	– LH	7	25
+ FSH	LH from Day 3		23
	LH from Day 0	8	41

appeared to increase the number of follicles only if it was present from Day O. The joint effect of the two hormones seemed to be additive.

4. Total numbers of follicles with three or more layers of granulosa cells. Table III shows the total number of follicles with three or more layers in each cultured ovary. Every ovary cultured in the presence of FSH had more such follicles than the corresponding one cultured without it. Similarly, three of the four ovaries cultured with LH from Day O had more follicles than the corresponding one cultured without it, the exception being the one cultured without FSH and fixed after three days.

TABLE III

Numbers of follicles with 3 or more layers of granulosa cells in cultured ovaries

		Duration of culture	
		3 days	6 days
– FSH	– LH	20	38
	LH from Day 3		18
	LH from Day 0	8	54
+ FSH	– LH	26	62
	LH from Day 3		63
	LH from Day 0	39	92

5. Differences between the number of follicles in cultured and control ovaries. Table IV shows the differences between the pairs of cultured and control ovaries with respect to the number of follicles having at least three layers of granulosa cells. All those cultured with FSH contained more follicles than their controls except for the one cultured for only three days in the absence of LH. The one cultured for six days with FSH, but without LH, had more follicles than its control and supplementing with LH from Day 3 appeared to have no additional effect. All those cultured in the absence of FSH contained fewer follicles than their controls except for the one cultured for six days with LH.

Three of the four ovaries cultured with LH from Day O contained more follicles than their controls, the exception being the one cultured for only three days without FSH. Addition of LH after three days made little difference. Altogether, five ovaries appeared to have lost follicles compared with their controls, the greatest loss occurring in the one

TABLE IV

Difference in numbers of follicles with 3 or more layers between cultured and paired control ovaries

		Duration of culture	
		3 days	6 days
– FSH	– LH	– 33	– 5
	LH from Day 3		– 2
	LH from Day 0	– 16	+ 15
+ FSH	– LH	– 22	+ 14
	LH from Day 3		+ 19
	LH from Day 0	+ 7	+ 30

with no gonadotrophin and cultured for only three days. In contrast the maximum increase occurred in the ovary cultured for six days with both FSH and LH.

Rough counts of follicles with one or two layers of granulosa cells suggested that losses of these small follicles were more drastic. Even in the ovary cultured for six days with both gonadotrophins, although the marked recruitment to larger categories was taken into account, many of the one and two-layered follicles seemed to have disappeared.

6. Antral spaces. No follicles with antral spaces were present in any control ovary. Only two of the cultured ovaries had any. The ovary grown with FSH, LH being added on Day 3, had one antral space in a follicle of Diameter Class (ii). The one grown with both hormones from Day O and fixed on Day 6 had four antral spaces, three in follicles of Class (v) and one in a follicle of Class (vii). Thus they were only present in ovaries exposed to both hormones and four of them were in particularly large follicles.

DISCUSSION

While, in the absence of replicate data, any conclusions drawn from these results must remain tentative, a comparison with the earlier study (Ryle, 1969 b) is of interest. In that experiment LH significantly stimulated the development of larger follicles while FSH did not. The present results similarly suggest that LH extends the range of follicle sizes, even if present in the medium for only the last three days of a six-day culture period. FSH once more does not appear to extend it.

In the earlier experiment FSH caused a significant increase, during a four-day culture period, in the number of follicles with four or more layers of granulosa cells. In this experiment ovaries treated with FSH consistently had more such follicles after six days of culture than did those grown without it. They also had more follicles with three or more layers after only three days.

In the earlier experiment LH did not significantly increase the number of follicles with four or more cell layers. Here, however, LH seems to have increased the number of follicles after six days but only if it was supplied from Day O. This contrast between the two experiments may arise from improvements in the culture technique permitting a more marked effect of LH.

The loss of follicles during the first three days is curious. It was greater when both FSH and LH were absent and was only reversed when both were present in the culture medium. The small follicles were mainly affected and, in the absence of FSH, LH only reversed the loss if supplied from the beginning to ovaries cultured for six days. This suggests that LH may be of importance for survival rather than for growth of small follicles and only stimulated growth when present during the last three days of culture. This interpretation is supported by evidence from ^{14}C-thymidine uptake experiments with similar cultures. FSH regularly stimulated a significant increase in thymidine uptake and increased deoxyribonucleic acid (DNA) formation above the control level during Days 2 to 4 of culture, whereas LH rarely caused a significant increase in uptake and did not increase DNA formation (Ryle, 1969a; Chaplin, How and Ryle, in preparation). Moreover, Lunenfeld and Eshkol (1968) found that DNA formation in the ovary was depressed within four hours of injecting human chorionic gonadotrophin into infantile mice. However, in a series of studies designed for other purposes the appearance of the tissue tended to be better in those ovaries cultured with LH than in those cultured without it (Ryle, unpublished). While this cannot be established quantitatively, it suggests that, if a positive effect on the survival of small follicles can be established, it may apply only *in vitro.* Possibly LH affects the permeability of cell membranes in the ovary, thus facilitating the diffusion of metabolites between the tissue and the medium.

The earlier study, on ovaries from 16- or 17-day-old mice, showed that antrum formation was stimulated significantly only by FSH. On the other hand, Lunenfeld and Eshkol (1968) found that, while human menopausal gonadotrophin induced antrum formation in the follicles of immature mice, highly purified FSH did not. In the present work on ovaries from 15-day-old mice antral spaces were seen in only two, namely in those cultured with both hormones simultaneously. It is possible that follicles must reach a certain size or maturity before they are capable of responding in this way to FSH, at least at the dose levels used here, and that LH was necessary for their maturation. Four of the five antral spaces were in follicles of Diameter Class (v) or more and these occurred only in ovaries treated with LH.

The slight reduction of cell size in the four-layered follicles of ovaries exposed to FSH is reminiscent of the significant decrease in the ratio of amino acid to thymidine taken up by ovaries treated with FSH in an earlier experiment (Ryle, 1969a). It suggests that cell division was proceeding so rapidly that cell growth was not quite keeping pace with it. It is curious that this did not occur in five-layered follicles.

ACKNOWLEDGEMENTS

I wish to thank Dr A. C. Crooke for his continued support, Dr W. R. Butt for the FSH, and Dr A. S. Hartree for the LH. Miss M. Wardill prepared the sections. The work was supported by the Ford Foundation.

GENERAL DISCUSSION

PARTICIPANTS: PEDERSEN, FRANCHI, ODELL, DONOVAN, PETERS LUNENFELD AND WEISZ.

1. Questions arose as to the significance of changes in follicle numbers indicated by the differences between cultured and paired control ovaries. Such comparisons indicated a net loss of follicles during the initial incubation period under various experimental conditions. This could arise specifically during the period when the tissue is adjusting to its new conditions. The increased number of follicles of various sizes counted after 6 days' culture may represent new follicles due to accelerated growth of small and medium ones.

2. The highest net loss occurred in the absence of gonadotrophins. Other studies show that in adult mice atresia is reduced by hypophysectomy, but this may not be true in the infantile stage.

3. Among other hormones which might support ovary cultures, insulin and thyroxine had been tried in a preliminary experiment and had had no positive effect. The limitations of cultures when the medium is not being changed were mentioned. Gonadotrophic hormones may lose their biological activity within the incubation period at 37°C, this needs to be tested. Metabolic products accumulating in the medium might function as inhibitors, although recovery of follicle numbers between Days 3 and 6 does not suggest this. Perfusion or superfusion culture methods would, however, probably provide more favourable conditions for the tissue.

REFERENCES

Lunenfeld, B. & Eshkol, A. (1968). In *Gonadotrophins 1968*, p. 197. Ed. Rosemberg, E. Los Altos, California: Geron-X.

Pedersen, T. & Peters, H. (1968). *J. Reprod. Fert.* **17**, 555.

Ryle, M. (1969a). *J. Reprod. Fert.* **19**, 87.

Ryle, M. (1969b). *J. Reprod. Fert.* **20**, 307.

Ryle, M. (1970). This volume, page 99.

Effect of Gonadotrophins on Protein Synthesis in the Ovary of the Mouse

S. Ben-Or and R. Broza

SUMMARY

The effects of pregnant mare serum gonadotrophin on protein synthesis in mouse ovaries were measured *in vivo* and *in vitro.* Actinomycin D was used in some experiments to inhibit the formation of new ribonucleic acid. The results of the experiments described show that the mechanism of PMSG action is complex and that one of its functions is related to protein biosynthesis. They suggest that activation of protein biosynthesis in the prepubertal ovary is reflected both at the DNA level, by inducing new transcriptional activity, and at the translational level, by activating the read-out of existing informational molecules.

INTRODUCTION

Recent studies on the mechanism of action of gonadotrophic hormones have shown that the remarkable increase in the size of the prepubertal ovary in response to their administration results from their stimulatory effect on ribonucleic acid (RNA) and protein synthesis (Blackburn & Kostyo, 1965; Ahren & Rubinstein, 1965; Civen *et al.*, 1966; Van Dyke & Katzman, 1968).

Ovarian response to pregnant mare serum gonadotrophins (PMSG) is already manifest several hours after the hormone injection but the full effect, which seems to be quite complex, occurs 50 to 70 hours later. The increased rate of the biosynthetic activities may result from various mechanisms operating simultaneously or consequentially at the different stages of the developing response. This study is limited to the early effects of PMSG on protein synthesis and attempts to clarify at what cellular level the hormone exerts its regulatory effect, at the transcriptional or translational level or at both levels of control. Protein synthesis was followed up in ovaries stimulated by PMSG, under normal conditions and under conditions where transcriptional activity was inhibited by Actinomycin D, and the results were compared to the rate of protein synthesis in the control, unstimulated ovary, under the same inhibitory treatment. Protein synthesis was studied *in vivo* and under *in vitro* conditions.

MATERIAL AND METHODS

Prepubertal female mice from the local albino strain were used at 18 at 22 days.

Gestyl PMSG, (Organon, OSS Holland), was given in a single i.p. injection of 20 i.u. per 8 to 10g mouse.

The radioactive tracers, L-Leucine-4,5,-T, 179 mC/mg, and Protein hydrolysate-C^{14}(U), 52mC/mA Carbon, were obtained from the Radiochemical Centre, Amersham.

In the *in vivo* incorporation experiments the labelled precursors were injected i.p. or i.v. (via the tail vein) as specified in the results and the ovary was removed for analysis 2 hours later. In the *in vitro* studies, the precursor L-Leucine was added to the medium at the concentration of 1 μC/ml and incorporation was studied during several hours in culture.

Culture conditions. Sterile conditions were maintained at all stages of the *in vitro* experiment. The ovary, freed from extraneous tissue was cut at the hilus under the dissecting microscope, and 'opened' with the aid of two needles, so as to allow spreading of the organ with minimum damage. This treatment permits direct contact of all parts of the tissue with the culture medium.

Four ovaries were put in a tightly sealed 10 ml Erlenmayer flask with 1 ml culture medium, in an atmosphere of 95 per cent air + 5 per cent CO_2. The flasks were rotated (70 r.p.m.) at 37°C on a gyratory shaker. The medium, Krebs-Ringer bicarbonate buffer pH 7.4, containing 5.5 mM glucose (Ahren and Rubinstein, 1965), was enriched with horse serum to a final concentration of 10 per cent, and with the vitamins and amino acid mixture of Eagle's Basal medium. Penicillin and Streptomycin, 500 u./ml of each, were also added. All the tissue culture materials were supplied by Microbiological Associates Inc. Bethesda.

Protein synthesis was determined by the incorporation of the labelled precursors into heat resistant trichloroacetic acid (TCA) precipitable counts. The cold 10 per cent TCA precipitate was washed with 5 per cent TCA, heated for 15 minutes at 90°C, cooled and centrifuged again. The precipitate was dissolved in NaOH to a final concentration of 0.1N, and the protein and radioactivity content of this solution were determined. The protein content was determined by the method of Lowry, *et al.* (1951). Aliquots from the same samples were counted in a Packard Tri Carb Liquid Scintillation Counter, in Bray's solution. The incorporation is expressed as counts/min/μg protein.

Actinomycin D (donated by Merck, Sharp and Dohme Research Laboratories) 3 μg/g body weight was injected i.p. one or two hours before the PMSG injection.

RESULTS AND DISCUSSION

The growth stimulating effect of PMSG. The stimulatory effect of PMSG, on the growth of the immature mouse's ovary, is one of the immediate and most remarkable effects of this hormone. The phenomenon is evident in the increased rate of protein synthesis, which begins to rise soon after PMSG administration (Table I) and in the increase in the total protein content of the ovary, which is already evident after 4 hours (Table II). The increase in protein biosynthetic activity, which is detected 1 hour after PMSG injection, continues to rise and reaches high levels after 4 hours (Table I).

The increased rate of protein synthesis, as measured by the incorporation of labelled amino acids, *in vivo,* may result from the greater availability of precursors to the stimulated ovary, due to hyperaemia caused by the hormone, rather than from activation of the protein biosynthetic apparatus in ovarian cells. The enhancement of blood flow through the ovary and thereby an enhancement in the supply of precursors is suggested by Reel and Gorski (1968) as one of the mechanisms for LH action.

Protein synthesis in the ovary in vitro. The *in vitro* incorporation experiments provide identical conditions for both the control and PMSG-stimulated ovaries, and may, therefore help to clarify the mechanism of the growth stimulating effect. The gonadotrophic stimulus was given to the ovaries *in situ* since preliminary experiments confirmed the fact that PMSG had no stimulatory effect when added to the culture medium (Blackburn and Kostyo, 1963; Ahren and Rubinstein, 1965; but see also Hamberger and Ahren, 1967). The ovaries were explanted 4 to 6 hours after PMSG administration since within this period the ovary became highly reactive. After a rapid, light operation (described in methods), the ovaries were transferred to the preincubated medium.

The results obtained in the incorporation studies *in vitro* varied. In five out of seven experiments the stimulatory effect of PMSG was evident during the first 4 hours in culture as shown in Figure 1; in one experiment there were no differences between the stimulated ovaries and their controls; and in another experimental group they were lower. The experimental conditions, were as far as possible the same. These differences in the results may be due to the different state of the stimulated ovaries at the time of explantation, on account of different kinetics in the developing response while still *in situ.* This assumption is supported by the results presented in Figure 2. In these experiments the incorporation *in vitro* was started at different time intervals after PMSG injection. It can be seen that the response as a function of time between PMSG administration and ovarian explantation varies, and the differences become more pronounced with time in culture.

The linear increase in specific activity of ovarian protein as a

TABLE I

The effect of PMSG treatment on the incorporation of amino acids into ovarian protein at different times after injection

Exp. No.	Treatment	Time after PMSG (hr.)	Average Protein Specific Activity Counts/min./μg Protein		
			CONTROL OVARY	STIMULATED OVARY	% CHANGE
1	5μc/mouse Algal hydrolysate C^{14} i.v.	1	3.90 (8)	4.62 (8)	+ 17.7
2		2	3.90 (8)	4.72 (8)	+ 21.0
3	10 μc/mouse H^3-Leucine i.v.	2	0.80 (2)	1.01 (2)	+ 26.0
4		4	0.80 (2)	2.07 (4)	+ 158.0
5		8	0.80 (2)	2.13 (2)	+ 166.0

Numbers in Brackets = Number of Animals

TABLE II

Effect of PMSG stimulation on the protein content of the ovary of the mouse

Exp. No.	CONTROL μg protein/ovary	PMSG μg protein/ovary			
		1 hr.	4 hr.	6 hr.	17 hr.
	120	133	133		
1	133	139	147		
			148		
2	107			139	
	104			139	
	145		180	215	311
	141		193	225	348
3	145		287	200	297
	169		304	218	272
					321

A pool of 8 ovaries was used for each determination

function of time in culture, indicates that the culture medium, though not optimal, is adequate to maintain the ovary for at least 4 hours.

However, comparison between the results of amino acid incorporation *in vivo* and *in vitro*, show that the stimulatory effect of PMSG on protein synthesis in the ovary, is much more pronounced *in vivo*. One obvious explanation for this difference is the greater availability of precursors to the stimulated ovary *in vivo*, but other still unknown factors may also be involved. It may result from the lack of the inducing agent in the culture medium, a requirement which may be essential for the full effect. While the greater availability of precursors may explain the difference between the stimulated ovary *in vivo* and *in vitro*, it cannot account for the difference between the stimulated ovary

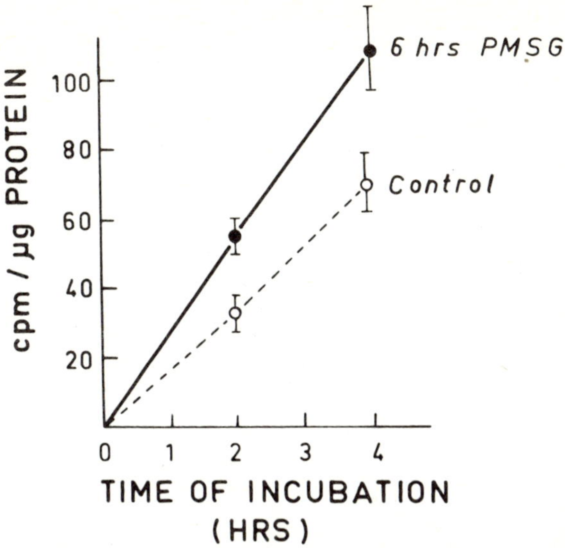

Fig. 1

The ovaries were explanted six hours after PMSG administration. Each point represents the mean of four observations. The standard errors are indicated by vertical lines.

Figure 1. Incorporation of amino acids into protein *in vitro* by control and hormone treated overies.

and its control *in vitro* which suggests that the mechanism of PMSG action involves activation of the protein biosynthetic apparatus in ovarian cells.

Dependency of the stimulating effect of PMSG on RNA synthesis. These experiments were intended to test whether the increase in protein synthesis caused by PMSG results from activation at the genetic level. Protein synthesis was therefore studied in animals in which RNA synthesis was inhibited by Actinomycin D. The effect of 3 μg/g body weight (24 μg/mouse) of Actinomycin D, injected i.p. was lethal. Many of the injected animals died within 8 to 16 hours although a few survived for about two days. Protein synthesis, in the drug treated animals, was therefore studied only during the first seven hours.

The effect of this antibiotic was followed in the PMSG-stimulated ovary and in the unstimulated control ovaries. The change in specific activity of ovarian protein, is calculated as a percentage of that in the control Actinomycin D-free ovaries. The time schedule of the experimental conditions, for the three somewhat different experiments, is indicated in the results summarized in Table III and Figure 3. Experiments 1 and 2 may be compared quantitatively since the animals received the same amount of the labelled precursor, 10 μc of ^{3}H-Leucine, and the difference in the response may be attributed to the

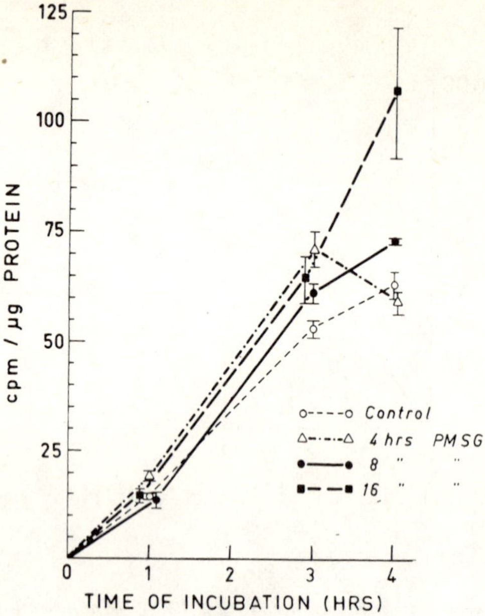

Fig. 2

Each point represents the mean of six or eight observations. The standard errors are indicated by vertical lines.

Figure 2. The effect of PMSG on the incorporation of amino acids into protein *in vitro*, at various periods after injection of PMSG.

difference in time between injection of Actinomycin D and injection of ^{3}H-Leucine. In Experiment 3 the precursor was ^{14}C-protein hydrolysate, 5 μc/mouse, and the results can be compared only qualitatively with the results of Experiment 2. The difference between them may be attributed to the difference in time between the administration of Actinomycin D and PMSG.

Several interesting data emerge from these results:

1. The early response of the ovary to Actinomycin D is shown as an increase in protein synthesis in both control and hormone treated ovaries, (Exp.1). The 76 per cent increase in the Actinomycin D-PMSG group, appears to be the sum of the effects produced by each of these agents alone.

2. This stimulation is only temporary, and declines with time, as indicated by the results of the Actinomycin D-Control group, in Experiment 2.

3. The 86 per cent increase, in the Actinomycin D-PMSG group, in Experiment 2, and the 27 per cent increase in the Actinomycin D-PMSG group, in Experiment 3, may therefore represent a stimulatory effect of PMSG which is independent of RNA synthesis. Since the total response to PMSG is increasing (Compare results of Experiment 2 with

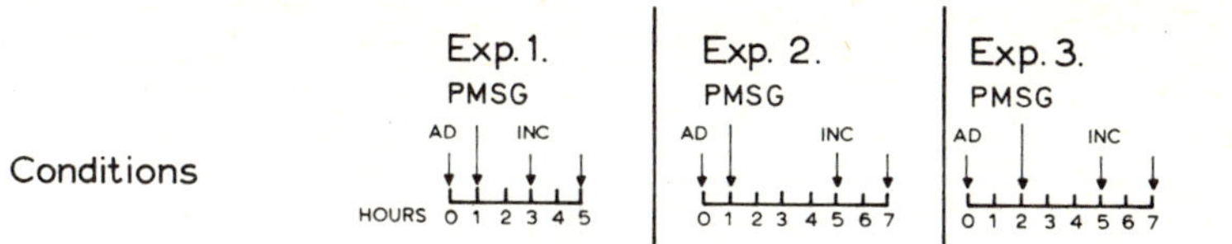

cpm / µg protein

		Exp. 1.	Exp. 2.	Exp. 3.
Controls	No Act.-D	0.8 0.8 av. 0.8	0.8 0.8 av. 0.8	3.38 3.67 3.00 av. 3.3
	Act.-D 3µg/g B.W.	1.20 1.04 av. 1.13	0.88 0.81 av. 0.85	
+PMSG	No Act.-D	1.04 0.96 av. 1.01	2.16 1.99 av. 2.07	8.03 5.13 av. 6.6
	Act.-D 3µg/g B.W.	1.31 1.51 av. 1.41	1.32 1.66 av. 1.49	3.3 4.4 4.9 av. 42

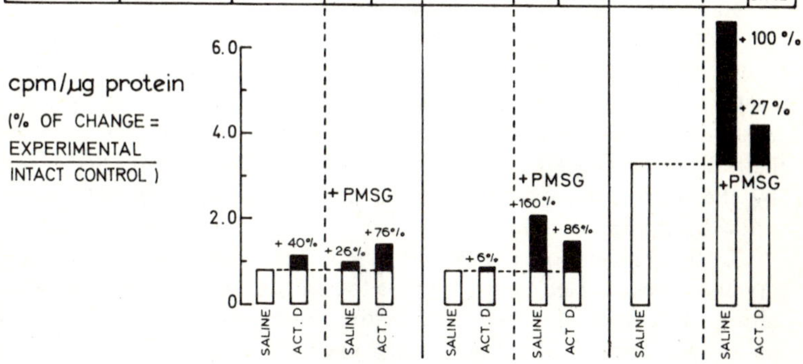

Fig. 3

Table III and Figure 3. Effect of pretreatment with actinomycin D on the amino acid incorporation into protein in control and PMSG-treated mice.

the results of Experiment 1), the fraction not dependent on RNA synthesis also represents an absolute increase.

4. The inhibitory effect of the drug becomes apparent 7 hours after its application in the PMSG treated ovaries. Inhibition of RNA synthesis reduces the PMSG effect by 50 per cent in Experiment 2 and by 75 per cent in Experiment 3, indicating that its predominant effect is reflected at the level of DNA activity.

The stimulatory effect of Actinomycin D on protein synthesis observed in the early stages of its action might be attributed to the inhibition of synthesis of a repressor protein. Inhibition of the production of m-RNA responsible for its synthesis would decrease the level of this repressor, and protein synthesis on other stable m-RNA molecules would thereby be increased (Garren *et al.* 1964; Papaconstantinou *et al.* 1966). The significance of these observations is under investigation.

ACKNOWLEDGEMENT

We would like to thank Merck Sharp and Dohme International for the donation of Actinomycin D.

GENERAL DISCUSSION

PARTICIPANTS: PETERS, ODELL, GONDOS, ESHKOL, WEISZ, RYLE, BRANDAU AND LLOYD.

1. Supplementary histological preparations of ovaries from Actinomycin D-FSH treated mice (fixed 52 hours after treatment) gave direct proof of the effect of Actinomycin D on the ovaries.

2. Regarding continued protein synthesis in the presence of Actinomycin D, other experiments indicated that the messenger RNA of these cells was much more stable than that of bacterial systems, with a half-life of 4 to 8 hours rather than of a few minutes. The effective half-life of Actinomycin D itself is not known. PMSG is excreted slowly by the mouse; the ovaries were probably exposed to it continuously until the animals were killed.

3. In follicles of animals injected with Actinomycin D many granulosa cells peeled away, fell into the antrum and later disintegrated, although neither the oocyte nor the zona pellucida seemed to be abnormal. Twenty-four hours after injection the large follicles were disintegrating but the smaller ones looked normal, the effect being more striking where many large follicles had developed in response to PMSG. Curiously, in spite of its great activity, the appearance of the theca was not altered by Actinomycin D.

4. Direct measurement of the effects of Actinomycin D on RNA synthesis had proved difficult. The incorporation of labelled uridine was variable and very low. This might have been due to a very large pool of endogenous nucleotides.

5. The present results suggest: (a) that the action of PMSG may involve utilization of already formed RNA for protein synthesis since it is effective even when RNA formation is inhibited and (b) that a repressor, with a high turnover rate, may be present, inhibition of its production by Actinomycin D leading to acceleration of protein synthesis. Nevertheless, the type of protein synthesized in response to PMSG stimulation need not be identical with that synthesized in response to release from the repressor.

6. Comparable studies are required to separate LH from FSH-type activity. Moreover, different types of response may be confined to different cell types, e.g. Actinomycin D selectively destroyed actively dividing granulosa cells. In any case, there is no firm evidence for a primary effect of the hormone on DNA- or RNA-controlled synthetic

activity. It might, for example, still act primarily on a membrane. In general one could perhaps view the development of the ovary as a series of preparations for events, each of which finally requires the removal of some inhibition to trigger its completion.

REFERENCES

Ahren, K. & Rubinstein, L. (1965). *Acta physiol scand.* **64,** 463.

Blackburn, N.M. & Kostyo, J.L. (1963). *Endocrinology,* **73,** 81.

Blackburn, N.M. & Kostyo, J.L. (1965). *Biochim. biophys. Acta,* **107,** 168.

Civen, M., Brown, C.B. & Hilliard, J. (1966). *Biochim. biophys. Acta,* **114,** 127.

Garren, L.D., Howell, R.R., Tomkin, G.M. & Crocco, R.M. (1964). *Proc. natn. Acad. Sci. U.S.A.* **52,** 1121.

Hamberger, L.A. & Ahren, K.E.B. (1967). *Endocrinology,* **81,** 93.

Lowry, O.H., Rosenbrough, N.J., Farr, A.L. & Randall, R.J. (1951). *J. biol. Chem.* **193,** 265.

Papaconstantinou, J., Stewart, J.A. & Koehn, P.K. (1966). *Biochim. biophys. Acta,* **114,** 428.

Reel, J.R. & Gorski, J. (1968). *Endocrinology,* **83,** 1083.

Van Dyke, K. & Katzman, P.A. (1968). *Endocrinology,* **83,** 107.

Changes in Rate of Nucleic Acid and Protein Synthesis in Juvenile Mouse Ovaries after FSH Stimulation

Eliza Eshkol, Brita Hardy and Clara Pariente-Coriat

INTRODUCTION

Ovarian development and the production of ovarian steroids are stimulated by pituitary gonadotrophic hormones. Growth of the follicle is in part due to proliferation of granulosa cells. The growth rate of the follicle is a function of the number of cells proliferating, the duration of the mitotic cycle and the extent of metabolic processes occurring in the follicle and its environment. The overall growth rate of the immature ovary depends on the number of such follicular units in the process of development.

Biologically 'pure' follicle-stimulating hormone (FSH) stimulated follicular growth in infant and immature mice even in the absence of circulating endogenous gonadotrophins (Eshkol and Lunenfeld, 1967; Lunenfeld and Eshkol, 1968). The aim of the present study was to investigate the sequence of some events following FSH stimulation which lead to proliferation of granulosa cells. An attempt was made to measure the rate of ribonucleic acid (RNA), deoxyribonucleic acid (DNA) and protein synthesis and to analyse changes in the RNA profiles following FSH stimulation.

MATERIALS AND METHODS

The animals used in these experiments were immature intact female mice of Swiss Albino Abu Kabir strain, aged 21 to 23 days and weighing from 12 to 18 g. Two preparations of human menopausal gonadotrophin (HMG) were used:

1. A crude HMG extract obtained from Ikapharm, Ramat Gan, with a potency of 0.25 i.u. FSH + 0.15 i.u. luteinizing hormone (LH) per mg. The hormone was obtained by kaolin extraction from human menopausal urine and precipitation with acetone (equivalent to HMG-K as described by Lunenfeld, 1964).

2. An HMG preparation (Pergonal-25 E 35 B) supplied by Istituto Farmacologico Serono, Rome, with a potency of 75 i.u. FSH + 75 i.u.

LH per ampoule. The hormone was obtained by purification of the kaolin extract from human menopausal urine by diethyl aminoethyl cellulose and chromatography on a Permutit column (equivalent to HMG as described by Donini *et al.* 1964).

A biologically 'pure' FSH preparation was obtained by adding appropriate amounts of an antiserum to human chorionic gonadotrophin (HCG) (which cross reacts with LH), in order to bind the LH present in the HMG preparations (Donini *et al.* 1966). Such an FSH preparation induces follicular growth similar to that observed with HMG containing both FSH and LH activities but it induces no increase in uterine weight (Eshkol and Lunenfeld, 1967).

Groups of 5 to 12 animals were given single or multiple subcutaneous injections of the hormones dissolved in 0.5 ml saline. Control animals received saline alone. The animals were sacrificed with ether and the ovaries were removed and transferred to the incubation flasks.

Pools of 5 or 6 ovaries were incubated in 0.6 ml Hank's Medium containing 10 μc/ml ^{3}H thymidine, ^{3}H L-leucine or ^{3}H uridine. The incubations were carried out in a Dubnoff metabolic incubator at 37°C for 30 minutes. At the end of the incubation period the isotopic incorporation was stopped with sodium azide (10^{-2}M final concentration). The material was either processed immediately or stored frozen at −20°C prior to homogenization. The Tables summarize a long series of estimations; each result shown is the mean counts per minute per 100 mg tissue obtained from two pools each consisting of five or six ovaries.

To measure the rate of tracer incorporation into DNA and protein aliquots of 0.4 ml were taken from each homogenate and 0.4 ml of cold 10 per cent trichloracetic acid (TCA) (w/v) was added. The precipitates were washed with medium containing unlabelled thymidine or L-leucine respectively. The filters were dried and counted with scintillation fluid (PPO and POPOP) in liquid scintillation counter (Packard). Protein concentration was determined by the Biuret method.

For extraction of RNA the ovaries were rinsed in cold medium prior to homogenisation. RNA was extracted by adding volume per volume of saturated phenol in acetate buffer, 0.1 M, pH 5.5. The mixture was shaken for 20 minutes, centrifuged and the upper layer was removed and dialysed overnight against the buffer. Ribosomes were prepared from the homogenate by freezing and thawing three times and centrifuging for 20 minutes at 12,000 g. The supernatant was removed and re-centrifuged for 90 minutes at 105,000 g. The sediment was collected and suspended in 0.01 M magnesium acetate. Subunits of ribosomes were prepared by dialysing the ribosomes overnight against 0.2 M EDTA. Profiles of RNA and ribosomes were determined by sedimentation patterns in sucrose gradient. The sucrose gradient of 5-25 per cent was prepared in 4.5 ml acetate buffer 0.2 M, pH 5.5, containing Mg-acetate 10^{-2} M. The tubes were stabilised for 3 hours at 4°C. 0.5 ml of the solution containing RNA or ribosomes was layered on top of the gradient and centrifuged for 6.5 hours at 39.000 r.p.m. in a Spinco centrifuge, model L-2, with an SW 39 rotor. Following centrifugation, fractions of two drops were collected. Optical density was determined at 260 mμ and each fraction was counted for radioactivity after TCA precipitation.

RESULTS

The rate of ^{3}H-thymidine incorporation was investigated at different times after a single injection of HMG. The data indicate an increase of 31 per cent above control at 15 minutes following the HMG (0.5 i.u. FSH + 0.3 i.u. LH) injection, 39 per cent between 30 minutes and 120 minutes and 27 per cent at 240 minutes ($p = 0.01$) (Table I). When the HMG dose was given in two injections the effect on the rate of DNA synthesis was significantly higher.

The effect of different doses of HMG on ^{3}H-thymidine incorporation into ovarian DNA was investigated by first injecting 0.5 i.u. FSH + 0.3 i.u. LH followed two hours later by a second injection of different quantities of FSH and LH. The animals were killed two hours following the last injection. A second dose of 0.12 i.u. FSH + 0.075 i.u. LH, of 0.25 i.u. FSH + 0.15 i.u. LH or of 0.5 i.u. FSH + 0.3 i.u. LH more than doubled the rate of DNA synthesis. No relationship between dose and increase in the rate of DNA synthesis was observed in this range but 1 i.u. FSH + 0.6 i.u. LH seemed less effective (Table II).

TABLE I

^{3}H-Thymidine incorporation at different times after a single injection of HMG (c.p.m./100 mg tissue)

	Time after injection (min)					
Controls	**15**	**30**	**60**	**90**	**120**	**240**
2500						
2730						
2570	3150	3370	3388	3315	3285	304[illegible]
1794						
Mean 2398				3339		
% change	31			39		27

TABLE II

^{3}H-Thymidine incorporation following stimulation with different doses of HMG* (c.p.m./100 mg tissue)

Controls	Dose					
	i.u. FSH	0.12	0.25	0.5	1.0
	i.u. LH	0.075	0.15	0.3	0.6
2768		7761	6750	6577	3857
2952		4125	6808	4834	2214

*Primary injection of 0.5 i.u. FSH + 0.3 i.u. LH given 2 hours before these graded doses.

The time interval between the two injections was then varied to study the relationship between it and the rate of ^{3}H-thymidine incorporation. The rate of DNA synthesis was measured two hours after the second injection. Groups of controls received the primary injection only and were killed at the same time as the experimental groups. The data show a significant increase in rate of DNA synthesis when the interval between the injections was 1½ or 12 hours. Before 1½ and after 12 hours there was no difference in ^{3}H-thymidine incorporation between ovaries from animals which received one and those which received two injections ($p = 0.01$) (Table III).

The duration of an increased rate of DNA synthesis following FSH stimulation was estimated by administering the second injection of FSH and LH two hours after the first and sacrificing the animals at different times following the second injection. The mean rate of ^{3}H-thymidine incorporation was increased by 80 per cent at 30 and 60 minutes after the second gonadotrophin injection, 100 per cent at 1½ to 2 hours and 40 per cent at 4, 8 and 12 hours ($p = 0.01$) (Table IV).

The rate of DNA synthesis was then investigated at different times following injection of a high dose of HMG (12 i.u. FSH + 12 i.u. LH). The data show that it was not increased (Table V). A single dose of 1 i.u. of biologically 'pure' FSH did not appear to increase the rate of ^{3}H-thymidine incorporation, although when the same dose was divided in two, four or eight injections stimulation was observed (Table VI). The rate of DNA synthesis rose 90 minutes after a single high dose of FSH (12 i.u.). This was followed by an increased rate of protein synthesis at 120 minutes. At 240 minutes ^{3}H-thymidine incorporation was augmented to 200 per cent and ^{3}H-leucine incorporation to 138 per cent of the corresponding control levels (Table VII).

The amount of ^{3}H-uridine incorporated into RNA was measured in control ovaries and at various times after hormone injection (Figure 1).

TABLE III

^{3}H-Thymidine incorporation two hours after second of two injections totalling 1 i.u. FSH + 0.6 i.u. LH (c.p.m./100 mg tissue)

Interval between injections (hrs.)	Experimental	Corresponding control
0.25	3190	2750
0.5	3052	2750
1	3053	2500
1.5	4810	2500
2	6577	2750
	4835	2500
	5821	2750
	5210	2500
	3120	1794
4	3250	2500
	2616	1633
	2392	1633
8	6455	2750
	4666	2500
12	4104	2500
16	2081	2500

RNA synthesis declined by 15 minutes after hormone injection, then rose by 30 minutes, reached a peak at 45 minutes and fell again to a steady state at 90 minutes.

Profiles of RNA synthesized at 45, 90 and 180 minutes following FSH administration as well as that being synthesized in the untreated animals were determined by running the RNA on sucrose gradients. Figure 2 shows that the RNA profile of the 180 minute sample was similar to that of the control. The breadth of the peak can be attributed to the presence of m-RNA. In the samples analysed at 45 minutes and 90 minutes (Figure 3) most of the label was confined to specific sharp peaks. At 45 minutes the peak seemed to correspond to

TABLE IV

^{3}H-Thymidine incorporation at different times after the last of two injections totalling 1 i.u. FSH + 0.6 i.u. LH (c.p.m./100 mg tissue)

Time after 2nd injection (Hours)	Experimental	Corresponding control
0.25	3585	2750
0.5	6184	2750
	4365	1794
1	4500	2500
	4885	2750
1.5	5828	2750
	3804	1794
2	6577	2750
	4835	2500
	5821	2750
	5210	2500
	3120	1794
4	3500	2500
8	4125	2500
12	3520	2500
16	2065	2500

TABLE VI

Effect of single and multiple injections of FSH on ^{3}H-Thymidine incorporation 240 and 120 minutes after the first injection

Time since 1st injection	Total dose i.u. FSH	No. of injections	Interval between injections (min.)	c.p.m./100mg tissue
240 min.	–	–	–	2652
	1	2	120	4446
	1	4	60	7956
	1	8	30	5902
120 min.	1	1	0	2917
	1	4	30	4057

TABLE V

Effect of a single injection of HMG (12 i.u. LH + 12 i.u. FSH) on ^{3}H-Thymidine incorporation (c.p.m./100 mg tissue)

	Time after injection (min)		
Control	90	120	340
2610	2827	2527	2520
% change from control	+7	–4	–4

TABLE VII

Effect of a single injection of 12 i.u. FSH on rate of DNA and protein synthesis

Time after FSH injection (min.)	^{3}H-Thymidine Incorporation c.p.m./100mg tissue	% change	^{3}H-Leucine Incorporation c.p.m./100mg tissue	% change
Control	1040		3874	
60	780	– 25	4420	17
90	1850	77	3068	– 26
120	1820	74	6838	77
240	3120	200	9230	138

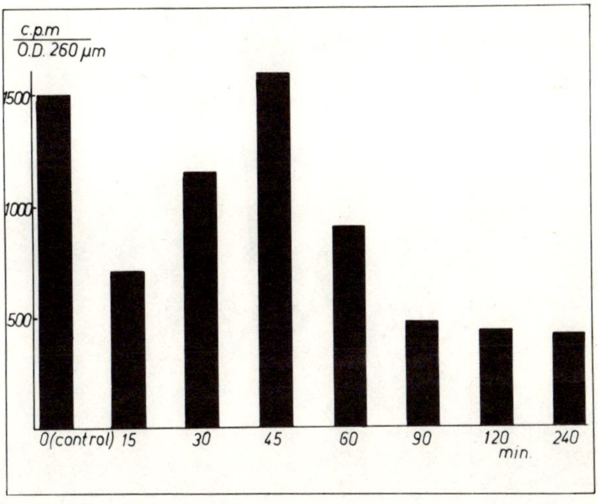

Figure 1. ^{3}H-uridine incorporation into RNA synthesized at various times after FSH injection.

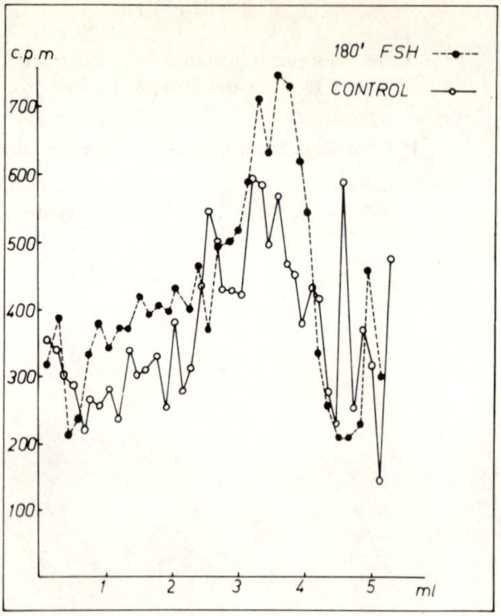

Figure 2. RNA profile 180 min. after FSH stimulation as compared to controls.

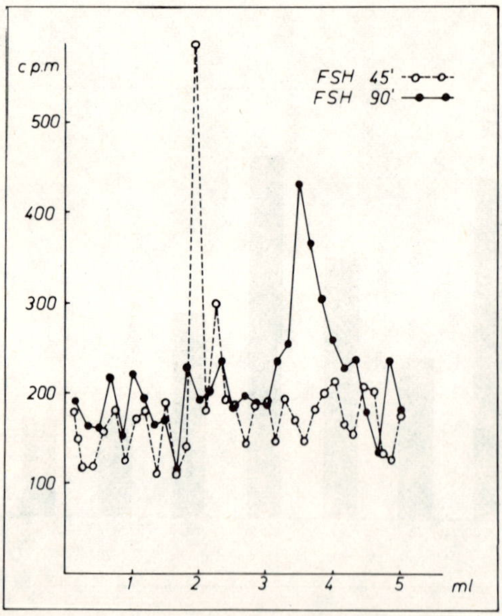

Figure 3. RNA profiles 45 and 90 min. after FSH stimulation.

the heavy ribosomal part, while at 90 minutes the peak seemed to correspond to the light part of the ribosomal subunit.

To confirm these findings, profiles of ribosomes from 45 minute samples (Figure 4) showed that radioactivity was mainly incorporated into the 60 S ribosome subunit and into the 78 S ribosomes. When the latter was split into its subunits by EDTA, most of the label was found in the 60 S subunit (Figure 5). Profiles of ribosomes obtained 90 minutes following hormone stimulation showed that the radioactivity was mainly incorporated into 40 S ribosome subunits (Figure 6). No significant incorporation of the label into 78 S ribosomes was observed at 90 minutes.

DISCUSSION

An increase in RNA synthesis before an increase in either total protein or DNA synthesis has been observed in other hormone dependent systems. The primary effect of thyroxine is increased 'de novo' synthesis of m-RNA and RNA polymerase activity (Tata, 1964). Pregnant mare serum gonadotrophin (PMSG) stimulates increased RNA polymerase activity due to the 'de novo' synthesis of this enzyme (Van Dyke, 1966). Stimulation of RNA synthesis by growth hormone precedes protein synthesis, cytoplasmic growth and cell division

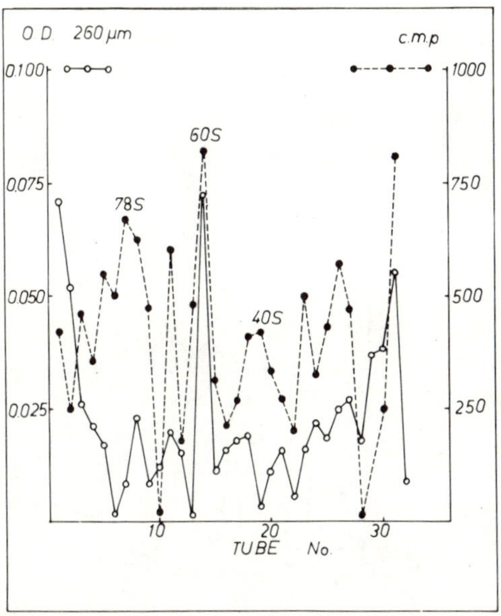

Figure 4. Profile of ribosomes 45 min. after FSH stimulation.

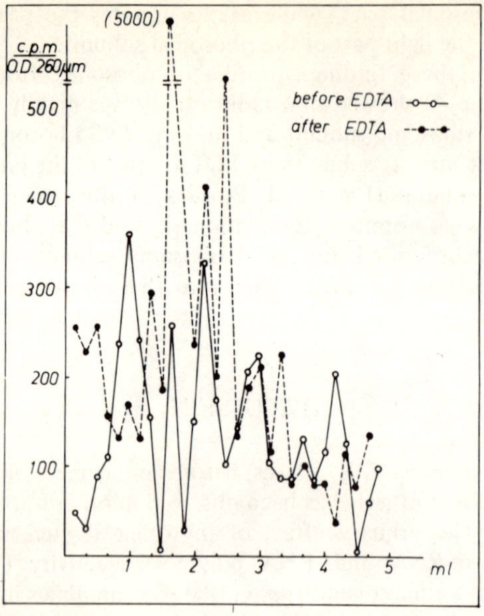

Figure 5. Ribosome profile 45 min. after FSH stimulation - prior to and after splitting with EDTA.

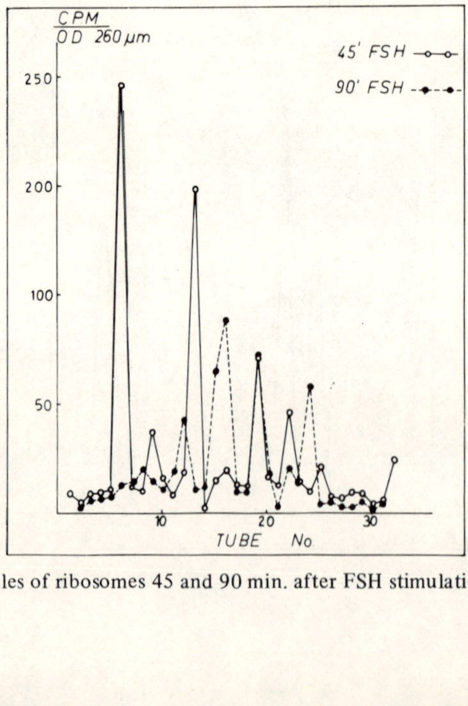

Figure 6. Profiles of ribosomes 45 and 90 min. after FSH stimulation.

(Korner, 1964). One of the early responses of uterine epithelium to oestrogen is RNA formation, followed by protein synthesis (Mueller, 1965). Gonadotrophic hormones act on the ovary and stimulate two co-ordinated although distinct activities, mitotic and steroidogenic. The effects of these hormones on the rates of DNA, protein and RNA synthesis were investigated as intermediate processes involved in ovarian metabolic regulation.

The rate of incorporation of ^{3}H-thymidine into DNA was increased by HMG. The effect was detectable 30 minutes after a single injection but was maximal after two injections separated by an interval of 2 hours. It lasted for 12 hours. Why was the effect of the hormone greater when the same amount was administered as two injections instead of one? Moreover, why did 12 i.u. of FSH in HMG fail to produce any effect while the same dose of 'pure' FSH stimulated DNA synthesis? One may conclude that the stimulation of DNA synthesis is due to the FSH component of HMG, but that at high doses LH interferes directly or indirectly with it. This does not however explain the greater effect of the total dose when divided between several injections.

While the rate of DNA synthesis was increasing at 90 minutes after hormone injection, the rate of protein synthesis was reduced at that time but increased exponentially at 120 minutes. This suggests that certain functional units at the sub-cellular level, necessary for rapid protein synthesis, must first accumulate. RNA synthesis had decreased by 15 minutes after injection but then rose to a peak at 45 minutes before declining to a steady state at 90 minutes. At 180 minutes when increased protein synthesis was taking place the RNA profile was similar to that of the control, and the breadth of the peak indicated m-RNA. However at 45 minutes most of the label was concentrated in a very sharp peak in the 28 S fraction which comprises the precursor for the heavier 60 S subunits of the ribosome. At 90 minutes the label was mostly in the 18 S fraction which appears to correspond to the lighter (40 S) subunit of the ribosome. (The sharpness of these peaks might indicate that relatively less m-RNA was being synthesized at 45 and 90 minutes).

Ribosomes isolated at 45 and 90 minutes after hormone injections confirmed these findings. The profiles at 45 minutes showed an increased incorporation of the tracer into the 78 S ribosomes and their 60 S subunit. Furthermore, after splitting the 78 S ribosomes (by EDTA) most of the label was found in the 60 S subunit. This indicated that the 78 S ribosomes were composed of labelled heavy subunits and non-labelled light subunits. Ribosome RNA profiles at 90 minutes showed an increased synthesis of the 40 S subunit.

These results suggest that FSH treatment initially induces inhibition of protein synthesis, but later results in an abrupt increase. *In vitro* experiments with bacterial ribosomes have shown that an excess of the 50 S fraction results in inhibition of protein synthesis (Pestka and Nirenberg, 1966). Perhaps this occurs here, the hormone stimulating differential increase of the heavy ribosomal particles which in turn inhibit protein synthesis. The increase in 40 S particles at 90 minutes may be necessary for increased protein synthesis. Thus ribosomes may act in two ways in the control of protein synthesis and cell division.

In conclusion, FSH increases the pool of proliferating granulosa cells by initiating some process leading to DNA replication. LH induces differentiation of these cells for other biochemical functions in such a way that their mitotic activity is reduced. These suggested effects of FSH and LH on the mitotic cycle of granulosa cells are shown diagrammatically in Figure 7.

ACKNOWLEGEMENTS

This study was supported by Ford Foundation Grant No. 67-470. Sections of this study were part of the M.Sc. theses (Bar-Ilan University) of C. Coriat, under the direction of B. Lunenfeld and Aliza Eshkol, and of B. Hardy, under the direction of I. Ashkenazi and B. Lunenfeld. The continued interest and guidance of B. Lunenfeld and I. Ashkenazi in the course of this study are greatly appreciated.

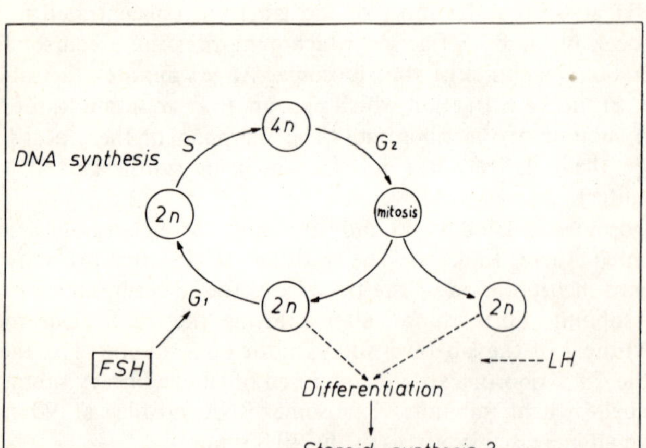

Figure 7.

GENERAL DISCUSSION

PARTICIPANTS: BEN-OR, LUNENFELD, PEDERSEN, RYLE AND PETERS.

1. Since fractions of whole ovaries were examined it is possible that protein synthesis was not occuring in the same cells as DNA synthesis. However, all available evidence indicates that in the mouse only granulosa cells respond to FSH. Autoradiography confirms that DNA formation is confined to this cell population and probably the increased protein synthesis is also. Interpretation of the results from RNA fractionation studies requires caution, in view of information available from other mammalian cell systems.

2. Synthesis of DNA persisted up to about 12 hours after injection of FSH, which has a short half-life in the circulation. DNA synthesis in response to FSH *in vitro* persists for between 9 and 12 hours (Ryle).

3. If FSH induces entry into the G1 phase of the mitotic cycle it could recruit previously inactive cells into the proliferating pool. In large follicles all granulosa cells are already proliferating so FSH would probably have no effect, but in medium follicles only a fraction are doing so and it may increase this fraction.

4. LH depressed DNA synthesis in these short incubations. It also does so in cultures of infantile ovaries maintained for several days (Ryle). Possibly cells which have been stimulated by LH are unable to respond to FSH, for DNA synthesis is generally arrested during cell differentiation.

5. The inhibitory effect of LH is probably indirect, depending on the synthesis of one or more steroids. Thus earlier papers in this conference showed that when the interstitial tissue of very young ovaries was strongly stimulated, the follicles failed to develop. However, when oestrogenic steroids are produced these may themselves induce mitosis. The larger follicles in *in vitro* cultures respond to LH by increased growth (Ryle).

REFERENCES

Donini, P., Montezemolo, R. & Puzzuoli, D. (1964). *Acta Endocr. Copenh.* **45**, 321.

Donini, P., Puzzuoli, D., D'Alessio, I., Lunenfeld, B., Eshkol, A. & Parlow, A.F. (1966). *Acta Endocr. Copenh.* **52**, 186.

Eshkol, A. & Lunenfeld, B. (1967). *Acta Endocr. Copenh.* **54**, 91.

Greep, R.O. (1959). *Recent Prog. Horm. Res.* **15**, 139.

Korner, A. (1964). *Biochem. J. 92,* 449.

Lunenfeld, B. (1964). Proc. of the 2nd Int. Congr. of Endocr. p. 828. *Excerpta Med. Int. Cong. Ser. No.* **83.**

Lunenfeld, B. & Eshkol, A. (1968). In *Gonadotropins,* p. 197. Ed. Rosemberg, E. Clinton, Los Altos, California: Geron-X, Inc.

Mueller, G.C. (1965). Proc. of the 6th Pan Am. Congr. of Endocrinology, P. 108. Excerpta Med. Int. Congr. Ser. No. 112.

Pestka, S. & Nirenberg, M. (1966). *Cold Spring Harbor Symp. quant. Biol.* **31**, 641.

Tata, J.R. (1964). Proc. of the 2nd Int. Congr. of Endocr. p. 46. *Excerpta Med. Int. Cong. Ser. No.* **83.**

Van Dyke, K. (1966). *Ph.D. Thesis.* Ann Arbor, Michigan: University Microfilms.

Histochemical Localization of Enzyme Activities in Normal and Gonadotrophin Stimulated Mouse Ovaries

Hartmut Brandau

INTRODUCTION

Eshkol and Lunenfeld (1967) have demonstrated the significance of gonadotrophins for the development of the mouse ovary during the infantile and early juvenile periods. Proliferation and organization of follicle cells is solely dependant on FSH, while LH initiates the development of interstitial tissue. Cell proliferation and induction of steroid synthesis dependent on gonadotrophins are consequences of the specific effects of these hormones on intermediary cell metabolism. Histochemical localization of the oxydoreductases involved in carbohydrate metabolism and of hydroxysteroid-dehydrogenases is particularly suited to the study of the cellular metabolism of the different functional structures of the ovary. Thus, comparison of the distributions of enzymes directly or indirectly related to steroid synthesis, such as of NADPH-producing enzymes and of hydroxysteroid-dehydrogenases, should help to shed some light on possible steroid biosynthesis in the individual structures.

METHODS

The stimulation experiments using gonadotrophins were made on infantile mice with pig follicle-stimulating hormone (FSH) (Sigma Chemical Corp.) and human chorionic gonadotrophin (HCG) (Organon). From their first day of life, the animals received daily subcutaneous injections of 2 i.u. FSH or 20 i.u. HCG. Untreated animals of the same litters served as controls. All animals were killed on the seventh day. Their ovaries were frozen in liquid air and sectioned in a cryostat. Oxydoreductases of the Embden-Meyerhof pathway, the citric acid cycle and the pentose phosphate shunt, as well as 3β-hydroxysteroid-dehydrogenase, were histochemically localized with the aid of the usual tetrazolium technique.

RESULTS

In the ovary of the untreated 7-day-old mouse, most of the follicles have one layer of cells (type III a + b; Pedersen and Peters, 1968), only a few having two layers (type IV). Only slight to moderately strong activity of the glycolytic enzymes, glyceraldehyde-phosphate dehydrogenase and lactic dehydrogenase, can be detected in the granulosa of any of these follicles (Figure 1). No difference in activity between the individual layers is evident. The interstitial cells which, on the seventh day, are scant, flat and sparsely scattered over the stroma, show markedly less activity due to these enzymes than do the granulosa cells, so they are difficult to locate in these preparations.

After FSH treatment the number of follicles with two layers is increased. Some follicles with three layers are always present and occasionally some with four layers (type V a + b). In these growing follicles, the activity of the glycolytic enzymes (Figure 2) is considerably raised, while that of the small follicles and of the interstitial cells remains as in the untreated ovary.

Treatment with HCG does not accelerate follicle maturation; most of the oocytes are surrounded with one layer of follicle cells as in the untreated ovary. The activity of the glycolytic enzymes in the follicles cannot be distinguished from that in the control ovary. On the other hand, the activity of these enzymes in the more numerous and well-rounded interstitial cells is greater, so that granulosa and interstitial cells now possess comparable activities.

Study of the distribution patterns of the NADPH-producing enzymes, glucose-6-phosphate-dehydrogenase and NADP-specific isocitrate dehydrogenase as well as NADP-diaphorase, shows that their activity behaviour is different to that of the glycolytic enzymes. While at seven days only traces of these three enzymes are detectable in the granulosa of the control ovary, the sparse interstitial cells possess markedly higher activities. FSH treatment leads to a very slight increase in activity of all NADP-specific enzymes in developing follicles while the activity of the interstitial cells remains unchanged (Figure 3). The administration of HCG, however, produces a conspicuous increase in activity of these enzymes in the interstitial cells. For instance the colour developed due to NADP-diaphorase (Figure 4) demonstrates the increases in number and size of the interstitial cells and in addition the formation of the theca interna layer which is also initiated by HCG. In the granulosa the activities of NADP-specific enzymes remain virtually unchanged, occasionally the activity seems to be even reduced in comparison with the control.

The enzymes of the citric acid cycle show completely different characteristics. In the control ovary there is low to moderately high activity in the granulosa. Succinate, isocitrate (NAD-specific) and malate are metabolized with decreasing vigour. The interstitial cells are

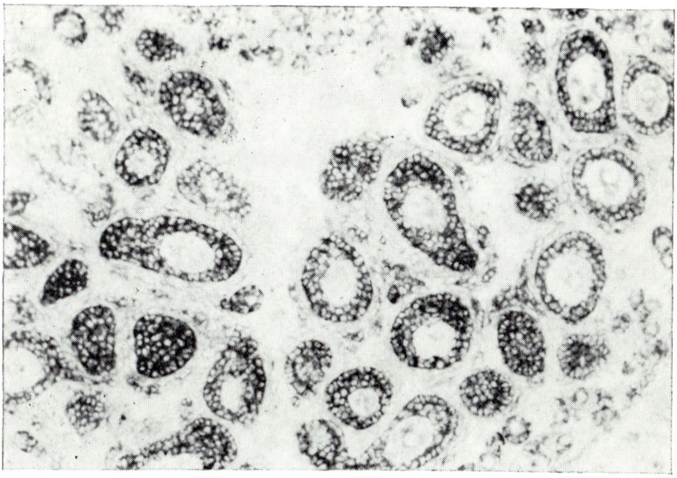

Figure 1. Localization of GAPDH in control ovary. Little to moderately strong activity in the granulosa cells. Interstitial cells only poorly active. ×200

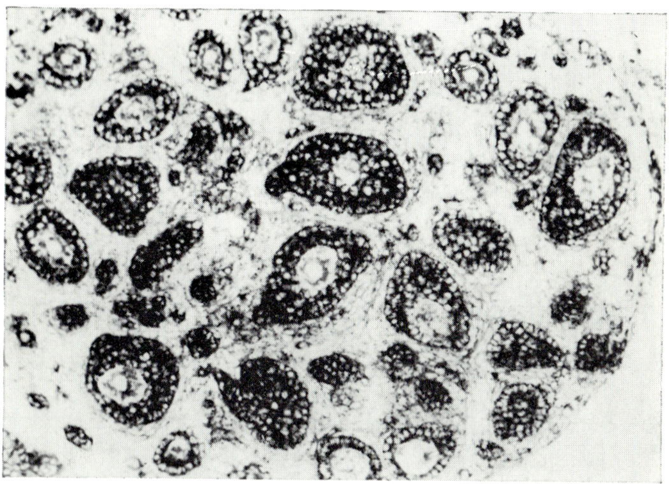

Figure 2. Localization of GAPDH after FSH stimulation. Considerably increased activity in the granulosa of growing follicles; interstitial cells only poorly active. ×200

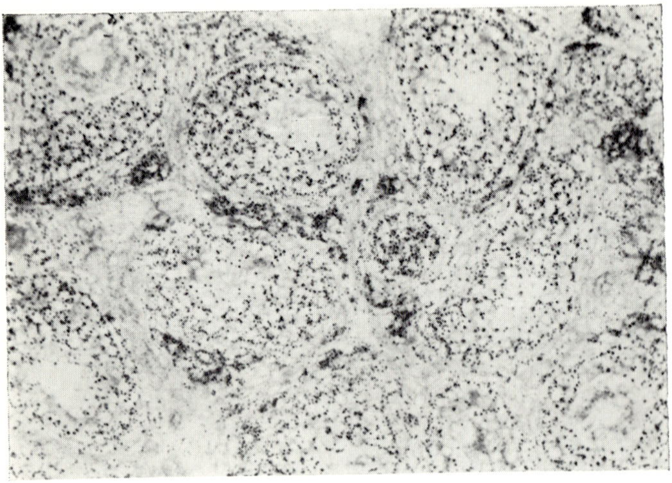

Figure 3. Localization of NADP-diaphorase after FSH stimulation. Activity in the granulosa of growing follicles slightly increased. Interstitial cells unaltered. ×320

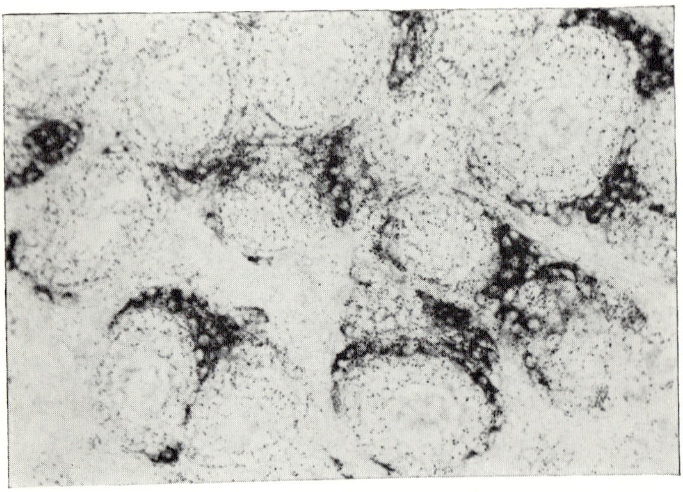

Figure 4. Localization of NADP-diaphorase after HCG stimulation. Interstitial cells show considerably increased activity. Granulosa occasionally shows less activity than control. ×320

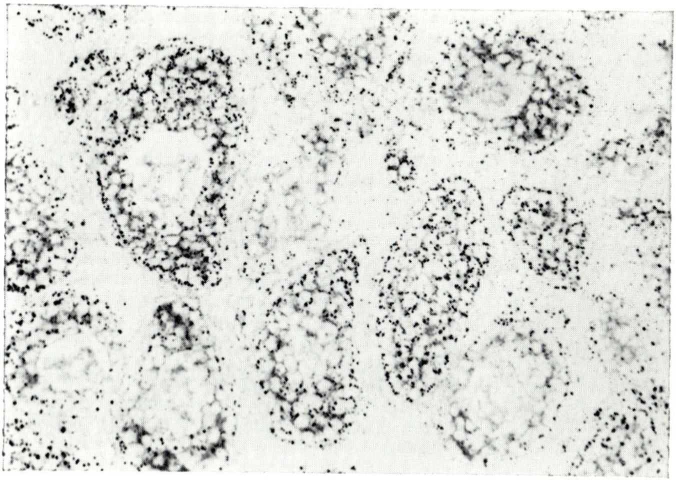

Figure 5. Localization of 3β-OHDH after FSH stimulation. Colour in granulosa of maturing follicles stronger and more granular. Interstitial cells sparse, scattered; activity as control. ×320

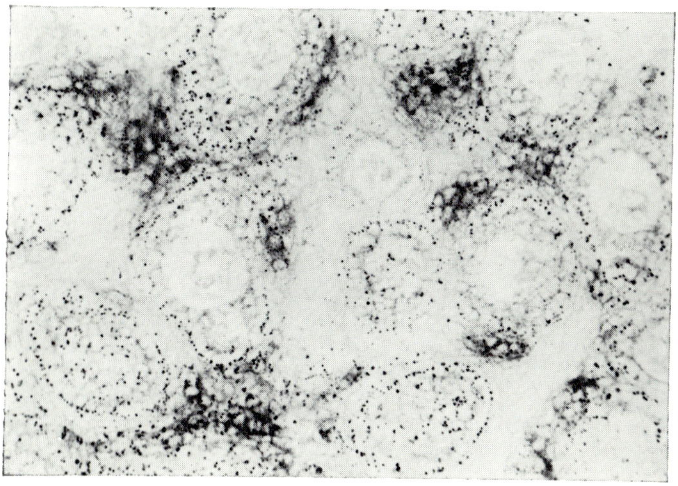

Figure 6. Localization of 3β-OHDH after HCG stimulation. Activity in the interstitial cells markedly increased. Activity in granulosa occasionally less than in control. ×320

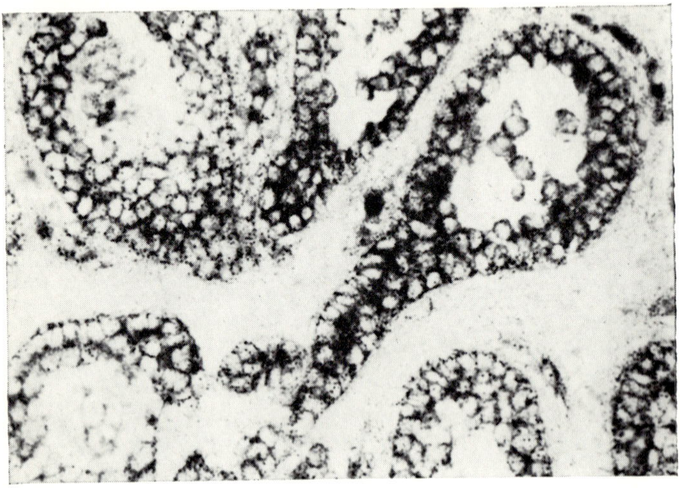

Figure 7

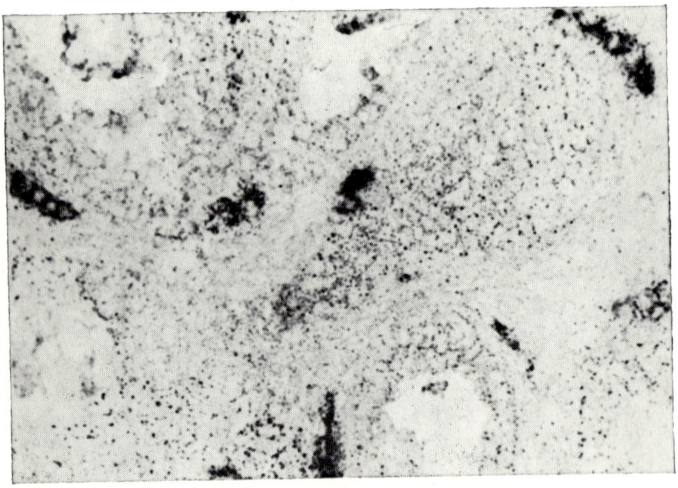

Figure 8

Figures 7 and 8. Serial sections of 7-day-old control ovary. In the centre of the picture ramifications of the granulosa cell cord. Figure 7 shows GAPDH localization and Figure 8 G6PDH localization. Distribution of both enzymes in the ramifications identical with that in the granulosa. The interstitial cells differentiating in the stroma show high activity of G6PDH also only poorly active. Stimulation with FSH or HCG does not alter the pattern. Granulosa and interstitial cells continue to show only low to moderately high activities.

Finally, the distribution pattern of 3β-hydroxysteroid-dehydrogenase requires description. As the investigations by Baillie *et al.* (1966) have shown, even the ovary of the foetal mouse has a little activity due to this enzyme. In the infantile ovary, activity is conspicuous in the granulosa of smaller follicles, while in the late juvenile and puberal ovary it is found solely in the granulosa of maturing tertiary follicles. In the control the activity of this enzyme is generally very low though, in multi-layered follicles, the colour of the innermost layer is somewhat stronger and less granular. The sparse interstitial cells are a little more strongly coloured than the granulosa. We did not find the colouring of the interfollicular stroma described by Baillie *et al.* (1966). After treatment with FSH the enzyme activity in the granulosa, in particular that of the innermost layers, is intensified (Figure 5). Otherwise the distribution is comparable to that in the control ovary. Stimulation with HCG leads - as in the case of NADPH-producing enzymes - to a pronounced increase of activity in the interstitial cells (Figure 6).

The development of the interstitial tissue, including the so-called primary interstitial cells, can be studied easily after stimulation with HCG. According to Dawson and McCabe (1951) and Rennels (1951) the primary interstitial cells in the rat ovary are derived from the granulosa cell layer. In the mouse ovary, too, protrusions or ramifications of the granulosa frequently occur. These cell groups, however, always show the enzyme pattern typical of the granulosa with moderately strong glycolytic enzyme activities (Figure 7) and, unlike interstitial cells, little activity of the NADPH-producing enzymes (Figure 8). Stimulation with HCG does not increase the activity of NADPH-producing enzymes or of 3β-hydroxysteroid-dehydrogenase in them. The typical primary interstitial cells, with high activities of NADPH-producing enzymes are, however, derived from the interfollicular stroma, individually and in groups. They adhere to the follicles to form the typical theca interna. Judging from the available facts growth of primary interstitial cells from the granulosa seems improbable.

DISCUSSION

When interpreting the effects of stimulation with gonadotrophins in intact animals, the possibility of complications due to 'feed back' mechanisms must be considered. It is remarkable, for instance, that the follicular development characteristic of the infantile period should be checked by HCG, so that the activities of some enzymes, such as

NADP-specific isocitrate-dehydrogenase and 3β-hydroxysteroid-dehydrogenase, are lower in the granulosa than in the control. HCG stimulates the secretion of pituitary FSH in the late juvenile period (Evans and Simpson, 1950) but the present results suggest that it curbs pituitary gonadotrophin secretion in the infantile period as it does in the adult, (Sawyer and Kawakami, 1959). FSH and HCG have very different effects on enzymatic and metabolic processes. The tendency to proliferate, induced by FSH, is accompanied by change in the enzyme pattern of the granulosa cell; the glycolytic enzymes are strongly activitated and there is a slight increase in the activity of NADPH-producing enzymes. It is difficult to explain the slight but clearly visible rise in activity of 3β-hydroxysteroid-dehydrogenase in the granulosa cells of small and medium follicles. The presence of this steroid dehydrogenase in the granulosa cell of the young follicle may perhaps be associated with a receptor for some action of steroid hormones. They are unlikely to synthesise steroids because they lack the NADPH-producing enzymes typical of the steroid secreting cell.

In the interstitial cell HCG produces a very marked rise in the activity of the NADPH-producing enzymes accompanied by a moderate rise in that of 3β-hydroxysteroid-dehydrogenase and the glycolytic enzymes. Thus, regarding the NADPH-producing enzymes, the interstitial cell of the infantile mouse ovary possesses the high sensitivity to LH or HCG characteristic of the steroid-forming cell. It is, however, questionable whether the activities of both enzyme systems are directly induced by HCG. It has been reported that HCG does not directly influence the activity of 3β-hydroxysteroid-dehydrogenase in the corpus luteum (Tamaoki and Pincus, 1961) or in the placenta (Venning and Sibulski, 1964). It seems probable that - as for example, has been shown in the case of the granulosa cell of the preovulatory follicle (Brandau *et al.*, 1967) - the NADPH-producing enzymes are directly induced by LH or HCG, while a substrate induction of 3β-hydroxysteroid-dehydrogenase occurs after initiation of steroid synthesis.

GENERAL DISCUSSION

PARTICIPANTS: BEN-OR, PRESL, STEGNER, FRANCHI, RYLE, GONDOS, WEISZ AND BYSKOV.

1. In this work pregnenolone was found to be the best substrate for 3β-hydroxysteroid-dehydrogenase (3βHSD). However, the optimal substrate varies with species and with age.

2. The increase in 3βHSD in the granulosa of small follicles in response to FSH is surprising. Other enzymes required for steroid biosynthesis are lacking and electron microscopy also indicates that these cells are not equipped for steroidogenesis. Steroid synthesis occurs only in the primary interstitial tissue at this age. Possibly the 3βHSD is associated with a receptor for stimulation by a steroid;

possibly it is concerned with protecting the young follicle from damage by some steroid. It is also found in granulosa cells in the foetal mouse. In the postnatal rat 3βHSD is never found in young granulosa cells but only in the theca, the interstitial tissue and in granulosa cells of some Graafian follicles just before ovulation and luteinization (Presl).

3. There is no direct correlation between the presence of lipid droplets and steroid synthesis. Lipid droplets are found in foetal ovaries in some species. In the new-born mouse the ovary contains neither 3βHSD nor lipid droplets. Lipid is present at 5 days in rat granulosa cells, which have no 3βHSD. It may be utilized in zona pellucida formation or in a variety of other metabolic processes. Having accumulated, it may be used in steroid synthesis if the appropriate stimulus initiates enzyme formation. One should probably distinguish between the different types of lipid.

4. The correlation between the level of a particular enzyme and a particular steroid product is usually not good. The overall enzyme pattern, including those involved in carbohydrate metabolism and NADPH production is more relevant. 3βHSD activity is not usually particularly sensitive to gonadotrophic stimulation and may not normally be a limiting factor for steroidogenesis. Once formed, it may be slow to disappear and therefore at times be irrelevant. Alternative pathways of synthesis may be available and the level of a particular enzyme may not therefore be critical. Moreover, the preferred pathway may differ at different stages of development. The steroids themselves may have different functions at different ages, e.g. they may be involved in the differentiation of young tissues.

REFERENCES

Baillie, A.H., Ferguson, M.M. & McK. Hart, D. (1966). *Developments in Steroid Histochemistry.* New York: Academic Press.

Brandau, H., Remmlinger, K. & Luh, W. (1967). *Acta endocr. Copnh.* **56**, 433.

Dawson, A.B. & McCabe, M. (1951). *J. Morph.* **88**, 543.

Eshkol, H. & Lunenfeld, B. (1967). *Acta endocr. Copnh.* **54**, 91.

Evans, M.H. & Simpson, M.E. (1950). In *The Hormones,* Vol. II. Ed. Pincus, G. & Thimann, K.V. New York: Academic Press.

Pedersen, T. & Peters, H. (1968). *J. Reprod. Fert.* **17**, 555.

Rennels, E.G. (1951). *Am. J. Anat.* **88**, 63.

Sawyer, C.H. & Kawakami, M. (1959). *Endocrinology,* **65**, 631.

Tamaoki, B. & Pincus, G. (1961). *Endocrinology,* **69**, 527.

Venning, E. & Sibulski, S. (1964). *In Biological and Clinical Aspects of Placental Steroidogenesis.* Milan, 1962. Ed. Polvani, F. & Bompigni, A. Baltimore: Williams and Wilkins.

Cell Population Kinetics of the Ovary of the Immature Mouse after FSH Stimulation

Torben Pedersen

INTRODUCTION

The effect of follicle stimulating hormone (FSH) on the ovary of the mouse and rat has been investigated both *in vivo* (Ben-Or, 1963; Eshkol and Lunenfeld, 1967) and *in vitro* (Pavic, 1963; Fainstat, 1968; Ryle, 1969). These investigations showed that FSH-treatment causes an increase in the growth of the follicles. The results were obtained by making differential counts of the follicles or by measuring diameters of the largest follicles and the size of the ovaries. But no attempt has been made to clarify the effect of FSH on the granulosa cell populations in follicles at different developmental stages, nor the effect on the proliferation rate of these populations.

Recently the growth of the granulosa cell populations and of the whole follicles has been determined in the ovary of the immature mouse (Pedersen, 1969) using autoradiography after pulse-labelling with tritiated thymidine. By this method it is possible to estimate the growth characteristics of granulosa cells, including their growth rates, and thereby also the growth rates of whole follicles.

It is the purpose of the present investigation to determine the effect of FSH treatment on the growth of granulosa cells and on whole follicles in ovaries of immature mice. The ages of 14, 21, and 28 days were chosen for the investigations.

The follicles constitute the major part of the immature ovaries, and a variation in their growth rates will be followed by variations in the volume of the whole ovaries. It was therefore decided also to determine the volume of the ovaries of untreated as well as FSH-treated animals. The results of the growth rates of follicles were correlated with the volume-measurements.

PRINCIPLES AND DEFINITIONS

The classification of follicles into types according to Pedersen and Peters (1968) has been used in the present investigation. It is based on the number of granulosa cells in the largest cross section of each follicle (see page 224).

The proliferation rate of the granulosa cells in a follicle is expressed as their doubling time (T_D), i.e. the time it takes a number of granulosa cells in a follicle to double their number. T_D can be calculated from the following parameters: (1) the labelling index (LI) of the granulosa cells in a follicle and (2) the duration of the DNA-synthesis phase (S-phase) of the individual granulosa cells. The LI is defined as the number of labelled granulosa cells expressed as a percentage of the total number of cells determined in autoradiographs prepared one hour after pulse-labelling with tritiated thymidine (^{3}HTdR). The duration of the S-phase of the single granulosa cells can be determined in autoradiographs prepared at varying time intervals after pulse-labelling with ^{3}HTdR (Quastler and Sherman, 1959).

From the total number of granulosa cells in the different types, which has been determined earlier (Pedersen, 1970), and the doubling times of the granulosa cells in these types, it is possible to calculate the growth rate of the follicles expressed as their transit times (T_F), i.e. the time it takes to grow from one type to another.

To be able to describe the kinetics of follicles in detail it is furthermore necessary to determine the number of follicles present of each different type (the differential counts) in treated as well as untreated mice. The growth rates of granulosa cells and follicles are determined by the values of the LI and the duration of the S-phases. By measuring these parameters and the differential counts in ovaries from treated mice and by comparing these with the values obtained in untreated mice it is then possible to determine accurately the effect of the treatment on the kinetics of the follicles.

Measurement of the volume of ovaries. In order to determine the volume of the ovary a grid was inserted into the ocular of the microscope. The number of squares covering every twentieth section of the ovaries were counted. As the area of each square was known, the volume (V) of the ovary could be calculated from: $V = NAf20$, where N is the total number of squares counted in every twentieth section, A = the area of the square in μ^2, and f = the section thickness.

MATERIAL AND METHODS

The effect of FSH-treatment was investigated in ovaries of Bagg mice at the ages of 14, 21, and 28 days. The FSH was made from HMG (Pergonal) by binding the LH-activity with an antiserum to HCG as described by Donini *et al.* (1966). It was dissolved in saline, every ml containing 3.0 i.u.

Two experiments were performed in order to determine (1) the differential counts and the LI of the follicles and (2) the S-phase of the granulosa cells.

Experiment 1. (LI and differential counts). The mice were treated as follows:

14-day-old mice. Four mice were given a single injection of 1.5 i.u. FSH. Two of these animals were pulse-labelled three hours later and the other two five hours after FSH. Six mice were injected once a day with 0.75 i.u. FSH. Two of these were pulse-labelled two hours after the second injection, two other animals were labelled after three, and the last two were pulse-labelled after four injections.

21-day-old animals. Four mice were injected once with 1.5 i.u. and pulse-labelled after three or five hours. Two mice were injected with 1.0 i.u. twice with an 8-hour interval between injections and then pulse-labelled three hours later. Six mice were injected every 12 hours with 0.75 i.u. FSH and pulse-labelled two hours after three, five or seven injections.

28-day-old mice. These were treated in the same way as 21-day-old mice.

Controls. Five mice served as controls at each of the ages: 14, 21 and 28 days, and two mice at each of the ages: 17, 24, and 31 days. All control mice were killed one hour after pulse-labelling with 5.0 μCi ^{3}HTdR/g body weight.

Experiment 2. (Duration of S-phase). Ten mice were treated at the age of 28 days with three injections of 1.0 i.u. FSH at 8-hour intervals. Two hours after the second injection of FSH all mice were pulse-labelled with 50 μCi ^{3}HTdR. They were killed at varying time intervals after pulse-labelling from one to 15 hours, autoradiographs were prepared and the curve of labelled mitoses determined.

Autoradiographic and counting procedures. These have been described elsewhere (Pedersen, 1970a).

RESULTS

Labelling index of follicles.

14-day-old mice. Neither after one injection nor after more injections of FSH could any influence on the LI of the follicles be detected (Table I). The LI of those types of follicles present at this age were of the same order of magnitude as in the controls. Even after daily injections of 0.75 i.u. FSH for three days no significant influence on the LI could be seen.

21-day-old mice. The LI of the follicles were measured in mice treated once and in mice treated several times with FSH (Figure 1). Three hours after a single injection of FSH an increase in the LI could already be found in some of the types (Type 3b, 5b, 6). Five hours after the first injection the LI of all types of follicles was increased. More injections of FSH were not followed by a further increase in the LI, but it remained on a higher level than in the controls as long as injections continued.

TABLE I

Mean labelling index of follicles in mice treated with FSH from the age of 14 days

Number of FSH injections	1	1	2	3	5	Untreated 14-day-old mice	Untreated 17-day-old mice
Time from first injection to pulse-labelling	3 h	5 h	26 h	50 h	74 h		
	%	%	%	%	%	%	%
Type 3b	11.5	11.9	7.5	10.0	11.8	13.2	13.8
Type 4	8.1	10.5	7.4	7.8	10.9	10.3	9.3
Type 5a	8.8	–	–	13.0	17.3	14.3	14.1

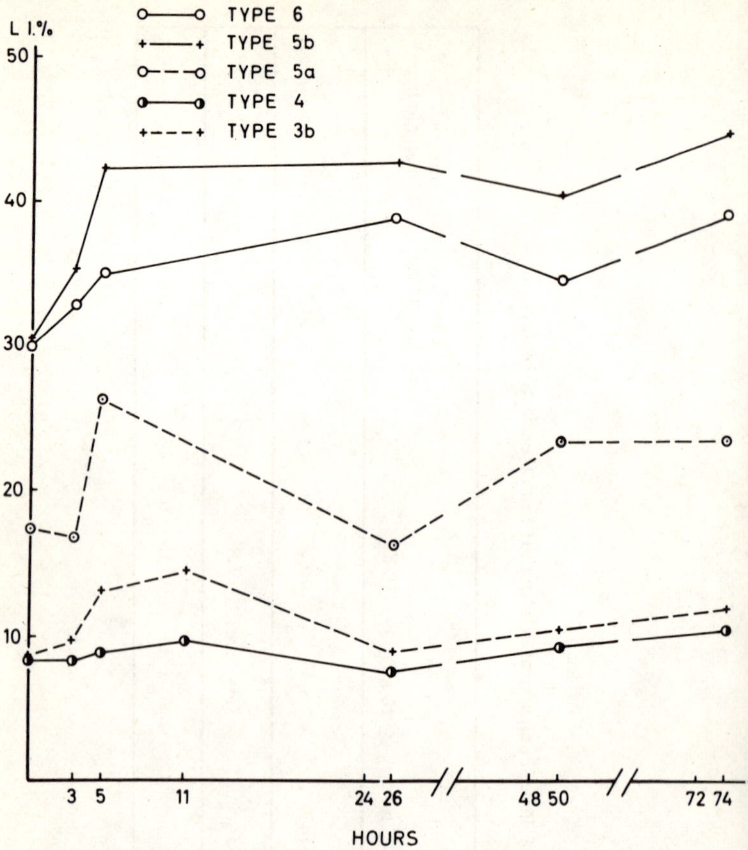

Figure 1.

28-day-old ovaries. Three hours after the first injection of FSH a slight increase in the LI of type 5a to 6 inclusive could be found (Figure 2). Five hours after a single injection the LI increased further and reached a maximum in the large follicles. After two injections with an 8-hour interval the LI of the large follicles decreased, while that of the medium follicles reached a maximum. In spite of further FSH-treatment every 12 hours the LI of all types of follicles decreased and already on the second day of treatment no difference between treated and controls was noted.

It can be concluded that the influence of FSH on the LI of the follicles depended on the age of the mouse: in the 14-day-old animals treatment did not alter the LI, i.e. the same proportion of granulosa cells synthesize DNA independently of FSH-treatment; in 21-day-old animals the LI remained elevated during treatment, while in 28-day-old mice the effect of FSH was only a temporary increase in the LI.

The duration of the DNA-synthesis phase. In order to be able to calculate the doubling times of granulosa cells it is necessary to determine the duration of their S-phases. This was done in an earlier investigation by the method of labelled mitosis (Pedersen, 1970) and it was shown that the S-phase varies considerably according to the type of follicle to which the granulosa cells belong, but within a given type there is no age-dependent variation. The same method was applied in this investigation to estimate if the FSH-treatment had any influence on the duration of the S-phase.

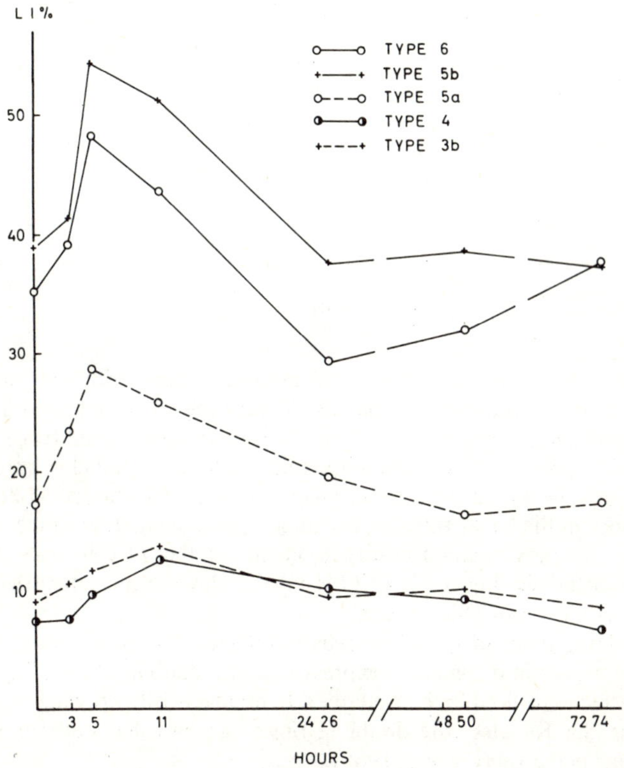

Figure 2.

The percentage of labelled mitoses was determined in follicles of different types, at varying time-intervals after pulse-labelling, in untreated mice as well as in treated 28-day-old animals (Figure 3). The results obtained in the controls are shown as the full line (the individual observations from which the curves are drawn are not shown) while the single observations shown represent the results obtained in the FSH-treated animals. There seems to be no difference between the

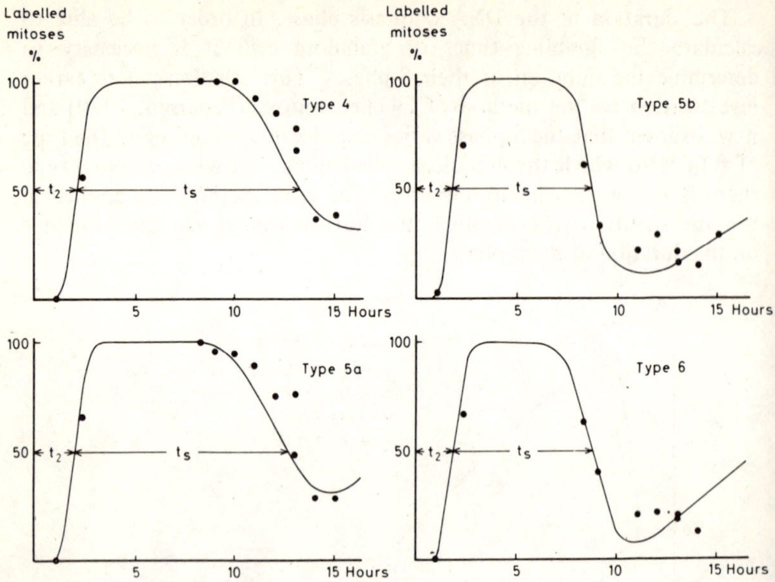

Figure 3.

duration of the S-phase before and after treatment with FSH, as the observations in the treated animals lie near the curve obtained in the untreated ones. This means that, in 28-day-old mice treated twice with FSH, the S-phase of the granulosa cells in type 3b follicles lasts 9.8 hours, in type 4 it lasts 11.2 hours and in type 5a 10.3 hours, while it is 6.8 hours in the larger follicles. It can be concluded that, at a time when FSH has its most marked influence on the number of cells in S-phase, namely after two injections of FSH into 28-day-old mice, no effect on the duration of this phase occurs.

Doubling times of granulosa cells in follicles. The growth rate of the granulosa cells in a follicle is expressed as the doubling time (T_D), the time it takes to double their number. In exponentially growing systems such as the follicles, the doubling times can be calculated from the following equation (Cleaver, 1965):

$$\text{L.I.} = \left(\exp.\frac{t_s \cdot \ln 2}{T_D} - 1\right)\left(\exp.\frac{t_2 \cdot \ln 2}{T_D}\right)$$

Assuming that the S-phase durations are unaltered by treatment with FSH the doubling times have been calculated (Table II).

Transit times of follicles. The transit time of a follicle type (T_F) is defined as the time it takes a follicle to grow through that type, i.e. to develop from one with a minimum number of cells (N_i) to one with a maximum number of cells (N_o) in that type of follicle.

TABLE II

Doubling times (in hours) of granulosa cells in follicles of different types in ovaries of untreated 14-, 21-and 28-day-old mice, and in ovaries of 21-and 28-day-old mice treated with FSH

Age of mouse	Number of FSH injections	Time to pulse-labelling	Medium follicles Type 3b	Type 4	Type 5a	Large follicles Type 5b	Type 6
14 days	untreated		54	80	55	–	–
21 days	untreated		84	99	46	19.0	19.1
21 days	1	3h	76	99	47	16.6	17.8
21 days	1	5h	57	92	32	14.2	16.8
21 days	2	11h	52	86	57	17.9	17.2
21 days	3	26h	81	106	48	14.0	15.3
21 days	5	50h	72	88	35	14.6	16.9
21 days	7	74h	66	79	35	13.5	15.1
28 days	untreated		76	110	46	15.4	17.0
28 days	1	3h	69	107	35	14.5	15.2
28 days	1	5h	62	84	29	11.6	12.8
28 days	2	11h	54	67	32	12.1	14.0
28 days	3	26h	77	82	41	15.8	19.5
28 days	5	50h	72	89	48	15.4	18.2
28 days	7	74h	85	117	45	16.0	15.8
35 days	untreated		66	101	48	16.7	17.8

T_F can be calculated from : $T_F = T_D \dfrac{\ln N_o - \ln N_i}{\ln 2}$

The transit times for the different types of follicles have been calculated in ovaries from untreated mice and after FSH treatment at the ages of 21 and 28 days (Table III).

The doubling and transit times of follicles in ovaries from 14 day old FSH-treated animals do not differ from the values in untreated animals as both parameters, LI and the S-phase length, which determine the doubling and transit times are unchanged during the FSH-treatment. Only the values obtained in 14-day-old untreated mice have therefore been listed in Tables II and III.

The variations in the doubling and the transit times follow the same trend as the variations in the LI: they are unchanged in the 14-day-old ovaries, they are shortened in the 21-day-old animals while under treatment, but are only temporarily shortened in the 28-day-old mice.

Differential counts. The number of follicles of each type present in one ovary in controls and after certain time intervals of treatment was determined. There was no significant variation in the differential counts from the start of the experiment to the end at any of the ages investigated.

Volumes of ovaries The measurements of the volume of the ovaries in controls and after 1, 2 and 3 days' treatment are shown in Figure 4. It is seen that if mice are treated from the age of 14 days with one daily injection of FSH there is no increase in the volume of the ovaries beyond that found at comparable ages in untreated mice. If the treatment with FSH starts when the mice are 21 days old, however, the volume increases considerably; this is already apparent 24 hours after the first injection of FSH, and after three days of treatment the ovaries are about 50 to 100 per cent larger than the controls. In mice treated from the age of 28 days the volume increases during the first two days, but thereafter decreases to normal values.

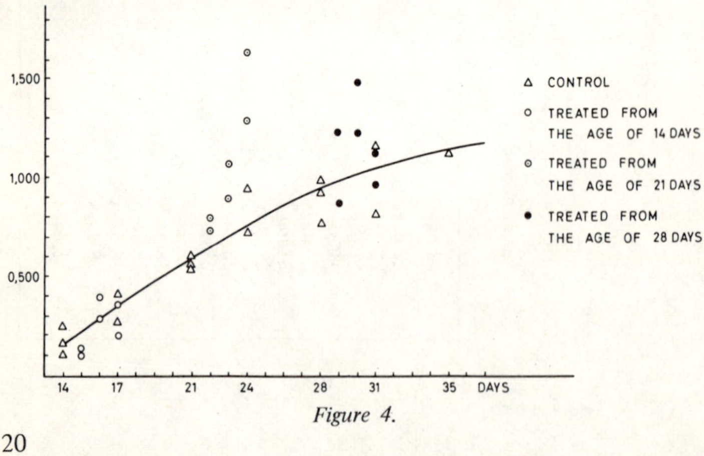

Figure 4.

TABLE III

Transit times (in hours) of follicles of different types in ovaries of untreated 14-, 21-, and 28-day-old mice, and in ovaries of 21- and 28-day-old mice treated with FSH.

Age of mouse	Number of FSH injections	Time to pulse-labelling	Medium follicles Type 3b	Type 4	Type 5a	Large follicles Type 5b	Type 6
14 days	untreated		119	91	67	–	–
21 days	untreated		185	113	56	29	15
21 days	1	3h	167	113	57	25	14
21 days	1	5h	125	105	39	21	14
21 days	2	11h	114	98	69	27	14
21 days	3	26h	178	121	58	21	12
21 days	5	50h	158	101	43	22	14
21 days	7	74h	145	89	43	20	12
28 days	untreated		167	126	56	23	14
28 days	1	3h	151	122	43	22	12
28 days	1	5h	136	96	35	17	10
28 days	2	11h	118	77	39	18	11
28 days	3	26h	169	94	50	24	16
28 days	5	50h	158	102	58	23	15
28 days	7	74h	181	134	55	24	13
35 days	untreated		145	115	58	25	14

DISCUSSION

The effect of the FSH-treatment on the growth of the follicles differs completely according to the age at which the treatment starts. In the 14-day-old mice FSH-treatment has no influence on the proliferation of the granulosa cells; they grow at the same speed as in comparable follicles in untreated mice. That FSH does not cause an increased growth of follicles is corroborated by the measurements of the volume of the ovaries which does not change more rapidly after FSH than in normal mice.

Unfortunately this age-group was not treated in the same way as the other groups, they only got one injection of FSH daily in contrast to the 21- and 28-day-old mice which were injected twice a day. Although the results are not directly comparable in the three age-groups it is most likely that FSH supplied to 14-day-old mice does not hasten the growth of the follicles.

In mice treated with FSH from the age of 21 days the results are quite different. The increase in the LI arises from shorter doubling times of the granulosa cells and shorter transit times of follicles. The decrease in the transit times after FSH is almost proportional for the different types and therefore the number of follicles entering a given type per hour is increased just as much as the number leaving that type per hour. This implies that although more follicles pass into the different types per hour the differential counts show no variations between treated and untreated animals. At the age of 21 days follicles do not develop further than to an early type 7 after which they degenerate. As more follicles reach this developmental stage after FSH than in normal mice, it means that more large follicles start to degenerate per hour in treated than in untreated animals. This increase in the number of degenerating large follicles is responsible for the increase in the volume of the whole organ. It can be concluded that at the age of 21 days treatment with FSH causes an increase in the growth rate of follicles which persists as long as the mice are treated.

In the 28-day-old mice the FSH has almost the same immediate effect on the proliferation of the granulosa cells as in 21-day-old animals: they grow faster in all follicle types. This effect is, however, only temporary and in spite of continuing treatment the growth rate of the granulosa cells and of the follicles falls to normal levels. The same trend in the variations was also found in the volume measurements: the volume increased in the first two days of treatment and then returned to normal levels. It can be concluded that at the age of 28 days FSH hastens follicle growth only temporarily in spite of continued treatment.

It has been reasoned earlier (Pedersen, 1970) that in medium follicles only a proportion of the granulosa cells are proliferating, i.e.

the growth fraction is below one, while in large follicles all granulosa cells proliferate, i.e. the growth fraction is equal to one. FSH caused an increase in the LI at the ages of 21 and 28 days, but had no influence on the durations of the S- or G_2-phases (cf. Figure 4). In the large follicles the increased LI can therefore only be accomplished by a shortening of the G_1-phase and thereby also a shortening of the generation time. In the medium follicles the same possibility exists, but the rise in the LI could furthermore be due to an increase in the growth fraction.

Other workers have shown that the effect of gonadotrophin treatment depends on the age of the animal (Pfeiffer and Hooker, 1942; Price and Ortiz, 1944). It was shown that before days 14 to 16 gonadotrophin treatment exerted no effect on the growth of the follicles, but thereafter the effect increased and reached a maximum about day 26 (in the rat) after which the effect again diminished. The present investigation shows the same trend, namely that the ovaries of 14-day-old mice do not respond to FSH, while 21-day-old ovaries show a marked effect and 28-day-old ovaries a somewhat less marked influence of the treatment.

The observations at the age of 14 days are, however, not in agreement with the observations of Eshkol *et al.* (1970) who showed that FSH is necessary in the first two weeks of life for normal development and growth of follicles. Ryle (1969) also concluded that FSH stimulates follicle growth in 15-day-old mouse ovaries. The negative result in this investigation may be due to too low a dose or too long an interval between the injections. Moreover differences in the degree of development in the different strains of mice used could be responsible for the different findings. And finally there is the possibility that, even with higher doses or shorter intervals of treatment, it is not possible to accelerate the growth of follicles with exogenous FSH at this age although endogenous FSH is necessary for normal growth.

One might explain the temporary effect of FSH injections in 28 day-old-mice by assuming that the ovary can, by this age, respond to it in such a way as to counteract the artificially raised level of the hormone. If, a week earlier, at the age of 21 days, the ovary had not reached that stage of development the raised FSH level would persist unopposed and continue to stimulate follicle growth.

ACKNOWLEDGEMENTS

The FSH was kindly prepared and supplied by Drs B. Lunenfeld and A. Eshkol, Institute of Endocrinology, Tel Hashomer Government Hospital, Israel.

GENERAL DISCUSSION

PARTICIPANTS: GREEP, BEN-OR, LUNENFELD, RYLE, LLOYD, WEISZ, ESHKOL AND DONOVAN.

1. One would expect the ovaries of 28-day-old mice to respond markedly to gonadotrophins by increasing in weight. They can readily be induced to superovulate at this age. Possibly in this work they were already almost maximally stimulated by endogenous hormone. In terms both of protein synthesis and of mitotic rate the ovaries of 18-day-old mice (of a different strain) show a relatively greater response to FSH than do those of 26-day-old animals (Ben-Or). In another strain the maximum response occurs at 21 days.

2. Here the lack of LH in the injected material may be important at 28 days, although endogenous LH was, of course, present. Any excess anti-HCG serum (used to neutralise the LH fraction of the HMG to produce biologically pure FSH) may be disregarded; it does not cross-react with mouse LH. Premature ovulation, like that induced with PMS, cannot be obtained with purified FSH. Perhaps it fails to produce steroids which may be necessary to induce release of sufficient LH for ovulation. Pure FSH, given for three days, does not cause the release of sufficient oestrogenic steroids to stimulate uterine hypertrophy.

3. Ovaries from 11- to 15-day-old CFW mice respond to purified FSH *in vitro* by increased thymidine uptake. It is therefore curious that, in the present work, no response was obtained in 14 day old animals. This may perhaps have been due to the mice receiving only one FSH injection daily and the plasma content of hormone therefore not being maintained at a sufficiently high level.

REFERENCES

Ben-Or, S. (1963). *J. Embryol. exp. Morph.* **11**, 1.

Cleaver, J.E. (1965). *Exp. cell Res.* **39**, 697.

Donini, P., Puzzuoli, D., D'Alessio, I., Lunenfeld, B., Eshkol, A. & Parlow, A.F. (1966). *Acta endocr. Copenh.* **52**, 186.

Eshkol, A. & Lunenfeld, B. (1967). *Acta endocr. Copenh.* **54**, 91.

Eshkol, A., Lunenfeld, B. & Peters, H. (1969). This volume, page 249.

Fainstat, T. (1968). *Fert. Steril.* **19**, 317.

Pavic, D. (1963). *J. Endocr.* **26**, 531.

Pedersen, T. (1970). *J. Reprod. Fert.* **21**, 81.

Pedersen, T. (1969). *Acta Endocr. Copenh.* **62**, 117.

Pedersen, T. & Peters, H. (1968). *J. Reprod. Fert.* **17**, 555.

Pfeiffer, C.A. & Hooker, C.W. (1942). *Anat. Rec.* **84**, 311.

Price, D. & Ortiz, E. (1944). *Endocrinology,* **34**, 215.

Quastler, H. & Sherman, F.G. (1959). *Exp. cell Res.* **17**, 420.

Ryle, M. (1969). *J. Reprod. Fert.* **20**, 307.

An In Vitro Test for Changes in Sensitivity to FSH of Infantile Mouse Ovaries

Margaret Ryle

SUMMARY

Pairs of ovaries from CFW mice aged 11, 13 and 15 days were grown *in vitro.* One of each pair received FSH; the other did not. Their uptake of ^{3}H-thymidine during three days' culture was measured. Every FSH-treated ovary contained more tritium than its corresponding control. The mean difference between paired treated and control ovaries did not alter significantly with age, nor did the mean ratio between them. The absence of such effects may have resulted from variation in control uptake between mice of different litters, for counts in FSH-treated ovaries varied significantly with age ($P < 1$ per cent), a sharp rise occurring between 13 and 15 days. Age had no significant effect on the variability within litters of any parameter.

INTRODUCTION

Human pituitary follicle-stimulating hormone (FSH) has been shown to increase follicular growth in the ovaries of 15- and 17-day-old mice cultured *in vitro* (Ryle, 1969c). It also made them take up more labelled amino acid and thymidine (Ryle, 1969a), nearly all of the latter being incorporated into desoxyribonucleic acid (DNA) (Chaplin, How and Ryle, in preparation). Preliminary experiments suggested that the ovaries of mice less than 15 days old or less than 5 g in weight were generally less sensitive and more variable in their response to FSH than those of older and heavier animals. Ben-Or (1963) also observed changes in sensitivity and in type of morphological response to FSH in mice around 12 to 15 days *in vivo.* The present work was undertaken to test more critically for a threshold of sensitivity in this age range.

The uptake of thymidine *in vitro* offers two particular advantages for this work. Firstly the response is quantitative and standard statistical procedures can be applied to the results. Secondly the two ovaries of each mouse, genetically identical and with effectively identical developmental histories, may be used to obtain matched experimental and control measurements. If the sensitivity to FSH increases with age, the difference in thymidine uptake between the FSH-treated ovary and the control ovary of each pair should rise. However, since the ovary grows, the number of cells able to respond may also increase with age and the ratio between each pair of measurements may be a more appropriate parameter to study than the simple difference.

The rate of change of sensitivity may not be constant at all ages. A period of rapid increase could be intercalated between periods of relatively constant sensitivity. If sensitivity should change rapidly over a brief phase of development, increased variability in response between mice at this age would be expected, because completely synchronous development is improbable even among animals of the same litter. In consequence the standard error of the mean response of a group of litter mates should rise during such a phase. Again a correction for the initial size of the ovaries is advisable because corresponding differences in total thymidine uptake are likely to occur. The standard error would be expressed more appropriately therefore as a percentage of the corresponding mean.

METHODS

Highly purified human pituitary FSH was prepared by the method of Butt *et al.* (1963). Its potency was estimated by the ovarian augmentation assay in mice (Steelman and Pohley, 1953; Brown, 1955). The luteinizing hormone (LH) in such preparations was estimated, by the rat ovarian ascorbic acid depletion assay (Parlow, 1961), to be less than 0.012 i.u. LH per 1 i.u. FSH. The FSH was used at a concentration of 1 i.u./ml of medium. This was approximately twice the concentration which normally produced a maximal response in the ovaries of 15-day-old mice cultured in comparable conditions.

Systematically random-bred mice, of the CFW strain, were reared from birth in foster litters of ten females. A group of five mice was selected from each of six such litters to give minimal differences of weight within each group (Table I). Two groups were used at 11 days after birth, two at 13 days and two at 15 days. The mice were killed abruptly and the ovaries dissected out rapidly and aseptically. One ovary of each mouse was explanted in a dish containing FSH, the other in a dish containing no hormone.

The ovaries were placed on strips of lens tissue supported on stainless steel grids. Each grid stood in a small petri dish containing 5 ml of a modified Eagle's medium (Vantsis and Wildy, 1962) supplemented with 5 per cent serum from a hypophysectomized sheep and, where appropriate, with FSH. Each set of ten dishes containing the

TABLE I

Weights of mice used

Litter	Age (days)	No. of mice weighing: 4.0	4.5	5.0	5.5	6.0	6.5	7.0g
A	11	–	–	–	5	–	–	–
B	11	5	–	–	–	–	–	–
C	13	–	–	–	–	1	2	2
D	13	–	–	–	5	–	–	–
E	15	–	–	–	–	4	1	–
F	15	–	–	–	–	5	–	–

ovaries from a particular group of mice was put in a McIntosh and Filde's bacteriological jar, which was gassed with 5 per cent CO_2 in air and sealed at 2 lb/sq. in. above atmospheric pressure.

The jars were incubated at 37°C overnight to permit diffusion of FSH from the medium into the tissue and to allow time for the completion of DNA formation due to stimulation by endogenous FSH. Earlier results had indicated that the stimulus due to human FSH *in vitro* was completed by 12 hours after removing the ovaries to fresh culture medium which lacked the hormone (Ryle, 1969b). The following day 1 μC of thymidine (methyl-^{3}H) (18.3 C/mM; Radiochemical Centre, Amersham), in 0.1 ml Eagle's medium, was added to each dish. The jars were then re-gassed, sealed and incubated as before for a further three days.

The method used to prepare the tissue for counting and the counting procedure are described elsewhere (Ryle, 1970).

RESULTS

Every ovary which was exposed to FSH contained more tritium than its corresponding paired control. The mean count for each group of five ovaries treated with FSH and the mean count for each untreated group are shown in Table II. The difference between the counts in the two ovaries from each mouse has been obtained, so also has the ratio of the count in the treated ovary to that in the untreated one. The means of these differences and of these ratios for every group of five mice are also presented in the table. The standard errors of all means are shown

TABLE II

Means and their standard errors. Counts/0.01 min. in FSH-stimulated and control ovaries. Differences and ratios between counts of paired stimulated and control ovaries.

Litter	Age (days)	+ FSH			– FSH			Diff.			Ratio		
		Mean	SE	SEx100/mean	Mean	SE	SEx100/mean	Mean	SE	SEx100/mean	Mean	SE	SEx100/mean
A	11	273	20	7.3	101	7	6.9	171	27	15.8	2.78	0.35	12.6
B	11	289	37	12.8	140	9	6.4	149	43	28.9	2.13	0.35	16.4
C	13	279	22	7.9	97	4	4.1	182	20	11.0	2.89	0.19	6.6
D	13	293	34	11.6	137	11	8.0	155	34	21.9	2.17	0.26	12.0
E	15	354	23	6.5	195	19	9.7	159	35	22.0	1.90	0.25	13.2
F	15	370	18	4.9	218	15	6.9	152	15	9.9	1.72	0.11	6.4

Counts/0.01 min.

and they are also expressed as percentages of their corresponding means.

An analysis of variance has been carried out on the figures in each column of Table II to test for effects due to the age of the mice. In only one case – that of mean counts in ovaries stimulated by FSH – were the differences between age groups significant. The stimulated ovaries of 15-day-old mice contained markedly more thymidine than did those of either 11- or 13-day-old mice ($P < 1$ per cent). Although one might expect differences in the basal rate of DNA synthesis associated with ovary size, the differences between the mean counts in unstimulated ovaries in fact just failed to reach the 5 per cent probability level (Table III). There was no significant effect of age either on the mean difference between pairs of treated and control ovaries or on the ratio between them. Nor was there any significant effect on the standard error of any of the means.

TABLE III

Analyses of Variance: Mean Counts/0.01 min. in FSH-treated and Control Ovaries.

		Mean squares	
Variation source	**d.f.**	**+ FSH**	**– FSH**
Age	2	4121*	5140
Residual	3	118	608

*Significant at 1% level

DISCUSSION

These results suggest no marked increase in the sensitivity of mouse ovaries to FSH between 11 and 15 days after birth. There is no consistent trend in the mean differences between treated and control ovaries with increasing age. The figures suggest that the ratio between the counts in treated ovaries and those in untreated ones may even decline during this period. On examining standard errors as percents of their corresponding means one finds no firm evidence for changing variability *within* litters, i.e. no indication of a phase of rapid alteration in the level of response.

On the other hand the mean uptake of thymidine in FSH-stimulated ovaries from all 11- and 13-day-old mice was remarkably constant and distinctly different from that in ovaries of 15-day-old mice. This is the only definite indication of any shift of response, but it does tend to confirm part of the impression gained from the preliminary experiments. It is not correlated with the weight of the mice: mice in Litter C were mostly heavier than those in either

Litter E or Litter F. The dose of FSH used was relatively large and not likely to be a limiting factor. If one assumes that all ovaries responded to the maximum of their capacity in the given conditions of culture, then there must be a rise of this capacity at about 14 days. (One might find increased within-litter variation at precisely that age). Despite this there was no significant effect of age on the mean difference between treated and control counts. This could have arisen if there had been proportional increases of uptake due to age (i.e. growth) in the unstimulated ovaries, but this was not the case. The absence of any other significant effects associated with the age of the mice appears rather to stem from inconsistencies between litters in the amount of thymidine taken up by the unstimulated ovaries of the younger mice.

The basal rate of DNA formation, indicated by thymidine uptake in unstimulated ovaries, possibly varies less between 15-day-old litters than between 13 or 11 day old ones. These variations are not due to the size of the mice, for at both 11 and 13 days the mean control uptake was lower in the ovaries from the larger mice. (Unfortunately it is not practicable to weigh the ovaries themselves before explantation). Possibly the effect of mouse hormone persists longer than that of human hormone. Possibly younger ovaries take longer to complete their response to it. Litters of the same age may differ at the 11 to 13 day period in the activity of their pituitaries rather than in the basal mitotic rate of their ovaries. If so, a longer interval between explantation and addition of labelled thymidine would reduce between-litter variations of uptake in control ovaries. This in turn would reduce between-litter variations in both the differences and the ratios between treated and control ovaries and facilitate the search for changes in ovarian response.

These results show that the measurement of thymidine uptake *in vitro,* using paired ovary controls, combined with careful mouse production, gives adequately reproducible results. Thymidine uptake was measured in 12 groups of only five ovaries each; in none did the standard error exceed 13 per cent of the corresponding mean and in nine groups it was less than 8 per cent. The method can of course give no information about the location of the newly formed DNA. It tells nothing about which cells are dividing nor the time between successive divisions. It could, however, supply a screening procedure, over a wider age range than that examined here, for identifying critical periods in which there occur changes in overall mitotic rates, including those due to altered sensitivity to gonadotrophins. Morphological and histochemical studies related to these periods might be particularly rewarding.

ACKNOWLEDGEMENTS

I am indebted to Dr A. C. Crooke for his constant encouragement, to Dr W. R. Butt for the FSH preparation and to Dr P. C. Leglise and Dr

Martinet (C.N.R.Z., Jouy-en-Josas, France) for the hypophysectomized sheep serum. Miss M. Wardill, Miss P. Connell and Miss M. Smith provided essential help. The work was supported by the Ford Foundation.

GENERAL DISCUSSION

PARTICIPANTS: BEN-OR AND ESHKOL.

1. The *in vitro* data correlate well with results of *in vivo* experiments. The change in response to FSH at 14 days coincides with various changes described in earlier papers. The shorter transit times of follicles at younger ages could account for the absence of significant variations in thymidine uptake with age in unstimulated ovaries.
2. TCA precipitable counts would be better than total counts, especially as the time of incorporation is so protracted.

REFERENCES

Ben-Or, S. (1963). *J. Embryol. exp. Morph.* **11**, 1.
Brown, P.S. (1955). *J. Endocr.* **13**, 159.
Butt, W.R., Crooke, A.C., Cunningham, F.J. & Wolf, A. (1963). *J. Endocr.* **25**, 541.
Parlow, A.F. (1961). In *Human Pituitary Gonadotropins,* p. 300. Ed. Albert, A. Springfield, Illinois: Thomas.
Ryle, M. (1969a). *J. Reprod. Fert.* **19**, 87.
Ryle, M. (1969b). *J. Reprod. Fert.* **19**, 349.
Ryle, M. (1969c). *J. Reprod. Fert.* **20**, 307.
Ryle, M. (1970). This volume, page 100.
Steelman, S.L. & Pohley, F.M. (1953). *Endocrinology,* **53**, 604.
Vantsis, J.T. & Wildy, P. (1962). *Virology,* **17**, 225.

Reactions of the Postnatal Ovary after Transplantation into Castrated Normal and Sterilised Rats

G. Mayer and A.J. Duluc

SUMMARY

One or two day old ovaries, grafted into castrated adult female rats, either normal or androgenized during postnatal life, and into castrated adult males, show neither follicular development nor corpora lutea, but a stimulation of the theca and interstitial gland. Ovaries aged 11 days show: (1) Intense stimulation of the follicular apparatus, without appearance of corpora lutea when transplanted into postnatally androgenized females, spayed when adult, or into castrated normal or oestrogenized males. (2) Intense luteinisation when transplanted into castrated adult normal females.

INTRODUCTION

Several studies have been made on the relationship between age and ovarian reactivity to gonadotrophins, particularly to various exogenous gonadotrophins (cf. Hisaw, 1947; Krohn, 1955; Picon,1956). One may explore reactivity to endogenous gonadotrophins on the other hand by grafting ovaries of different ages into hosts of the same species. Results obtained by these two methods argue in favour of progressive acquisition by the ovary of reactivity to gonadotrophins (Corey, 1928; Aron, 1933; Collip, 1935; Selye and Thompson, 1935; Saunders and Cole, 1936; Price and Ortiz, 1944).

We have re-examined this problem using rats, by transplanting the ovaries of new-born animals into adult castrated animals, some of which have been injected with gonadal hormones during the post-natal period. It is known that post-natal androgenisation of females prevents ovulation and the development of corpora lutea, and provokes the appearance of persistent oestrus. This syndrome is due to modifications in hypothalamo-pituitary functions, resulting in the disappearance of the cyclical discharge of LH by the hypophysis (cf. Pfeiffer, 1936; Mayer *et al.* 1965; Harris, 1964; Barraclough, 1967). In the male rat post-natal oestrogenisation produces important testicular lesions, in particular in the seminal tubules (Mayer *et al.* 1965; Duluc and Mayer, 1967). We have explored the reactivity of the ovaries, as a function of their age, to these different hormonal conditions.

MATERIALS AND METHODS

Ovaries for transplant have been taken from rats aged 1, 2 and 11 days. The host animals were female and male rats, aged 4 to 5 months, castrated some 40 days before transplantation. Some of them had been treated on the fourth day after birth by a single dose of steroids : 2.5 mg of testosterone propionate for the females and 500 μg of oestradiol benzoate for the males.

One ovary was transplanted under the capsule of the left kidney of the host. The grafted animals were isolated and kept at a constant temperature. They were sacrificed 17 to 20 days after the ovarian transplantation.

RESULTS

Transplants of 11-day ovaries. All the grafted ovaries were well developed, but their histological structure varied according to the hormonal conditions of the host animal. What is striking in the case of ovaries grafted into normal castrated female rats is the presence of numerous and voluminous corpora lutea. The vaginal epithelium was more or less cornified. The uterine mucosa was developed and the endometrial epithelium prismatic. The mammary gland was not developed. The ovaries grafted into the postnatally androgenized female rats and those grafted into the normal castrated males or into the postnatally oestrogenised castrated males, were similar. Corpora lutea were not found, but there were numerous follicles at different stages of maturation, as well as atretic follicles. The vaginal and endometrial mucosae were of the same type as in the preceding series. In certain regions of the uterine mucosa there were some areas of squamous metaplasia. When the ovary was grafted into males, the seminal vesicles of the hosts were not stimulated to any appreciable degree.

Transplants of 1- or 2-day ovaries. The results were similar, no matter what the host animal : normal spayed female, androgenized spayed female, or castrated male. The grafts took less well than the 11-day grafted ovaries. Their histological structure reveals little or no follicular development and corpora lutea are absent. The ovary contains essentially primordial and primary follicles. The stroma is abundant. and the theco-interstitial tissue is often highly developed. The vaginal epithelium is in general of a squamous stratified type, but sometimes it appears like that of a castrated animal. The uterine endometrium and the mammary gland remain atrophic and the seminal vesicles of the grafted males are not developed.

DISCUSSION

1. The hormonal state of the host animal is different according to the experimental conditions, judging from the reactions of the 11-day-old ovarian grafts. In the normal female rat, castration provokes a marked release of LH, as is shown by the formation of numerous corpora lutea in the graft. This does not seem to occur after castration in the female rat androgenized at birth, in the normal male rat, or in the male rat oestrogenized during the post-natal period. In these animals the action of FSH seems to be predominant, because the grafts contain only follicles and no corpora lutea.

2. The sensitivity of the ovary to gonadotrophic hormones varies with the age of the gonad. In grafts of 1 or 2 days of age, the follicles were not stimulated, since one finds in fact no secondary but only primordial and primary follicles. On the other hand the theco-interstitial tissue shows a precocious sensitivity to gonadotrophins liberated by the hypophysis of male or female type. These conclusions agree with results obtained after injecting hormones (Corey, 1928; Aron, 1933; Smith *et al.* 1934; Saunders and Cole, 1936; Price and Ortiz, 1944; Picon, 1956) (For normal development of the prepuberal ovary cf. Peters, 1969). The ovary of 11 days of age has acquired its sensitivity to gonadotrophins, since the follicles can evolve towards complete maturity and luteinization. What makes the follicular apparatus able to respond to gonadotrophins is not yet clear (cf. Hisaw, 1947). In this regard, it would be interesting to explore the reactivity of 1-or 2-day-old ovaries grafted into an adult after some days' culture *in vitro,* to discover whether this ability can be acquired outside the organism (Bousquet, 1969).

3. Stimulated ovaries secrete oestrogens, as is shown by the reactions of the vaginal mucosa. The secretion of androgens by the grafts has not been demonstrated by the techniques employed. (The seminal vesicles of the host animals were not stimulated.)

4. In interpreting the results the progressive establishment of feed-back regulating mechanisms between the transplanted ovary and the hypophysis of the host animal must be taken into consideration. The absence of luteinization of the ovaries grafted into the kidney in castrated males or castrated postnatally androgenized females contrasts with the results of Kempf (1950) and Desclin (1950), who report the presence of numerous corpora lutea in intrasplenic ovarian grafts in the castrated male rat. Likewise, the ovaries of females androgenized at birth are not capable of luteinization *in situ* but these same ovaries

grafted into the spleen of the same animal develop corpora lutea (Mayer and Thevenot-Duluc, 1962). This difference seems to be due to inhibition of LH release by the oestrogen produced in the ovary, which, in the case of an intrasplenic graft, is inactivated by the liver. In our observations, the absence of corpora lutea in ovaries grafted into males and castrated androgenized females, could be explained by the influence on the hypophysis of the hormones secreted by the graft.

GENERAL DISCUSSION

PARTICIPANTS: BEN-OR, LUNENFELD, PETERS, ODELL, GREEP, LLOYD, ESHKOL, DONOVAN AND STEGNER.

1. It was surprising that there was no immune response to homologous ovarian grafts beneath the kidney capsule in these heterozygous rats. In another strain intraocular grafts were essential if severe responses were to be avoided (Ben-Or).

2. The present paper agrees with another in this conference (Ben-Or, p.226) suggesting preferential sensitivity of the interstitial tissue to gonadotrophic stimulation in very young ovaries and, at a later age, a follicular tissue response.

3. It was suggested that 1- or 2-day grafts in non-androgenized female hosts might have developed follicles and corpora lutea if left in place longer. Prolonged oestrus in the presence of intra-ocular grafts has eventually been followed by the establishment of ovarian cycles, as though the various hormones ultimately adjusted to a normal balance.

4. The question was asked: were the corpora lutea in the grafted ovaries truly corpora lutea, i.e. had their ova been released? No trace of ovum or of zona pellucida was found in serial sections though what had happened to them was not known. If an adult ovary is grafted into the eye of an intact male the follicles luteinize without rupture. In other contexts the differentiation between true corpora lutea derived from ruptured follicles and luteinized unruptured follicles is important. A specific name for the latter is needed. The definition of a true corpus luteum should include absence of remnants of the zona pellucida, which persists and stains clearly with PAS, although the oocyte itself may rapidly disintegrate and disappear. Secretion of progesterone is not a satisfactory criterion; for example, in the highly PMS-HCG stimulated animal abnormally small follicles may luteinize and produce progesterone. Capillary invasion and other such diagnostic morphological features do not appear immediately.

5. The distinction between the two structures may reflect, under experimental conditions, differences in the amount and/or timing of LH secretion. In the normal animal it may reflect differences in the sensitivity of particular follicles. However, before any sort of gross response is evident follicles may be changing under hormonal influences

in ways requiring more subtle methods of detection. Completion of these preparations may be essential for a fully normal response to gonadotrophic stimulation. Cells may, for example, have to undergo a certain definite number of mitoses before they can respond to a specific stimulus in a fully appropriate manner.

REFERENCES

Aron, M. (1933). *Archs. Anat. Histol. Embryol.* **15**, 237-423.

Barraclough, C.A. (1936). *Rec. Prog. Horm. Res.* **22**, 503-539.

Barraclough, C.A. (1967). In *Neuroendocrinology*. Ed. Martini, L. & Ganong, W.F. Vol. 1, p. 62-99. New York: Academic Press.

Bousquet, J. (1969). Histophysiologie des gonades du Rat impubère. Etude expérimentale en culture organotypique et par la méthode des greffes. *Thèse Doct. Sciences Bordeaux,* **No. 245.**

Collip, J.B. (1935), *J. Am. med. Assoc.* **104,** 556-558.

Corey, E.L. (1928). *Anat. Rec.* **41,** (abstract).

Desclin, L. (1950). *Annls. Endocr.* **11,** 656-659.

Duluc, A.J. & Mayer, G. (1967). *C.r. Séanc. Soc. Biol.* **161,** 774-776.

Harris, G.W. (1964). *Endocrinology,* **75,** 629-648.

Hisaw, F.L. (1947). *Physiol. Rev.* **27,** 95-119.

Kempf, R. (1950). *Archs. Biol., Liège,* **61,** 501-594.

Krohn, P.L. (1955). *Ciba Fdn. Colloq. Ageing,* **1,** 141-161.

Krohn, P.L. (1962). In *The Ovary*. Ed. Zuckerman. S. Vol. 2, p. 435-462. New York: Academic Press.

Mayer, G. & Duluc, A.J. (1967). In *Postnatal Development of Phenotype,* Symposium Czechoslovak Academy of Sciences, Liblice, September 18-22 (in press).

Mayer, G. & Thevenot-Duluc, A.J. (1962). *C.r. Seanc. Soc. Biol.* **156,** 377-379.

Mayer, G., Thevenot-Duluc, A.J. & Burin, P. (1965). *Path. Biol., Paris,* **12,** 989-1002.

Peters, H. (1969). *Acta Endocr. Copenh.* **62,** 98-116.

Pfeiffer, C.A. (1936). *Am. J. Anat.* **58,** 195-225.

Picon, L. (1956). *Archs. Anat. Micros. Morphol. exp.* **45,** 311-341.

Price, D. & Ortiz, E. (1944). *Endocrinology,* **34,** 215-239.

Saunders, F.J. & Cole, H.H. (1936). *Proc. Soc. exp. Biol. Med.* **33,** 504-505.

Selye, H. & Thompson, D.L. (1935). *Proc. Soc. exp. Biol. Med.* **32,** 800-803.

Smith, P.E., Engle E. & Tyndale H.H. (1934). *Proc. Soc. exp. Biol. Med.* **31,** 745-746.

Part III
Interactions Between the Hypothalamus, the Hypophysis and the Ovary

Pituitary Gonadotrophins and Circulating LH in Immature Rats - A Comparison between Normal Females and Males and Females Treated with Testosterone in Neonatal Life

Judith Weisz and Michael Ferin

Our studies of the hypothalamic-pituitary-ovarian relationship in the prepuberal rat have been motivated primarily by our interest in the polycystic ovary syndrome; more specificially by the polycystic ovary syndrome as a manifestation of a neuroendocrine disorder. As is well known, female rats injected with testosterone propionate as neonates develop a syndrome, the most obvious components of which are sterility, follicular cystic ovaries and constant vaginal cornification (Harris, 1964; Barraclough, 1966). The putative primary site of action of the testosterone is the hypothalamus. The sterility, the polycystic ovaries and the constant vaginal cornification are late manifestations of the abnormality induced by the testosterone. They are not specific to the androgen-treated rat and can be caused by a variety of other aetiological factors (Takewaki, 1962; Everett, 1964). The common denominator in all of these situations in which polycystic ovaries develop is that some error is introduced into the hypothalamic-pituitary ovarian circuit (Lloyd and Weisz, 1966b).

To understand how the introduction of an error into the circuit might lead to cessation of ovulation we were interested in establishing in a chronological sequence the changes that occur in different components of the circuit following the administration of testosterone. In some of our studies we have included a comparison of the androgen-treated female with normal males as well as normal females because it has been suggested that the testosterone acts by 'masculinizing' the hypothalamic gonadotrophin regulating centres of the female. Such a comparison we think might be of value not only in understanding the pathology of the testosterone treated rat (TTR) but also the physiology of sexual maturation.

Bioassay of pituitary gonadotrophins. Our first studies involved the measurement of the LH and total gonadotrophins in the pituitaries of normal and testosterone treated females (Matsuyama *et al.* 1966). We have supplemented these data by measurements of LH, FSH and total gonadotrophins in the male, and FSH in the female. LH was measured by the ovarian ascorbic acid depletion method of Parlow (1961); FSH by the Steelman Pohley assay (1953); total gonadotrophins by the mouse uterine weight method (Segaloff, 1962). In each study a single batch of animals, born on the same day, was used. With respect to LH, the sequence of changes in the three groups was very similar (Figure 1). In all three the concentrations of LH first rose steeply to reach a maximum before puberty and then fell precipitously, reaching minimal values around the time of vaginal opening in the female and the time of appearance of spermatids in the lumen of the seminiferous tubules of the male. The differences between the three groups lay in the ages at which the maximum and minimum values were found and in the

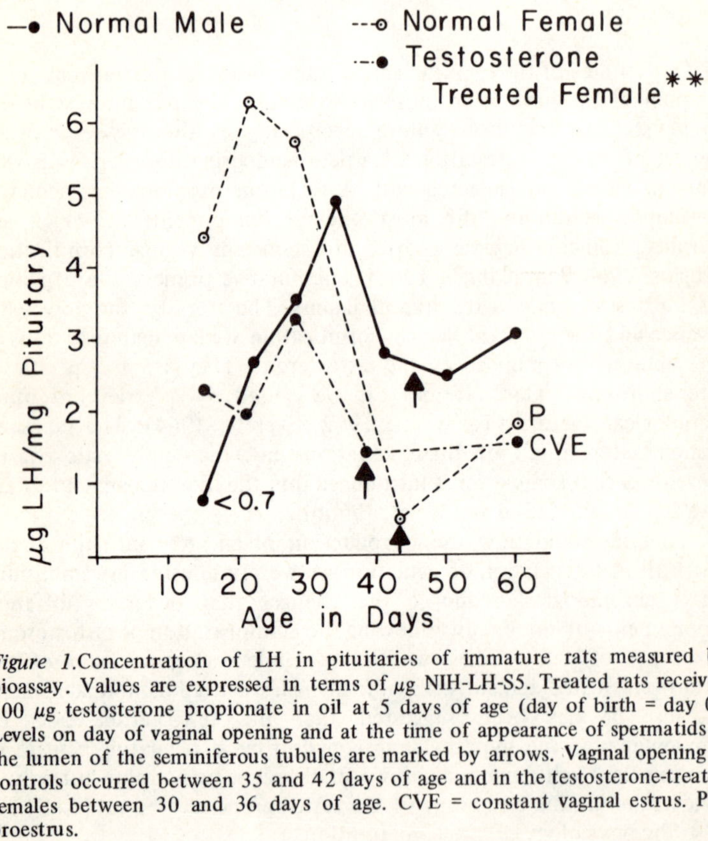

*Figure 1.*Concentration of LH in pituitaries of immature rats measured by bioassay. Values are expressed in terms of μg NIH-LH-S5. Treated rats received 100 μg testosterone propionate in oil at 5 days of age (day of birth = day 0). Levels on day of vaginal opening and at the time of appearance of spermatids in the lumen of the seminiferous tubules are marked by arrows. Vaginal opening in controls occurred between 35 and 42 days of age and in the testosterone-treated females between 30 and 36 days of age. CVE = constant vaginal estrus. P = proestrus.

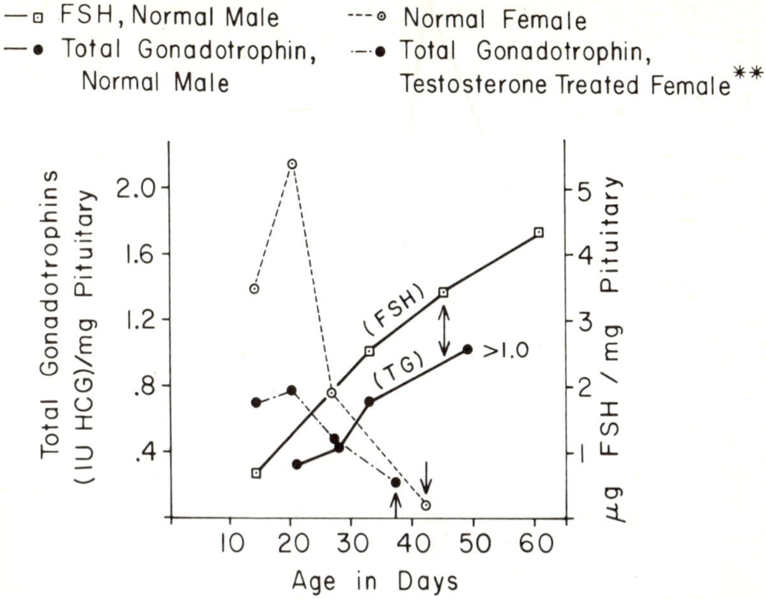

Figure 2. Concentration of total gonadotrophins in pituitaries of immature female and male rats and of FSH in the pituitaries of immature male rats measured by bioassay. FSH is expressed in terms of NIH-FSH-S-1. Values on the day of vaginal opening in the female and appearance of spermatids in the lumen of the seminiferous tubule in the male are marked by arrows.

absolute values. At its peak the LH was highest in the normal female, lowest in the female that had received testosterone propionate on the fifth day of life and intermediate between the two in the normal male. In constrast to LH, the patterns of pituitary FSH and of total gonadotrophins during sexual development were quite distinct in the two sexes (Figure 2). In the female they paralleled LH (Figure 3). In the male they continued to rise until 60 days of age (the oldest age group studied). At that time the values were similar to those reported by others in the adult male rat, suggesting that after 60 days of age a plateau is reached. Total gonadotrophins paralleled LH in the female and FSH in the male.

Bioassay data on immature rats reported by other workers are essentially in agreement with those reported here (Lauson *et al.* 1939; Corbin and Daniels, 1967; Kragt and Ganong, 1967; Lisk, 1968).

Radioimmunoassay of circulating LH. With the availability of radioimmunoassays for rat gonadotrophins it is now possible to measure blood levels of these hormones from individual animals. We are in the process of examining the changes in circulating gonadotrophins and correlating these with the changes in pituitary concentrations. Thus far we have measured the LH concentration in the serum of normal

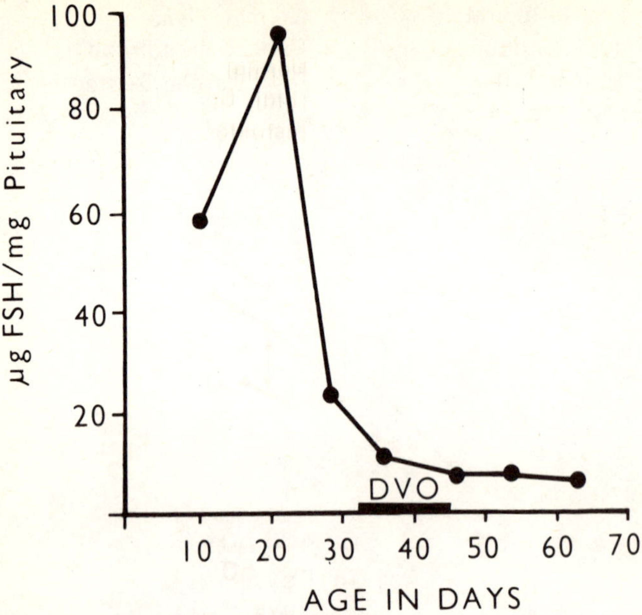

Figure 3. Concentration of FSH in pituitaries of normal immature female rats measured by bioassay and expressed in terms of μg NIH-LH-S-1. DVO = Day of vaginal opening. (A. Watanabe and E. Matsuyama, *unpublished data.)*

females, testosterone treated females and normal males from five days of age throughout the period of sexual maturation. Rat LH was measured by an adaptation of the technique developed by Odell *et al.* (1967) for human LH using an antibody (#15) supplied by Doctors Niswender and Midgley (Niswender *et al.* 1968). Rat pituitary LH preparations for standards were supplied by Dr Reichert, LER 1056 C2 for iodination and LER 1213 A for reference standard. Iodination with ^{125}I was carried out by the method of Greenwood (Greenwood *et al.* 1963). All animals were of the Sprague Dawley strain (Charles River CD) and were from a single batch born on the same day. Blood was collected in the morning under deep anaesthesia. The assay was carried out on bloods from individual animals except in the 5-day-old group in which bloods from two animals were pooled. The results are expressed in terms of ng LER 1213A, the standard used in the assay. We feel, as do some other investigators (Gay and Midgley, 1969) that until a suitable, uniform LH preparation becomes universally available it is probably better to view the results as patterns of change observed under a set of physiological conditions rather than as absolute values (Figures 4, 5 and 6).

In males the LH is detectable at five days of age, in females at seven days of age. We suspect that there are detectable levels in the female at

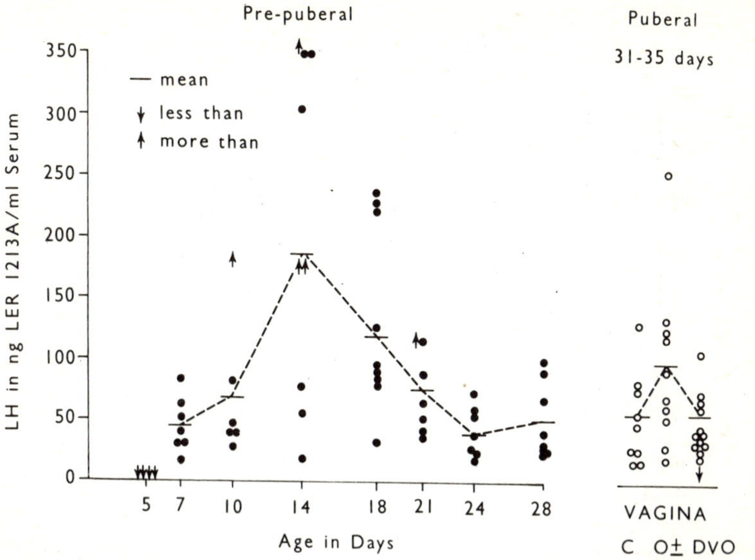

Figure 4. Concentration of LH in peripheral venous blood (serum) of immature female rats measured by radioimmunoassay and expressed in terms of ng reference standard LER 1213-A (15.83 ng LER, 1213-A = 1.0 ng NIH-LH-S-12 measured by radioimmunoassay). C = Vagina closed. O ± = Incipient vaginal opening. DVO = Day of vaginal opening.

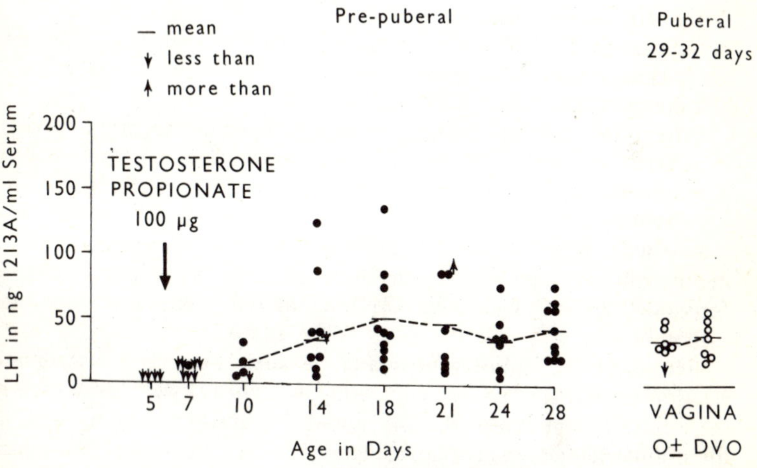

Figure 5. Concentration of LH in peripheral venous blood (serum) of immature female rats following an injection of 100 μg testosterone propionate in oil at 5 days of age. LH was measured by radioimmunoassay. The levels are expressed in terms of ng reference standard LER 1213-A. O ± = Incipient vaginal opening. DVO = Day of vaginal opening.

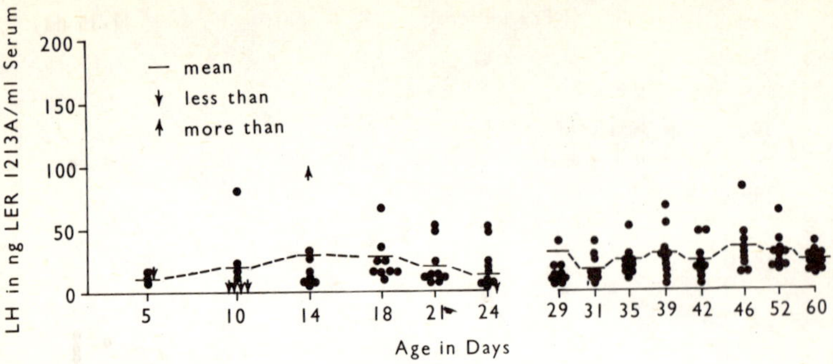

Figure 6. Concentration of LH in peripheral venous blood (serum) of male rats from 5 to 60 days of age. LH was measured by radioimmunoassay and the levels are expressed in terms of ng reference standard LER 1213-A. Spermatids were seen in the lumen of the seminiferous tubules around 45 days of age.

five days of age but that it would require pooling of blood from three to four animals.

In the normal female there was an increase after 7 days of age (Figure 4). Highest levels were seen in some of the animals in the 14-day-old group. Levels were also high in some of the 18-day-old rats. By 24 days of age the values were lower and similar to those in the 7-day-old group. Between 24 days and the time when signs of vaginal opening were first observed levels remained relatively low. In terms of the mean values there was a peak at 14 days. The individual differences in the 10, 14 and 18-day-old animals were striking. This may reflect differences in physiological age or stage of maturity at a time when rapid changes in LH levels are taking place. In animals of the same chronological age, in the fourth week of life the mean levels were lower and the individual variations were less.

After 28 days of age the animals were grouped not according to their chronological age but rather according to the stage of their vaginal development. There was no difference between rats over 28 days of age with vaginas closed and rats on the day of full vaginal opening. The group with incipient vaginal opening consisted of animals with a depression over the vagina. In some of these the vagina could be forceably opened. The mean LH level in this group was somewhat higher.

Serum LH in testosterone-treated females. Following the administration of testosterone, the circulating levels of LH were lower than in the normals in all except the 24-day-old rats, at which time the level in the normal female was also low (Figure 5). As in the normal, the values were higher in the 14, 18 and 21-day groups.

Serum LH in normal males. There was surprisingly little change in circulating LH between 5 and 60 days of age (Figure 6). An occasional animal had a higher level of LH but in the majority the values were very similar to those in the adult.

In spite of the similarity of the pattern of changes in LH in pituitaries of male and female rats before puberty, the patterns of circulating LH are strikingly different. In the female there is a peak at or just after 14 days of age, followed by a trough and then by another minor peak at the time of vaginal opening. In the male, essentially adult levels are reached by 10 days of age and these are maintained throughout sexual maturation. The correlation between stage of sexual maturation and gonadotrophins in the male may be restricted to FSH or perhaps LH/FSH ratios. In the female the period of sexual maturation would seem to have three phases:

Second and third week of life. The increase in circulating LH around 14 days of age occurs at a time when pituitary LH concentrations are rising. There have been indications from a number of different sources that around 14 days of age components of the hypothalamico-pituitary-ovarian circuit undergo rapid changes. At this time there is in the anterior pituitary a sudden increase in the number of basophils, designated as gonadotrophs (Siperstein *et al.* 1954). This was reported to be most pronounced between 11 and 14 days of age.

There is evidence for an increase in oestrogen content of ovaries at 14 days (Cieciorowska and Russfield, 1968). Uterine weight increases relative to body size between the fourteenth and eighteenth days and the effects of prior castration on uterine weight become evident at 14 days of age (Price, 1947). We can clearly recognize primary interstitial gland tissue in the ovaries of these rats by 7 days of age. It contains Δ^5-3β-ol steroid dehydrogenase and lipids stainable with oil red-O, indicating at least a potential for steroid synthesis. Thereafter the amount of interstitial tissue increases rapidly and with it the amounts of enzyme and lipids (Nadkarni, V., Weisz, J. and Lloyd, C.W., unpublished data). The development of primary interstitial gland tissue in our strain of rats follows closely that described by Dawson and McCabe in 1951. They found that the full range of staining characteristics for lipids and steroids is acquired by the primary interstitial gland tissue between 10 and 12 days of age. Significant antral development also begins at about this time.

The organization of the stalk-median eminence region along adult lines also becomes evident at about this time (Glydon, 1957; Floresheim and Rudko, 1968; Kobayashi *et al.* 1968). In the median eminence on the ninth to tenth days of life, loops from the supratuberal vessels, large enough to be recognizable by using large colloid particles, are seen penetrating the external layer forming the capillaries of the primary portal plexus (Glydon, 1957; Floresheim and Rudko, 1968).

In the adult, monoaminergic nerve terminals are seen in the

neurovascular contact region in the external layer of the median eminence and the stalk. These fibres, we believe, are in some way involved in the regulation of gonadotrophin release (Kuwabara *et al.* 1967). Some monoamine-containing fibres have been observed in the median eminence of the mouse by the seventh day of life (Bjorklund *et al.* 1968). Within the next two weeks they develop and by day 21 the full adult picture is seen. In the strain of rats used in our studies, we have observed by fluorescence microscopy numerous monoaminergic fluorescent fibres in these regions by 12 to 14 days of age.

Fourth week of life. The decrease in circulating LH after 21 days of age coincides with the period of high LH concentrations in the pituitary. During this period the concentrations of LH in the pituitary are higher than at any other time in the life of the animal. A great increase in basophilic granules in the cells designated as gonadotrophs has been described in the pituitary of female rats during the fourth week of life (Siperstein *et al.* 1954). This was found to reach its maximum at 28 days of age and to be greater than in the male. This period, then, seems to represent one of gonadotrophin storage prior to the first ovulation.

Puberty. The lack of any very dramatic changes around the time of vaginal opening is probably due to two factors; our use of vaginal opening as the basis of grouping the animals and, more fundamentally, the method of sampling.

Vaginal opening is clearly a very imprecise index of sexual maturation (Lloyd *et al.* 1946) and its exact relationship to the first ovulation is somewhat variable. The vaginal smears on that day vary. The mode of sampling is probably also not suitable for showing abrupt and transient changes in circulating levels of gonadotrophins such as we believe might occur when the first ovulation is triggered. Sequential determinations on individual animals would be required to reveal such events clearly. This should be possible by cannulation of the jugular vein, a technique we shall try to adapt to the puberal animal.

Testosterone treated rats. In the testosterone treated rats, the lower levels of circulating LH paralleled the lowered levels of pituitary gonadotrophins. The changes in LH in blood were seen by 7 days of age, within two days of the injection of testosterone. Abnormalities in the ovaries can also be recognized by histochemical techniques within two days of the testosterone injection. In the interstitial gland tissue there is a reduction in the enzymes reactions for Δ^5-3β-ol steroid dehydrogenase and in the amount of oil-red-0 stainable lipids (Nadkarni, Weisz and Lloyd, unpublished data). After 18 days of age the amounts and the distribution of these enzymes and of the lipids in any given structure are indistinguishable from the normal. The incorporation of labelled progesterone into labelled oestradiol and oestrone *in vitro* by the ovaries of testosterone-treated rat at 21 and 28 days of age is no different from that of the normal (Weisz and Lloyd, 1965). However ovarian weight in the testosterone treated rat is reduced from 7 days of age.

It is not possible to deduce anything from these observations about the nature of the change induced by testosterone in the central nervous system in neonatal life and little about causal relationships. The reduction in circulating LH seen two days after the injection of testosterone propionate could be one of the earliest expressions of a change in the course of differentiation of the gonadotrophin regulating centres. However, before accepting this interpretation, it will be necessary to establish how long the testosterone remains in circulation and that it does not, in the first instance, lower LH through its negative feedback effect. The metabolism of testosterone may be considerably slower in the immature rat than in the adult. This has been suggested to be the case for oestradiol (Wooley *et al.* 1969).

Feedback from ovaries must come into play at a very early age, as an important determinant of the level of gonadotrophin secretion (Donovan and van der Werff ten Bosch, 1965). The steroid output of the testosterone treated rat may well be quantitatively, if not qualitatively, different during the prepuberal period and before the development of polycystic ovaries. Reduction in the mass of functional ovarian tissue in the testosterone treated rat is evident from 7 days of age until vaginal opening. Therefore, the nature of the change in the central nervous system induced by the testosterone has to be studied in the gonadectomized animal. So far we have made two observations on castrated immature rats.

We have obtained evidence that the precocious vaginal opening is a direct effect of the testosterone on the central nervous system. At 4 days of age 20 female rats were gonadectomized. The next day half received 100 μg testosterone propionate. At 28 days of age they all received two ovarian transplants from intact litter mate controls. The vaginas of the testosterone treated rats opened between 30 and 32 days of age; those in the oil injected controls after 32 days of age, corresponding to the times of vaginal opening of intact controls.

The response of the gonadotrophin regulating centres to the withdrawal of all gonadal hormones can also be altered by the testosterone. This is evident in the immature but not in the mature animal (Lloyd and Weisz, 1966a). In the rat gonadectomized at 4 days of age, pituitary LH levels are higher than normal at 21 days of age. The injection of testosterone propionate on the fifth day of life suppresses this response. In the adult, 14 days after castration, pituitary LH concentrations were identical in the testosterone treated and normal rats.

The hypothalamic-pituitary-ovarian circuit in the immature rat. There can be little doubt that the hypothalamic-pituitary-ovarian circuit becomes established as a functional unit at a very early age. Gonadotrophins, steroid hormones and releasing factors can be elaborated and secreted by the very young. With the help of the newer micro-methods we should soon know the pattern of these three types of hormones throughout the period of sexual maturation. What initiates the functioning of the circuit remains an intriguing question. The timing of the onset of puberty is controlled primarily by the central nervous system and appears to involve a change in the sensitivity of the hypothalamus to gonadal hormones (Donovan and van der Werff ten Bosch, 1965). What is the mechanism which determines the time of onset of sexual maturation? In the female rat could this involve a change in the sensitivity of the ovaries to gonadotrophins? In this context, the question of the reported insensitivity of the ovary of the newborn rat to gonadotrophins deserves re-examination (Price and Ortiz, 1944; Donovan and van der Werff ten Bosch, 1965).

POSTSCRIPT

Since this manuscript was completed we have measured, by radioimmunoassay, the LH content of pituitaries of the rats whose blood LH levels are reported in this publication. In the normal females there was a sharp decrease in LH content between 14 and 18 days of age. The mean LH content on day 10 was 12.30 μg. It rose to 70.76 μg by day 14 and dropped to 5.10 μg on day 18 (values expressed in terms of μg 1213 A LH standard). The corresponding values in the testosterone treated rats were 5.37 μg, 11.60 μg and 15.52 μg respectively. Thereafter levels rose again reaching their highest value on day 21–95.8 μg in the controls and 45.1 μg in the testosterone-treated rats.

This suggests that the prepuberal LH peak in the female is normally biphasic. This was not noticed previously because samples were obtained from fewer age groups.

ACKNOWLEDGEMENTS

Dr Ferin was a Fellow, Training Program in the Physiology of Reproduction, supported by a grant from the Ford Foundation.

We wish to acknowledge the skillful and dedicated technical assistance of Mrs P. Gunsalus and of Miss A. Belpulsi. This work was supported in part by grant AM-08184 of the National Institute of Arthritis and Metabolic Diseases, National Institutes of Health, Bethesda, Maryland.

GENERAL DISCUSSION

PARTICIPANTS: LUNENFELD, PETERS, RYLE, ODELL, LLOYD, PRESL, BEN-OR AND DONOVAN.

1. Studies on another strain of rats showed that the 5-day-old ovary

contained no Δ^5-3β-hydroxysteroid dehydrogenase and did not respond to PMSG by releasing oestrogen. A significant rise in plasma oestrogen occurred, however, in the uninjected animal, at about 10 days, when antrum formation, differentiation of the theca interna, the formation of the enzyme and responsiveness to PMSG were also beginning. Subsequently there was a gradual increase in response to the same dose of PMSG, indicating increasing ovarian sensitivity. In the control animals plasma oestrogen concentrations then remained more or less constant up to and including the period of vaginal opening at about 50 days (Presl).

2. The decline in rat pituitary LH content (presumably due to release) between 14 and 18 days reported in the paper under discussion coincided with, and probably caused, the appearance of (a) Δ^5-3β-hydroxysteroid dehydrogenase in the abundant primary interstitial tissue, and (b) considerable amounts of oestrogen in the plasma. Vascularization of the theca begins at 15-18 days. Possibly the switch of steroid production to secondary interstitial tissue, theca and granulosa begins at this time. The mouse ovary contains more medium and large follicles at 14 to 18 days than at any other time and many of the larger ones are regressing and transforming to secondary interstitial tissue. At 13 or 14 days the mouse theca first shows positive lipid and Schultz histochemical reactions.

3. The particularly marked variability between individual rats with respect to (a) plasma LH concentration at 14 days and (b) pituitary LH content at 24 days suggests periods of rapid change with incomplete synchronization of individual developmental patterns. At 24 days there is a rapid increase in the storage granules of the pituitary basophils, believed to be the site of gonadotrophin production.

4. Experimental evidence indicates that the precocious puberty seen in androgen treated rats seems to be due to some direct imprinting on the nervous system; neither gonadotrophin levels nor ovarian structure are primarily responsible. There is much less variation with respect to the age of vaginal opening in the treated animals than in the controls.

5. It is probable that small cyclic LH fluctuations begin some time before puberty, the oscillations becoming progressively more marked. The smaller number of offspring characteristic of first litters in the mouse may indicate that fully mature cycles are still not yet established.

REFERENCES

Barraclough, C.A. (1966). *Recent Prog. Horm. Res.* **22,** 503.

Bjorklund, A., Enemar, A. & Falck, B. (1968). *Z. Zellforsch. mikrosk. Anat.* **89,** 590.

Cieciorowska, A. & Russfield, A.B. (1968). *Archs Path.* **85,** 658.

Corbin, A. & Daniels, E.L. (1967). *Neuroendocrinology,* **2,** 304.

Dawson, A.B. & McCabe, M. (1951). *J. Morph.* **88,** 543.

Donovan, B.T. & van der Werff ten Bosch, J.J. (1965). *Physiology of Puberty*, Baltimore: Williams & Wilkins Co.

Everett, J. (1964). *Physiol. Rev.* **44**, 374.

Floresheim, W.H. & Rudko, P. (1968). *Neuroendocrinology*, **3**, 89.

Gay, V.L. & Midgley, A.R. Jr. (1969). *Endocrinology*, **84**, 1359.

Glydon, R. St. J. (1957). *J. Anat.* **91**, 237.

Greenwood, F.C., Hunter, W.M. & Glover, J.C. (1963). *Biochem. J.* **89**, 114.

Harris, G.W. (1964). *Endocrinology*, **75**, 627.

Kobayashi, T., Kobayashi, T., Yamamoto, K., Kaibara, M. & Asika, K. (1968). *Endocr. jap.* **15**, 337.

Kragt, C.L. & Ganong, W.F. (1967). *Fed. Proc.* **26**, 534.

Kuwabara, T., Weisz, J. & Lloyd, C.W. (1967). *Program 49th Meeting Endocrine Society, Miami, Florida* (Abstract).

Lauson, H.D., Golden, J.B. & Severinghaus, E.L. (1939). *Am. J. Physiol.* **125**, 396.

Lisk, R.D. (1968). *Neuroendocrinology*, **3**, 18.

Lloyd, C.W., Rogers, W.F. & Williams, R.H. (1946). *Endocrinology*, **39**, 256.

Lloyd, C.W. & Weisz, J. (1966a). *Proc. Second Internat. Congr. Hormonal Steroids*, p. 917.Ed. Martini, L., Fraschini, F. & Motta, M. *Exerpta Med. Int. Congr. Ser.* No. 132.

Lloyd, C.W. & Weisz, J. (1966b). *Proc. Fifth World Congr. on Fertility and Sterility*, p.21. Ed. Westin, B. & Wiqvist, N. *Exerpta Med. Int. Congr. Ser.* No. 133.

Matsuyama, E., Weisz, J. & Lloyd, C.W. (1966). *Endocrinology*, **79**, 261.

Niswender, G.D., Midgley, A.R. Jr., Monroe, S.E. & Reichert, L.E. Jr. (1968). *Proc. Soc. exp. Biol. Med.* **128**, 807.

Odell, W.D., Ross, G.R. & Rayford, P.L. (1967). *J. clin. Invest.* **46**, 248.

Parlow, A.F. (1961). In *Human Pituitary Gonadotropins*, p. 300. Ed. Albert, A. Springfield, Illinois: Thomas.

Price, D. (1947). *Anat. Rec.* **97**, 519.

Price, D. & Ortiz, E. (1944). *Endocrinology*, **34**, 215.

Segaloff, A. (1962). In *Methods in Hormone Research*. Vol. II, p. 603. Ed. Dorfman, R.I. New York: Academic Press.

Siperstein, E., Nichols, C.W. Jr., Griesbach, W.E. & Chaikoff, I.L. (1954). *Anat. Rec.* **118**, 593.

Steelman, S.L. & Pohley, F.M. (1953). *Endocrinology*, **53**, 604.

Takewaki, K. (1962). *Experientia*, **18**, 578.

Weisz, J. & Lloyd, C.W. (1965). *Endocrinology*, **77**, 735.

Wooley, D.E., Holinka, C.E. & Timiras, P.S. (1969). *Endocrinology*, **84**, 157.

The Development of the Mouse Ovary after Testosterone Propionate Injection on Day 5

Hanna Peters, Ivan Nørbaek Sørensen, Anne Grete Byskov, Torben Pedersen and Torben Krarup

A single dose of androgen given early in life has a far reaching effect on the reproductive organs of rodents. Rats injected with testosterone propionate in the neonatal period have no oestrous cycle and their ovaries have no corpora lutea (for review of literature see Harris, 1964). The ovaries of androgenized animals contain follicles in various stages of development capable of ovulation and corpus luteum development after injection of pregnant mare serum gonadotrophin and human chorionic gonadotrophin (Brown-Grant *et al.*, 1964). Barraclough and Leathem (1954) tested the fertility of mice after injection of 1 mg of testosterone propionate on day 5 and found them to be sterile. The treated females were paired when they were 3 months old; their reproductive behaviour before this age was not investigated. The histology of ovaries of 40-day-old animals (35 days after androgen injection) showed an increase in follicular atresia, an increase in stromal connective tissue and the absence of corpora lutea (Barraclough, 1961). Self (1966) studied morphological changes in the ovaries of testosterone injected rats and reported a 60 per cent decrease in the granulosa cell component and a 200 per cent increase in the 'interstitial gland' component in 21-day-old animals. With increasing age (up to 6 months) the areas of the granulosa cell, interstitial gland and theca interna components increased by 300 to 400 per cent. A direct effect of the androgen on the ovary of 5 day old animals was suggested by Jacobsohn (1964) who stated that the growth of the organ as defined by its weight was impaired from the time the treatment was given.

In order to gain more insight into the early development of ovaries in mice after androgen treatment during the neonatal period, morphological investigations of ovarian development during infancy and the juvenile period were undertaken. Furthermore, it was thought to be of interest to determine whether follicle growth during the juvenile period after androgen treatment in infancy was normal or altered.

From the morphological studies we hoped to learn more about the intermediate developmental stages from the ovary of a 5-day-old mouse to that of the adult with follicular cysts, follicles in several stages of development, and no corpora lutea. We also wanted to investigate whether oocyte atresia might be influenced by the androgen treatment because, if androgen injection on day 5 induces an acyclic release of gonadotrophin (Barraclough, 1961), it is possible that oocyte atresia is accelerated. In that case the ovary might age earlier than in the normal animal, since a reduction in the oocyte population, whatever its cause, is accompanied by morphological changes leading to premature aging and abnormal morphological development (Russell and Fekete, 1958; Thung, 1958; Krarup, 1967; Peters, 1969a). The speed of follicle growth was examined because studies on a group of androgenized mice showed that although the majority conceived at least once, they produced their first litter at a considerably later age than the controls (unpublished data). This suggested that follicle development during the juvenile period might be slowed down in the androgenized animals.

METHODS

Five-day-old Bagg mice were injected subcutaneously with 1 mg testosterone propionate (TP). To study the morphological development of the ovary, groups of 3 to 10 animals were sacrificed at weekly intervals from 7 to 49 days. The ovaries were fixed in Bouin's aqueous solution and serially sectioned at 5 μ. Alternate sections were stained with haematoxylin-eosin and Heidenhain's azan. To test for lipids, oil red O and sudan black B stains were used on fresh frozen as well as formalin fixed material. To detect the activity of 3-β-hydroxysteroid dehydrogenase, dehydroepiandrosterone and DPN were used as substrates on freshly frozen tissue in the staining procedure.

The number of small oocytes (those with a cell diameter of less than 20 μ) was determined by the method described by Mandl and Zuckerman (1951). Differential counts of follicles in different stages of development were made according to the classification previously reported (Pedersen and Peters, 1968).

To judge follicle growth during the juvenile period, cell kinetic studies on androgenized animals at ages 7, 14 and 28 days were carried out. To determine the labelling index of follicles (i.e. the percentage of labelled granulosa cells in a follicle) two animals in each age group were injected with 25 μCi, 40 μCi or 100 μCi tritiated thymidine (^{3}HTdR) respectively and sacrificed one hour later. Autoradiographs of the ovaries were prepared by exposing for two weeks after dipping in Ilford K2 liquid emulsion.

To determine the proportion of labelled mitoses in follicles, 15 androgenized mice, 28 days old, received 100 μCi ^{3}HTdR. They were killed at intervals 1 to 16 hours after pulse labelling and auto-

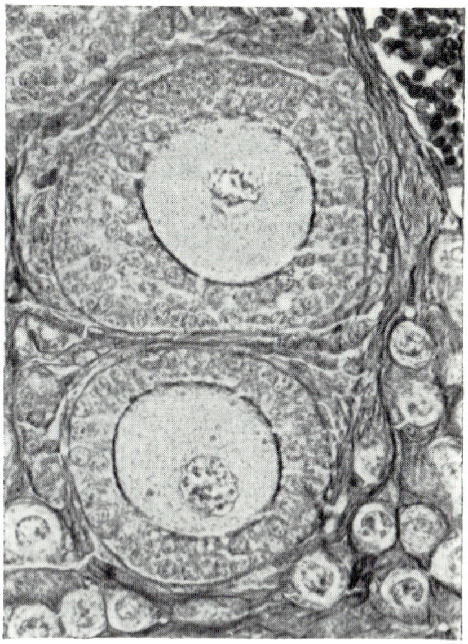

Figure 2. Ovary of a normal 14-day-old mouse. The granulosa layer is regular and surrounded by a well defined basement membrane (Azan stain)

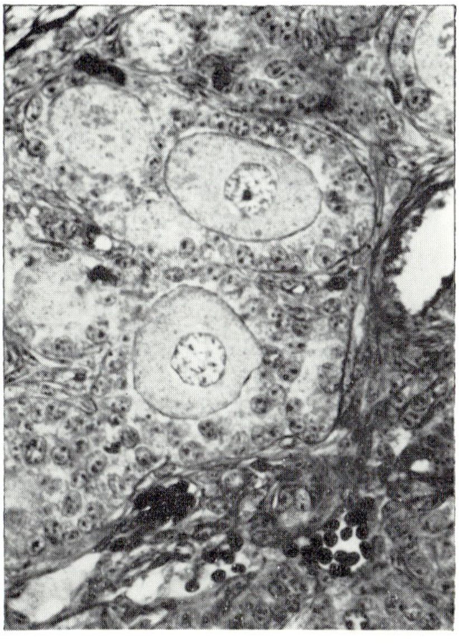

Figure 3. Ovary of a 10-day-old animal that had received 1 mg testosterone propionate on day 5. Follicles with irregular granulosa layer and faulty basement membrane are common

facing p. 352

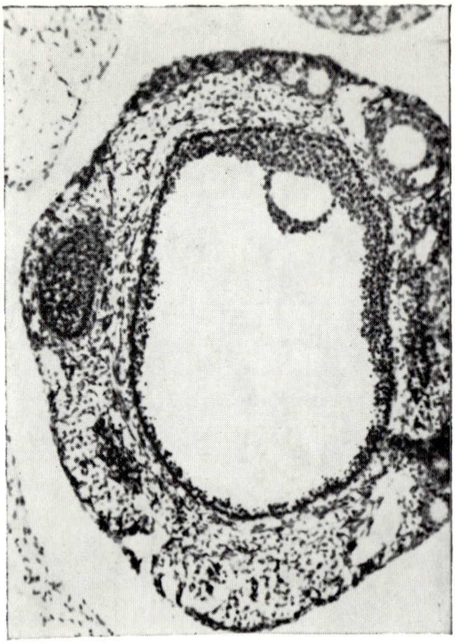

Figure 4. Ovary of a 35-day-old androgenized mouse. Large, cystic follicles with uneven walls are characteristic. The stroma is 'rarified'

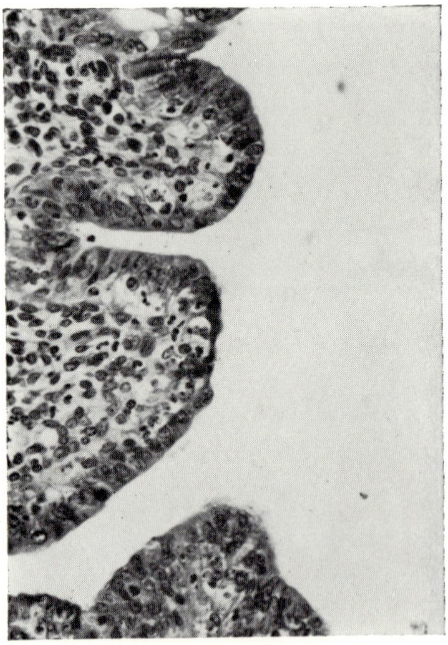

Figure 5. Uterus of a 42-day-old androgenized mouse. The endometrium shows many vacuoles and leucocyte invasion

radiographs on their ovaries were prepared. Curves of labelled mitoses for the different follicle types were determined. (For details of methods and principles in cell kinetic studies of follicles in immature mice see Pedersen, 1969).

RESULTS

Morphological studies. In order to follow the development of the ovaries in detail, quantitative as well as qualitative evaluations were made.

Quantitative evaluation. The number of small oocytes in normal Bagg animals between the ages of 1 week and 12 months has been determined and recorded (Figure 1). The small oocytes, the majority by

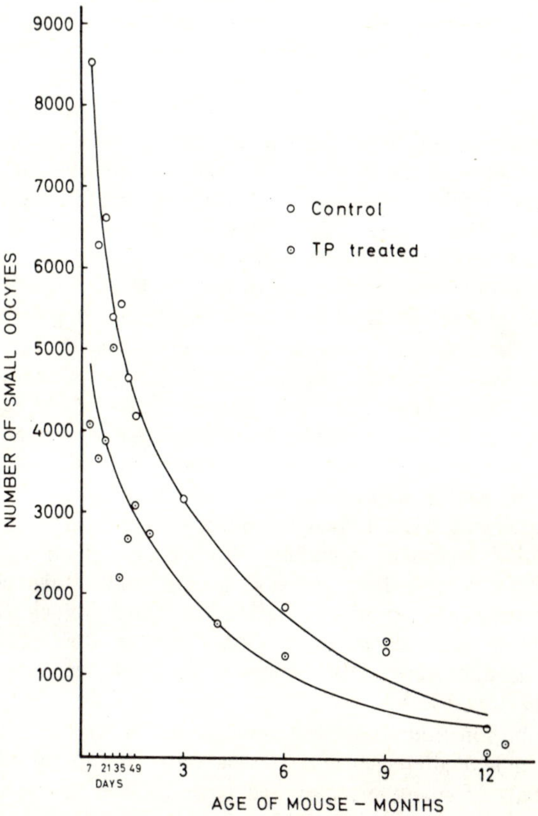

Figure 1. Number of small oocytes in relation to age in control and androgenized mice.

far of the total oocyte population (95 per cent at 1 month and 75 per cent at 12 months of age), represent the pool from which all subsequent follicles develop. TP treatment during the neonatal period reduces this pool immediately. Only about half the number of small oocytes are left in the ovary (4,000 small oocytes in TP mice as compared to 8,500 in normal mice) at the age of 7 days, i.e. two days after TP injection. The loss of small oocytes immediately after the injection of TP is considerable, however the atresia rate with advancing age is similar to that in the normal animals (note the slope of the curve in Figure 1).

Qualitative studies. The morphological development of the ovary was observed in ovaries from animals sacrificed at weekly intervals from 7 to 49 days.

The marked immediate reduction in the number of small oocytes, which becomes obvious when the oocytes are counted, is not always evident morphologically, but in some specimens examined two days after androgen injection the outer cortex seems less densely populated with small oocytes than normally and a histiocyte invasion may be present.

The morphological appearance of ovaries in the 7-, 10- and 14-day-old androgenized animals is not normal, but the abnormalities are subtle. With the ordinary haematoxylin-eosin stain the changes are hard to identify but with Heidenhain's azan stain, they become well defined. The arrangement of the follicle envelope seems faulty in many cases. In a normal one or two week old mouse the follicular envelope consists of cells which usually lie in a regular and orderly pattern around the oocyte (Figure 2). After androgen treatment many follicles are irregular and look disordered. The granulosa cells seem randomly placed. The contact between the oocyte and the granulosa cells may be incomplete, and large spaces and 'holes' in the granulosa layer are common. Furthermore the basement membrane of the granulosa layer may be faulty or even missing in some places between two adjacent follicles (Figure 3). In several cases two, three, and even four oocytes were found enclosed in a single follicle, a phenomenon which is most likely due to faulty basement membrane formation failing to enclose granulosa cells that 'belong' to one oocyte only. Bulging of the basement membrane and the formation of bizarre follicles are seen more frequently as well as for a longer time in the androgenized animals. In normal ovaries this phenomenon is rarely seen after the age of 10 days (Peters, 1969b).

The distribution of lipids, as seen in oil red and sudan black B stained sections, seems unaltered in the stroma cells, but increased in the granulosa cells of many medium and large follicles throughout the juvenile period. The reaction for 3-β-hydroxysteroid dehydrogenase was found to be severely altered during the infant and juvenile ages. In control animals activity of the dehydrogenase was present in all specimens tested between the ages of 6 days and 2 months. After TP treatment no activity of the enzyme could be detected before the age of 28 days. Already at 18 hours after treatment the dehydrogenase activity was absent from the ovary. In 28 day old animals enzyme activity seemed normal in distribution and amount.

The effect of hormonal imbalance on the ovary becomes clearly evident in 1-month-old androgenized animals. At that age the ovaries contain large (Type 7) follicles with walls of uneven thickness, which a week later (at the age of 35 days) have often already become follicle cysts (Figure 4). The stroma takes on a characteristic appearance: the nuclei, small compared to the large amount of cytoplasm, are eccentrically placed giving the impression of a 'rarified stroma'. Many (though not all) ovaries of animals between the ages of 28 and 42 days look abnormal and develop the three characteristics of bizarre follicles with abnormal basement membrane and theca development, the presence of follicle cysts and rarified stroma.

Some of the animals ovulate and develop corpora lutea.

The development of the uterus in the treated animals is of interest. In those showing abnormal follicle development and cysts the endometrium was abnormal. At the age of 21 days subendometrial oedema was noted. In animals aged one month or more the endometrium showed marked folding, metaplasia, vacuole formation in the endometrium and leucocytic infiltration and transmigration (Figure 5).

Follicle kinetics. In order to investigate whether androgen in infancy will influence the number of follicles which begin to grow during the juvenile period or their speed of growth, it was necessary to determine (1) the differential count of follicle types at different ages, (2) the growth rate of follicles expressed as their transit time and (3) the number of follicles which start their development as type 3b follicles per hour (inflow into type 3b).

Differential counts. The number of medium and large follicles in 7-, 14- and 28-day-old mice was determined (Table 1). No difference in the number and distribution of follicles was found in the TP treated and control mice on day 7 or 14. However, at the age of 28 days the ovaries of the TP treated animals contained only about half the number of medium and large follicles. The reduction in number is found not only among the large but also in all types of medium follicles.

Growth rate of follicles expressed as their transit times. The transit time of a follicle type is defined as the time it takes a certain type of follicle to develop from one containing the minimum number of cells to one containing the maximum number of cells within this type. This can be calculated from the doubling times of granulosa cells and the total number of granulosa cells in the different follicles.

The doubling time is calculated from the DNA synthesis (S) phase of

TABLE I

Differential counts of medium and large follicles in ovaries of androgenized (TP) and untreated mice

Age (days)		Medium and large follicles					
		type 3a	type 3b	type 4	type 5a	type 5b	type 6
7	TP-treated	442	90	–	–	–	–
	controls	375 ± 65	141 ± 45				
14	TP-treated	480	132	147	23	–	–
	controls	479 ± 19	155 ± 9	125 ± 0	31 ± 8	–	–
28	TP-treated	208 ± 24	98 ± 10	41 ± 5	33 ± 5	11 ± 2	1
	controls	504 ± 54	142 ± 20	94 ± 13	60 ± 7	22 ± 4	19 ± 9

the granulosa cells and the labelling index (LI) of the follicles (Pedersen, 1969).

In a previous investigation the S-phase of granulosa cells was determined in normal mice and found to be constant within the juvenile period for cells of follicles in the same stage of development. This is also the case after TP treatment. The curves of labelled mitoses (from which the duration of the S-phase can be read directly) of granulosa cells of treated animals are identical with the controls for type 3b follicles (S-phase = 9.8 hr.), for type 4 (S-phase = 11.2 hr.), and for type 5a follicles (S-phase = 10.3 hr.). In the large follicles (Type 5b and 6) the duration of the S-phase was somewhat longer in TP treated animals (about 8 hr.) than in controls (6.8 hr.). (The material is as yet too limited for statistical evaluation and it cannot therefore be decided whether this is significant or not). The labelling indices of follicles were determined at different ages after TP-treatment. They did not vary significantly from the controls.

The calculated transit times are listed in Table II. The transit times of the medium follicles are considerably longer than those of the large follicles in control as well as TP-treated mice. Furthermore, the transit times of the medium follicles in 7- and 14-day-old mice are shorter than in 28-day-old animals. But there is no difference between the transit times, i.e. the growth rate of follicles, in treated and untreated mice at the same ages.

Inflow into type 3b. The number of follicles which start their development as type 3b follicles at different ages during the infant and juvenile period in normal mice has previously been determined (Pedersen, 1969). (This number can be expressed as the number of type 3b follicles present in the ovary divided by the transit time of type 3b follicles). It was shown that in a 7-day-old mouse the number of follicles that begin their development is about twice as large (1.7 follicles per hr.) as in a 28-day-old mouse (0.9 follicles per hr.) (Table III). In TP-treated animals there are also more follicles which start their development at the end of the first week of life than at later ages; however there are fewer follicles that start their development after androgen treatment than in normal animals.

DISCUSSION AND CONCLUSIONS

The results suggest that testosterone propionate treatment given soon after birth already influences the development of the ovary during the infant and juvenile period. The altered morphological development seems to be due not only to the effect of the androgen on the central nervous system and the resulting altered release of gonadotrophins (Barraclough, 1961; Harris, 1964) but also to a direct effect of the androgen on the infant organ itself. The fact that the number of small

TABLE II

Transit times of follicles at different ages in the immature period in ovaries of mice treated with testosterone propionate (TP) and in untreated mice (hours).

		Medium follicles			Large follicles	
Age (days)		type 3b	type 4	type 5a	type 5b	type 6
7	TP-treated	92	–	–	–	–
	controls	83	–	–	–	–
14	TP-treated	119	94	78	–	–
	controls	119	91	67		
28	TP-treated	164	89	36	23	
	controls	167	126	56	23	14

TABLE III

The number of follicles beginning development as type 3b per hour at different ages in mice treated with testosterone propionate (TP) and in untreated mice.

	7 days	14 days	28 days
TP-treated	1.0	1.1	0.6
controls	1.7	1.3	0.9

oocytes is reduced 2 days after treatment by about half suggests an immediate destruction of the small oocytes. This reduction of the pool of small oocytes, from which all subsequent follicles develop, causes in itself changes in the subsequent development of the organ. The reduction of the pool is followed by a decrease in the number of follicles which start to grow; this is already recognizable on day 7 but more pronounced later in the infantile period. The speed with which these follicles grow is not influenced by the treatment. As the number of medium and large follicles in the ovary of two week old animals is not yet significantly altered, but is reduced to half in 1-month-old mice, it can be concluded that those follicles which had started to develop when the TP was injected, remained unaffected by the treatment; they continued to grow, and grew at a normal speed. However, TP had an influence on the number of follicles which subsequently started on their development. That TP might have a direct influence on the signal which is necessary for a follicle to begin its growth is unlikely. Rather the reduced size of the small oocyte pool in itself is most probably responsible for the fewer follicles that begin to grow (Krarup *et al.*, 1969). Thus the effect of TP on the quantitative development of the ovary in the first month of life is characterized by a direct effect, reducing the number of small oocytes, as a consequence of which fewer follicles start on their development, leaving the ovary of a 1-month-old androgenized mouse deficient in the number of hormone producing follicles.

The reduction of the oocyte pool is not the only change in ovarian development following TP treatment. The lack of the reaction for 3-β-hydroxysteroid dehydrogenase before the age of 28 days suggests an abnormal enzyme production which might lead to an altered steroid synthesis. The faulty formation of the basement membrane which surrounds the granulosa cell layer, as well as the abnormal arrangement and cohesion of the granulosa cells and the changed relationship between the granulosa cells and the oocyte in some of the developing follicles suggest a severe disturbance of intraovarian cellular relationships.

It is not yet possible to distinguish clearly between (1) the direct action of TP and the direct consequence of this action on the juvenile ovary and (2) the changes in the ovary caused by imbalance in the production and/or release of gonadotrophin initiated by the TP. At what time this imbalance begins to influence the juvenile ovary has not yet been determined. It is possible that the low number of hormone producing follicles in the 1-month-old androgenized mouse further influences the already altered gonadotrophin production and/or release.

GENERAL DISCUSSION

PARTICIPANTS: LUNENFELD, BRANDAU, PRESL, STEGNER, WEISZ, ODELL, LLOYD, MAYER AND RYLE.

1. Female rats injected with androgen at 5 days develop the constant oestrus syndrome but, if mated on the day of vaginal opening, will become pregnant and give birth to young, though litters tend to be small and labour prolonged. If repeated pregnancies are allowed they seem to be protected against the consequences of the androgen treatment, although the non-pregnant animal rapidly develops ovarian cysts and constant vaginal oestrus and is sterile. Mice of the Bagg strain, on the other hand, become pregnant but the earlier litters fail to implant. The first vaginal plug is seen around Day 35 but the first litter is born about Day 75, after a 20-day pregnancy, suggesting initial uterine inadequacy which is gradually overcome. It is not known whether androgenized female Bagg mice separated from males until 2 to 3 months old will, like rats, prove sterile (Peters).

2. Sensitivity to androgen is probably greater in animals less than 5 days old, but if they are injected at, for example, one day after birth the genitalia develop abnormally and they are therefore less useful experimentally.

3. The levels of those enzymes involved in steroid biosynthesis appear, histochemically, to be lower than normal in the ovaries of androgenized animals. However, the *in vitro* rate of conversion of progesterone to testosterone, androstenedione and oestrogens by such ovaries is higher than normal and *in vivo* plasma levels of at least testosterone tend to be high.

4. Because of its rapid effect it seems likely that testosterone acts directly on very small follicles to reduce the oocyte pool. A reduced pool size, and hence an altered follicle recruitment rate might influence absolute or relative production of different steroids and consequently act back on the pituitary. On the other hand, specific localized hypothalamic lesions can lead to ovarian changes like those due to

androgen. Procedures for disentangling primary from secondary effects of androgen injection are needed.

REFERENCES

Barraclough, C.A. (1961). *Endocrinology,* **68,** 62.

Barraclough, C.A. & Leathem, J.H. (1954). *Proc. Soc. exp. Biol. Med.* **85,** 673.

Brown-Grant, K., Quinn, D.L. & Zarrow, M.X. (1964). *Endocrinology,* **74,** 811.

Harris, G.W. (1964). *Endocrinology,* **75,** 627.

Jacobsohn, D. (1964). *Acta endocr. Copenh.* **45,** 402.

Krarup, T. (1967). *Acta path. microbiol. scand.* **70,** 241.

Krarup, T., Pedersen, T. & Faber, M. (1969). *Nature, Lond.* **224,** 187.

Mandl, A. & Zuckerman, S. (1951). *J. Endocr.* **7,** 190.

Pedersen, T. (1969). *Acta Endocr. Copenh.* **62,** 117.

Pedersen, T. & Peters, H. (1968). *J. Reprod. Fertil.* **17,** 555.

Peters, H. (1969a). *Advances in Reproductive Physiology.* Vol. 4. Logos Press.

Peters, H. (1969b). *Acta Endocr. Copenh.* **62,** 98.

Russell, E.S. & Fekete, E. (1958). *J. natn. Cancer Inst.* **25,** 365,

Self, L.W. (1966). Proceedings of the 1965 Laurentian Hormone Conference. *Recent Progr. Horm. Res.* **22,** 533.

Thung, P.J. (1958). *Ovaria van onde Muizen.* Thesis. University of Leyden.

Ultrastructural Changes in the Immature Mouse Ovary after Injection of Androgen in the Neonatal Period

H. -E. Stegner and M. Poelmann

Treatment of female rats and mice with androgen in the neonatal period leads to persistent defects of their reproductive function (for review of literature see Harris, 1964). At the age of 40 days the ovaries of androgenized animals show a reduced number of follicles and an increase of connective tissue, formation of follicle cysts and lack of corpora lutea. Most of the females injected on day 5 of postnatal life become sterile (Barraclough and Leathem, 1954). The Early Androgen Syndrome (EAS) is considered to be the result of a disturbance of the hypothalamic system followed by a derangement of the neural control of hypophyseal activities (Barraclough and Gorski, 1961; Takewaki, 1962; Harris, 1964; Swanson and van der Werff ten Bosch, 1964; Harris and Levine, 1965). In addition, the process may be complicated by local interaction of the androgen with different cellular components of the developing ovary. This is concluded from morphological changes in the ovaries of animals to be found as early as 18 hours after androgen treatment. Peters *et al.* (1970), using differential follicle counts, found an immediate reduction in the number of small oocytes. Their follow-up studies showed that the morphological appearance of the ovaries in the 7-, 10- and 14-day-old androgenized animals was not normal, but the abnormalities were subtle and hard to define.

In order to get a better insight into local morphological effects we have studied, by means of electron microscopy, the ovaries of 11- and 13-day-old mice injected on day 5 with 1 mg testosterone propionate (TP).

METHODS

Female mice aged 11 and 13 days from the Bagg strain, pretreated on day 5 with 1 mg testosterone propionate, were sacrificed. The ovaries were removed immediately and fixed in 4 per cent glutaraldehyde in 0.1 M phosphate buffer for 2 hours, washed in phosphate buffer containing 0.2 M sucrose, and postfixed in 1 per cent osmium tetroxide. The tissues were dehydrated in a graded series of acetone-water mixtures, passed through two changes of 100 per cent acetone (dried over $CuSO_4$) and embedded in Vestopal W.

Thin sections were cut on a Reichert OmU2 Ultra-Microtome and stained with lead citrate (Reynolds). The sections were examined and photographed with a Zeiss EM 9 electron microscope. For purposes of orientation and survey sections 0.5 to 1.0 μ in thickness were cut from the same block with a Porter-Blum Ultra-Microtome and stained with Toluidine Blue.

RESULTS

The abnormalities of the ovarian structure in androgenized mice already described in recent light microscope studies (Peters *et al.* 1970) are more pronounced at the submicroscopic level. Both the follicle apparatus and the interstitial cell system are affected by the TP treatment. The classification of oocytes and follicles of Pedersen and Peters (1968) is employed in the following description.

Follicle apparatus. Among the individual groups the medium sized follicles display the most obvious alterations. The changes can be defined as a loss of normal cell contacts and interrelationships within the granulosa epithelium. In normal medium sized follicles up to stage 5a the follicular envelope consists of cells regularly placed around the oocyte (Figure 1). As the number of concentric granulosa cell layers grows the intercellular spaces become extended but the cells remain connected by cytoplasmic processes. In the treated animals granulosa cell association is evidently disturbed in as early as stage 2 follicles. In several areas the desmosomal contacts are loosened and lakes of varying sizes are formed between the cells, filled by cloudy material which is presumably follicular fluid (Figure 2). The disturbance of cell contacts is not restricted to the follicle cell interconnections but extends to the follicle cell-oocyte relationship. Normally in stage 2 follicles the young follicle cells are closely adherent to the oocyte, the surface membranes lie parallel and are held by membrane interdigitations and zonulae adhaerentes. The disorganization of the granulosa epithelium in the treated animals is not accompanied by oocyte atresia. In nearly all cases of follicular disorder the oocytes are morphologically intact and do not exhibit any signs of regression as do the granulosa cell elements.

It cannot be decided whether the abnormal cell dissociation is due to a loss of physiological membrane coherence or to an extreme expansion of the intercellular spaces by an increased, premature secretion of follicular fluid. In the normal follicle there must be regional differences in the degree of cell coherence. Antrum formation does not commence normally either in the cells of the corona radiata or in the most peripheral layer of the growing follicle. This suggests that the cell interconnections are more stabilized between follicle cells of the corona radiata and between those of the basal layer of the granulosa epithelium. The abnormal cell dissociation in the treated animals is not confined to special areas but may arise anywhere in the follicle except the most peripheral layer of the granulosa epithelium, where at least the basal portions of the cells remain connected (Figure 3).

Histological findings suggested that the arrangement of the follicular envelope may be faulty and in some cases distended or disrupted (Peters *et al.* 1970). In our material we constantly found continuous electron microscopic basement membranes surrounding the follicles of all stages but the perifollicular spaces were often found to be broader than in follicles of the controls.

Interstitial system. The electron micrographs reveal abnormalities in the interstitial system both in the undifferentiated stromal cells and in the specialized interstitial cells which are presumably involved in steroid metabolism. The specialized interstitial cell is characterized by a typical assortment of organelles. Like the steroid producing cells of the adrenal cortex and the Leydig cells of the testis they contain big mitochondria of the tubular type, a highly developed endoplasmic system of smooth membranes and accumulations of lipid droplets (interstitial cells of testis: Fawcett and Burgos, 1960; Crabo, 1963; Cristensen, 1965. Steroid cells of adrenal cortex: Lever, 1956; Sabatini and Robertis, 1961. Interstitial cells of mouse ovary: Beltermann and Stegner, 1968). Completely differentiated interstitial cells of the ovary are not found in juvenile mice before day 7 of postnatal life. They are situated predominantly in the outer concentric layers of the theca folliculi and in clusters within the ovarian stroma. In the androgenized animals most of the interstitial cells and in some places many of the undifferentiated stromal cells are severely altered (Figure 4). The endoplasmic lamellae and smooth vesicles are extremely distended and ruptured. The outer membrane of the nuclear envelope is elevated and the nuclear material seems to be condensed. The lipid droplets show lamellar deformation. The mitochondria, which are obviously reduced in number, show different stages of degeneration. In some areas the degenerating stromal cells exhibit an accumulation of intracytoplasmic fibrillar material. In other regions, especially in the stroma cells underlying the surface epithelium of the ovary, the neogenesis of collagenous material seems to be stimulated.

DISCUSSION

The e.m. observations in the ovaries of 11 and 13 day old androgenized mice can be summarized as follows. Compared with controls both the granulosa epithelium and the interstitial cell system

Figures 1-4 are of sections from ovaries of 11-day-old mice. TP = treated with 1 mg testosterone propionate on day 5

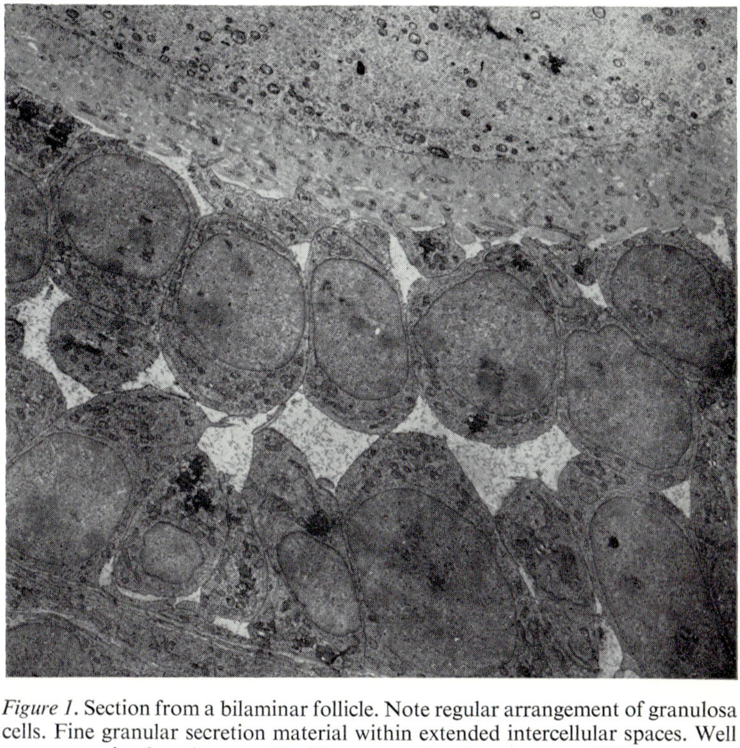

Figure 1. Section from a bilaminar follicle. Note regular arrangement of granulosa cells. Fine granular secretion material within extended intercellular spaces. Well developed zona pellucida surrounding the oocyte. ×3500

facing p. 364

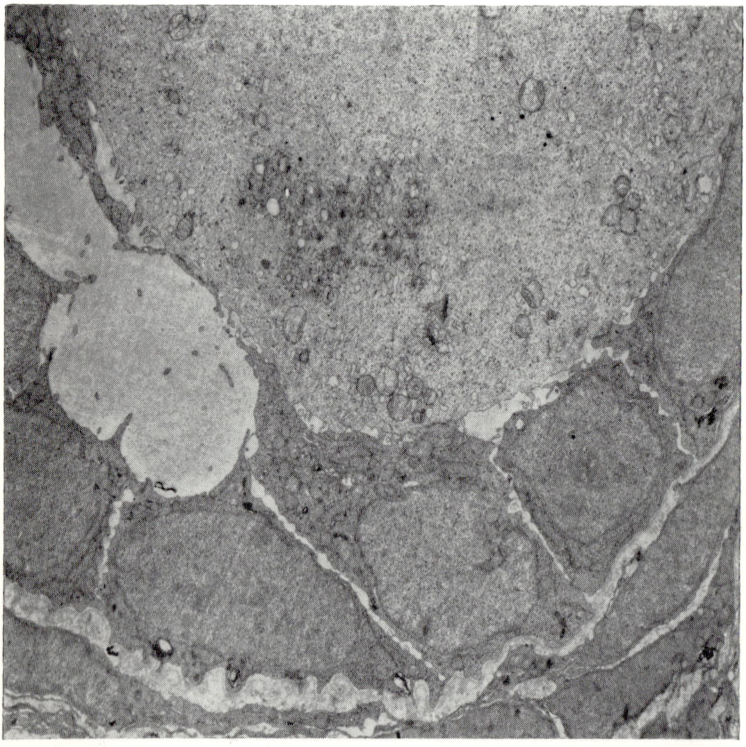

Figure 2. TP. Section from a type 3 follicle. Note disorder of cell interrelationship. The granulosa cells are retracted from the oocyte's surface membrane. 'Lakes' of various sizes are formed between the cells. No secretion of zona-like substance. ×6300

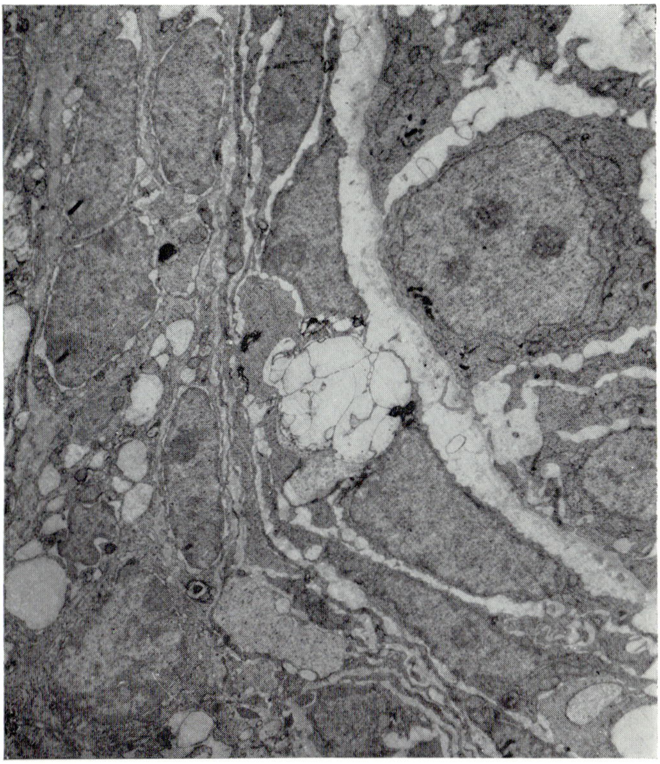

Figure 3. TP. Section from the basal layer of a bilaminar follicle. Note distended perifollicular space and marked degenerative changes of both thecal and stromal cells. Extension and rupture of endoplasmic vesicles. Lamellar disorganization of lipid droplets. ×6300

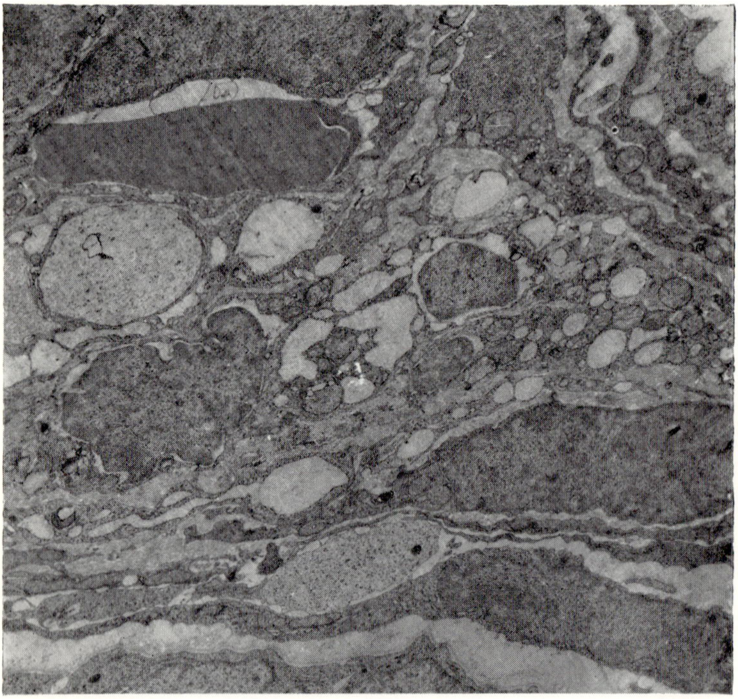

Figure 4. TP. Section from interstitial tissue. Note vacuolization of both stromal and interstitial cells. ×8820

facing p. 364

exhibit marked structural deviations. The most obvious local effects of TP treatment are (1) disturbance of the cell interconnections within the granulosa epithelium (2) degeneration of interstitial and of stromal cells (fibrillar degeneration) and (3) increase of collagenous material.

Histological findings suggest that the altered structural development of the ovary is due not only to the effect of the androgens on the hypothalamic centres but also to a direct action on the immature organ itself. Ultramicroscopical observations give further support to this suggestion. The severe alteration of the interstitial cell structure found by e.m. studies coincides with abnormalities of the cellular enzyme pattern. According to Peters *et al.* (1970) 3-β-hydroxysteroid-dehydrogenase, normally already present in 6-day-old specimens, is absent from the ovaries of androgenized animals as early as 18 hours after treatment. Raig *et al.* (1968) noted diminished activity of almost the entire set of enzymes in testosterone treated rats. Although the way in which mechanisms of local cellular interactions are influenced is still unknown, the results of several experiments indicate a direct effect of androgen on the immature ovary. The onset of local cellular destruction (oocyte atresia, interstitial cell degeneration) within 18 hours after androgen treatment and the dose-dependance of the consecutive structural abnormalities (see Swanson and van der Werff ten Bosch, 1964) can be interpreted more easily as a direct action of the androgen than as an indirect effect via hypothalamic-hypophysial control. Thus in our opinion the characteristic morphological and functional condition of the androgenized animal is a consequence of interacting central and gonadal disturbances in the developing juvenile organism.

ACKNOWLEDGEMENT

This work has been supported by a grant from the Deutsche Forschungsgemeinschaft.

GENERAL DISCUSSION

PARTICIPANTS: ODELL, WEISZ, ESHKOL, LLOYD, FRANCHI AND PETERS.

1. Changes occur in the granulosa and in the interstitial cells of androgenized animals. Changes also occur in both these tissues after anti-gonadotrophin serum treatment. However, the types of change are distinct. After antiserum treatment the granulosa cells are crowded and stuck together. In androgenized animals there is a loss of cell contact. In the latter interstitial cells develop but their shape and ultrastructure are abnormal. They show lamellar degeneration and contain more fibrillar material than those seen in antiserum treated animals. It is therefore unlikely that a large dose of testosterone acts by depressing

gonadotrophin secretion. It is not known if the changes seen at 11 and 13 days persist into the adult stage.

2. There are probably three important consequences arising from large doses of androgen at 5 days: (a) a direct action on the ovary; (b) an alteration in the hypothalamic regulation of gonadotrophin production, which may become effective very rapidly; and (c) the development of potentially normal but functionally disturbed adult ovaries producing an abnormal spectrum of steroids which may in turn provide an abnormal hypothalamic feedback. These same ovaries can change to normal structure and function when grafted into a normal environment. Trials with smaller doses of testosterone (10-100 μg is sufficient in the rat) might help sort out these components. Supplementary gonadotrophin might also help, but it must be recalled that, although steroids suppress gonadotrophin release in the castrate, they do not always do so in the intact animal. Oestrogen can stimulate LH secretion in normal women on a chronic basis and may also do so in prepubertal animals (Odell).

3. How testosterone might act directly on the oocyte is unknown. This could be studied *in vitro* or by direct androgen implantation into the ovary of a larger species. The ultrastructural picture does not indicate typical atresia or granulosa cell degeneration. The individual cells resemble the controls but their mutual relationships are disturbed. It is not known if testosterone directly harms the oocytes of older animals, when the hypothalamus is no longer susceptible, but some conversion to oestrogen would, in any case, be expected in these.

4. It is likely that in many species androgen sterilization can only be effected by injection of the foetus. The rat and mouse are exceptional in the persistence of hypothalamic susceptibility into the post-natal period.

5. The ovaries of androgenized mice show morphological features typical of the aging animal earlier than normal, probably in the same manner as do those of animals in which the oocyte pool has been depleted by other procedures (Peters).

REFERENCES

Barraclough, C.A. & Gorski, A.R. (1961). *Endocrinology,* **68,** 68.

Barraclough, C.A. & Leathem, J.H. (1954). *Proc. Soc. exp. Biol. Med.* **85,** 673.

Beltermann, R. & Stegner, H.-E. (1968). *Acta endocr. Copenh.* **57,** 279.

Christensen, A.J. (1965). *J. cell. Biol.* **26,** 911.

Crabo, B. (1963). *Z. Zellforsch. mikrosk. Anat.* **61,** 587.

Fawcett, D.W. & Burgos, M.H. (1960). *Am. J. Anat.* **107,** 245.

Harris, G.W. (1964). *Endocrinology,* **74,** 627.

Harris, G.W. & Levine, S. (1965). *J. Physiol., Lond.* **181,** 379.

Lever, J.D. (1956). *J. biophys. biochem. Cytol.* Suppl. **2,** 313.

Pedersen, T. & Peters, H. (1968). *J. Reprod. Fert.* **17,** 555.

Peters, H., Sørensen, J., Byskov, A.G., Pedersen, T. & Krarup, T. (1970). This volume, page **355.**

Raig, H., Brandau, H. & Luh, W. (1968). *Arch. Gynäk.* *206,* 154.
Sabatini, D.D. & Robertis, E.D.P. (1961). *J. biophys. biochem. Cytol.* **9**, 105.
Swanson, H.G. & van der Werff ten Bosch, J.J. (1962). *Acta endocr. Copenh.* **45**, 1.
Takewaki, K. (1962). *Experientia,* **18**, 1.

Radioactive Oestrogen Uptake by the Brain, Pituitary, Uterus and Skeletal Muscle of the Rat during Postnatal Development

Jiri Presl

ABSTRACT

Target tissues for oestrogen include the median eminence and the anterior hypothalamus. The uptake of ^{3}H-oestradiol was measured in the median eminence and in anterior and posterior portions of the hypothalamus and also in the pituitary, uterus, cerebral cortex, cerebellum and muscle of female rats ranging in age from 5 to 50 days.

Uptake per milligram by all brain tissues declined from the fifth or tenth day after birth, the greatest fall occurring between the tenth and fifteenth day. Subsequent values were relatively stable and low. The highest uptake coincided with the period of neuronal competence, during which oestrogen injections can induce the 'early oestrogen syndrome'. High uptake might have resulted from the absence of the haemato-encephalic barrier which has not yet developed at this age. However, as skeletal muscle showed a similar relationship between age and uptake, this was probably not the cause. In both brain and skeletal muscle the weight of tissue increases immediately postnatally without cell proliferation i.e. the number of cells per unit weight decreases. The parallel decline in oestrogen uptake by the two tissues could have arisen from this morphological change.

Both pituitary and uterus took up several times as much ^{3}H-oestradiol as the hypothalamus, even in 5 day old rats. They showed similar age variations, without, however, the permanent decrease in uptake of radioactivity following the fifth and/or tenth day after birth. Since the rat uterus grows intensively and continuously in response to ovarian oestrogen, from the second postnatal week, comparison between it and skeletal muscle was of decisive importance. In spite of rapid growth the uterus took up more ^{3}H-oestrogen per milligram than the skeletal muscle (or c.n.s.) and the rate of uptake did not remain low from the fifteenth day onwards. Therefore the pattern of oestrogen uptake in other tissues, especially skeletal muscle, probably simply reflected the high initial, and subsequently declining concentration in the blood and did not result from changes in cell number per unit weight.

From 15 days onwards the median eminence took up significantly more ^{3}H-oestradiol per mg than the cerebral cortex. The anterior hypothalamus took up significantly more from 20 days onwards. Thus (1) preferential uptake by the median eminence occurred only after the negative oestrogen feedback had begun, and (2) preferential uptake by the anterior hypothalamus coincided with establishment of the positive oestrogen feedback. This suggests separate development of the two feedback mechanisms. At no time did the posterior hypothalamus take up significantly more ^{3}H-oestradiol than the cerebral cortex.

The number of ganglionic cells in the rat c.n.s. is constant after birth. Therefore increases with age in the number of oestrogen binding sites per unit weight of median eminence and anterior hypothalamus, compared to the uptake of radioactivity by the cerebral cortex, probably arose from either (1) qualitative or quantitative changes in the binding sites themselves at the subcellular level, or (2) relatively slower development of tissue components other than the ganglionic cells in these areas.

Finally it should be noted that the 'early oestrogen syndrome', due to inhibition of the development of the positive feedback, is induced by oestrogen injections at a time when, judging from these results, no preferential uptake by the anterior hypothalamus is demonstrable.

A full report of this work has appeared elsewhere (Presl, 1970, Endocrinology **86**, 899).

GENERAL DISCUSSION

PARTICIPANTS: LUNENFELD, ESHKOL, LLOYD, DONOVAN, RYLE, WEISZ, PETERS, ODELL, AND FRANCHI.

1. The maximum variation between rats with respect to counts in the anterior hypothalamus occurred on day 20. This recalls the maximum variation in pituitary LH levels on Day 24 described earlier. Induction of endogenous LH release, e.g. by oestradiol injection, is first possible around 20 to 23 days, which seems to be a period of rapid change.

2. The use of labelled hormones involves some methodological problems. If, in reacting with the target site, the hormone molecule is destroyed, or at least the label is removed, one might find: (a) no radioactivity in the target organ in spite of a positive biological response; and (b) radioactivity elsewhere. Labelled oestrogen injected into a spayed cat accumulates rapidly in the hypothalamus, but the label disappears from this organ long before the consequent mating behaviour develops. Similar situations with much shorter time scales are conceivable.

On the other hand, if labelled oestrogen is injected into rats, the uterine counts remain high over a prolonged period although the initially marked content of non-target tissues declines rapidly. This is of importance for selecting the best time for comparisons (Presl). However, such a situation may not be true for other hormones and other target tissues.

The degree of vascularity could also affect the apparent content of an organ and probably varies with age.

3. Immature animals may metabolise steroids more slowly than adults and their livers may act differently on injected material. The infant rat's liver destroys oestradiol but its ovary produces some oestrogenic material not susceptible to such degradation. Thus the products which reach the hypothalamus need not be identical at different ages. Identification of labelled material in the target tissue, by extraction in the presence of excess carrier and recrystallization, is needed.

Studies of Hypothalamic–Pituitary–Gonadal Interrelations in Prepubertal Cattle

W. D. Odell, M. A. Hescox and C. A. Kiddy

INTRODUCTION

The mechanisms of transformation of the 'immature', infertile hypothalamic - pituitary - gonadal unit of the young animal into the 'mature', fertile unit of the adult animal are unknown. In 1929 Kallas reported that when two immature rats were united by parabiosis and one castrated, the other underwent precocious puberty. He postulated that pituitary secretions were under the reciprocal control of gonadal secretions prior to puberty in rats. Byrnes and Meyer (1951), studying the same parabiotic animal preparation, showed that tiny doses of oestrogens administered to the castrated immature parabiont could prevent the precocious puberty. Studies from our laboratory (Odell and Ross, 1966; Odell *et al.* 1967), and that of Bagshawe *et al.* (1966), revealed that, by radioimmunoassay, luteinizing hormone (LH) was detectable in serum from most children over the age of one year. However, prepubertal girls with ovarian agenesis had blood LH concentrations within the range found in normal children. Based upon the observations in rodents and children, we suggested in 1966 that a dynamic hypothalamic - pituitary - gonadal interrelationship exists prior to puberty in children. According to our hypothesis, a delicate active feedback exists between gonadal steroids and gonadotrophins in children. Castration would result in small elevations of circulating gonadotrophins, but they would not increase to the levels observed in castrated adults. All fluctuations would occur within a narrow range, lower than observed in adults. A series of publications designed to identify gonadotrophin in urine of children by bioassay and to support or refute this concept followed from our laboratories and those of others (Kulin *et al.* 1967; Rifkind *et al.* 1967; Johanson *et al.* 1969; Kulin *et al.* 1969; Raiti *et al.* 1969).

The difficulty of adequately studying these interrelationships in children led us to turn to other mammalian species for study. For several reasons, we decided to study these interrelationships in prepubertal cattle: (1) highly purified preparations of bovine gonadotrophins were available for radioiodination, (2) prepubertal cattle are sufficiently large that repeated, generous-sized blood samples may be obtained for assay of both steroidal and polypeptide hormones, (3) castration or other manipulations designed to perturb the postulated hypothalamic - pituitary - gonadal interrelations can be performed. We present herein our studies of LH in prepubertal cattle. A portion of these studies were presented in January, 1969 (Odell and Kiddy, 1969).

METHODS

A radioimmunoassay for bovine LH was developed using the procedures previously described from our laboratory. In brief, bovine LH (NIH-LH-B4) was used to prepare antiBLH in rabbits. All antisera were screened for titre, affinity, and specificity as previously described (Odell *et al.* 1969). One was selected for additional study. Studies of specificity were performed using (1) other purified pituitary hormones, (2) physiological measurements in cattle under conditions where LH changes were predictable, and (3) quantification of LH in several preparations of bovine origin by simultaneous bioassay and radioimmunoassay. The bioassay procedure was the ventral prostate weight increment in hypophysectomized immature male rats (Greep *et al.* 1941). The procedure for radioimmunoassay of bovine LH was as we have previously described for human LH and FSH, (Odell *et al.* 1967; 1968). All samples from a single animal were determined in a single assay.

Serum samples were obtained by needle puncture of the jugular vein, usually between 9.00 and 10.00 a.m. NIH-LH-B5 was used as a reference preparation in all assays. The mean intra-assay coefficient of variation was 6 per cent; the interassay coefficient of variation was considerably greater. Thus, relative values of samples compared in a single assay were reliably determined, but absolute values may be in considerable error. These statistical evaluations of the BLH radioimmunoassay are considerably worse than other assays we have reported. For example, human LH, TSH, and FSH have intra-assay coefficients of variation of 2.5 per cent and interassay coefficients of variation of 18 per cent. The reasons for this difference are not readily apparent and considerable efforts to improve precision in the BLH assay are in progress.

The first oestrus cycle in heifers usually occurs at 12 to 13 months of age; sexual maturation is usually considered complete in both sexes by this age. We began studies of prepubertal animals therefore at 1 to 5 months of age.

RESULTS

Figure 1 depicts a dose-response curve for bovine LH in the

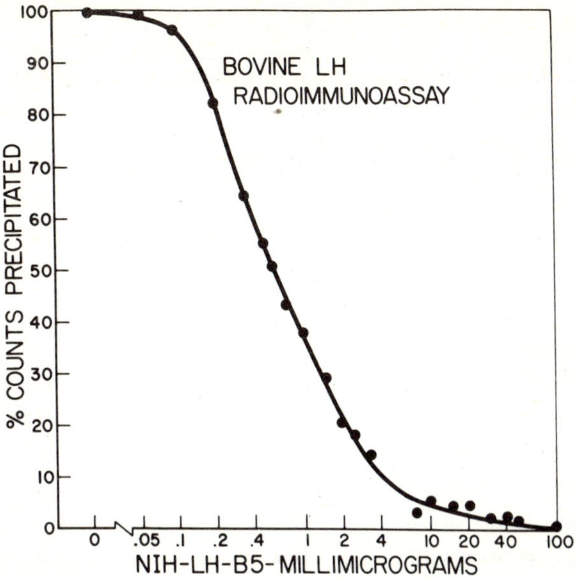

Figure 1. Dose-response curve for NIH-LH-B5 in the radioimmunoassay for bovine LH. Useful portions of the dose-response curve range from 0.2 to 3.0 mμg in one ml of incubation mixture, which corresponds to 0.5 to 7.5 mμg/ml of serum if 400λ of serum were assayed. As we have previously described, the percent counts precipitated are defined as 100 percent in control tubes containing no added unlabelled hormone.

radioimmunoassay. Table 1 presents bioassay and immunoassay potency estimates for several purified and non-purified bovine hormone preparations. Potency estimates by bioassay did not differ significantly from those obtained by immunoassay. Figure 2 illustrates the fluctuations in LH during an oestrus cycle in a normal heifer in which the ovulation time was determined by rectal palpation. This figure is illustrative of a large series of such studies reported separately and designed to determine the time between the LH ovulatory peak and ovulation (Kiddy and Odell, 1969). LH concentrations were relatively low and constant prior to the ovulatory peak, behavioral oestrus usually coincided with the LH ovulatory peak, and ovulation occurred at an average of 21.9 hours following the peak. LH concentrations following ovulation were generally lower and more constant than those prior to ovulation. These studies assist in supporting the specificity of the bovine LH assay used in the remaining studies reported herein.

Daily blood samples were obtained from 5-month-old cattle, three female and two male. At least 10 samples were obtained prior to and 10 or more after castration. The mean and 95 per cent confidence limits of the mean were calculated pre- and post-castration. Figure 3 depicts a study from one of these animals. There was considerable day-to-day

TABLE I

Potency estimates of bovine LH as determined by bioassay√[1] and immunoassay

Preparation	Bioassay μg/ml	Immunoassay μg/mg
BGH–B–13	1.7	5.6
BLTH–PB–2	<0.4	0.2
BTSH (Condliffe) √[3]	290	300
Pituitary preparation #1	259	348
Pituitary preparation #2	307	170
Pituitary preparation #3	127	196
Pituitary preparation #4	149	144
Pituitary preparation #5	204	107
Pituitary preparation #6	393	239
Pituitary preparation #7	216	261
Pituitary preparation #8	99	239

Immunoassay/Bioassay √[2] = 1.14
(0.59 – 1.69)

√[1] Prostate weight in immature, hypophysectomized male rats

√[2] For pituitary preparations

√[3] Containing 20 usp units per mg of TSH

(Table I reproduced from Kiddy, C.A., & Odell, W.D. (1969).

variability in LH both prior to and after castration. Mean LH for 12 days prior to castration was 4.2 ± 1.1 (standard error of the mean) millimicrograms per millilitre (mμg/ml). Mean LH calculated for the 20 days immediately following castration was 8.9 ± 1.0 mμg/ml. In all five animals, mean LH following castration was higher than mean LH prior to castration. Figure 4 depicts these changes.

Similar studies were then performed in 11 additional cattle ranging in age from one to three months. Figures 5 to 10 illustrate these studies. In all but one of these studies, LH concentrations after castration were higher than those prior to castration. Figure 11 shows mean pre- and post- castration concentrations in all 11 animals.

The marked variation from day to day within each animal (Figures 5 to 10) made it difficult to compare responses to castration at various ages. In addition, interassay variability made subtle evaluation of absolute BLH concentrations difficult. Random single day samples were obtained from adult and pre-pubertal male and female animals. These

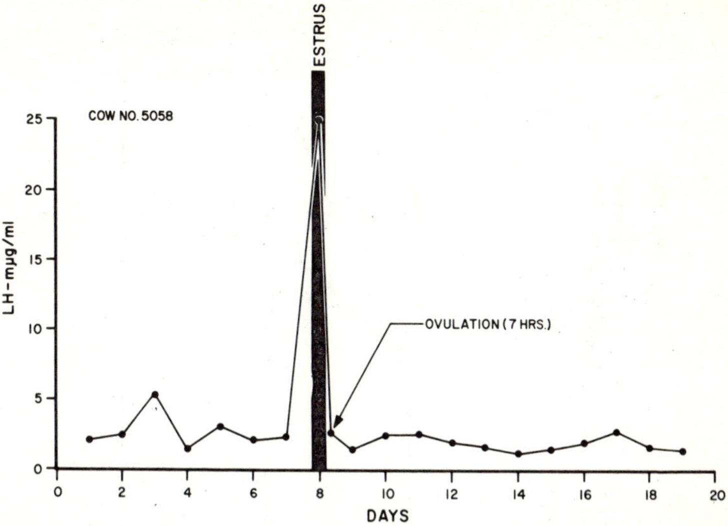

Figure 2. Serum LH measured daily during the oestrus cycle of a sexually mature cow. Note that LH concentrations appear slightly higher prior to the ovulatory peak than they do following this peak. Time of ovulation was determined by rectal palpation and in this animal occurred seven hours after the LH peak. Time of behavioral oestrus is shown in the black bar and coincided with the elevation of serum LH. This sequence is typical of a series of more than 20 such studies performed to determine the time interval between LH ovulatory peak and ovulation. The studies are being reported separately.

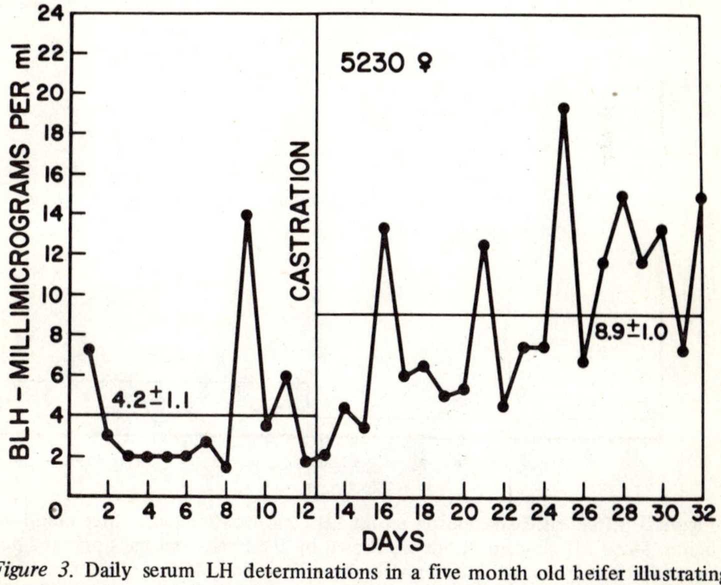

Figure 3. Daily serum LH determinations in a five month old heifer illustrating the day-to-day variation observed and the increase in mean LH following castration ($P < 0.05$).

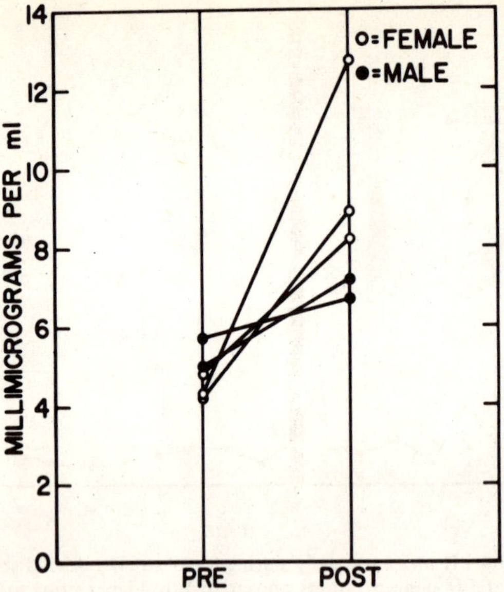

Figure 4. Mean pre and post castration LH concentrations in three female and two male cattle at five months of age. Each point represents the mean of at least 10 consecutive daily determinations. Mean LH increased in all five animals after castration; this increase was significantly different from precastration concentrations in all but one animal ($P < 0.05$, using a paired 't' test). All samples were determined in a single assay.

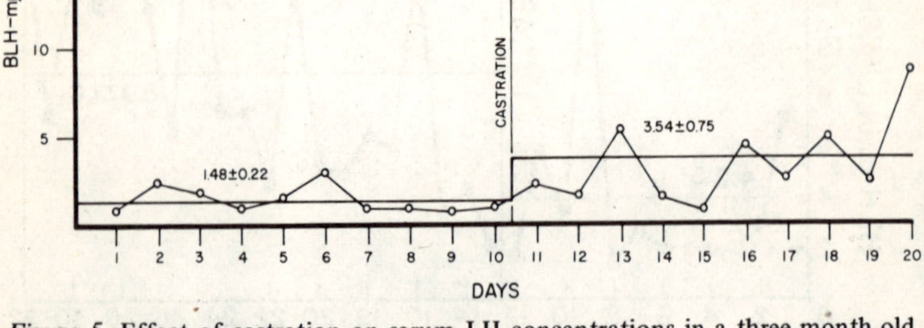

Figure 5. Effect of castration on serum LH concentrations in a three month old heifer. Mean LH concentrations are shown by the horizontal lines pre- and post-castration and mean ± SE are printed above these lines. Mean LH after castration is higher than before castration ($P < 0.05$).

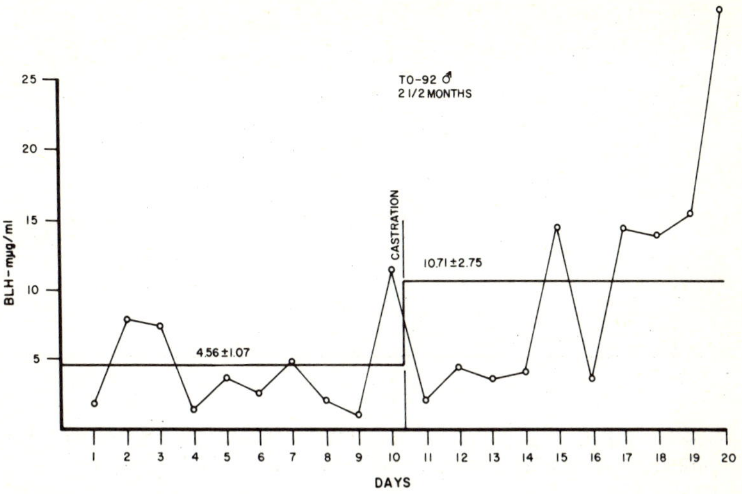

Figure 6. A study identical to that shown in Figure 5 performed in a 2½-month-old male. Mean LH is greater after castration than prior to castration ($P < 0.05$).

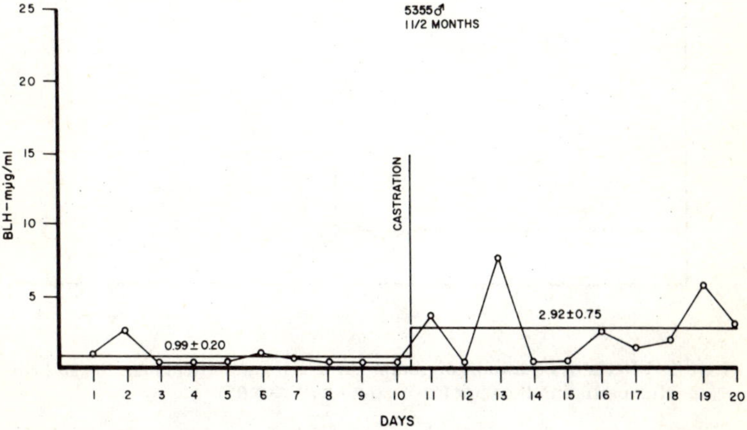

Figure 7. Effects of castration on serum LH concentrations in a 1½-month-old male. Mean LH is significantly greater after castration than prior to castration ($P < 0.05$).

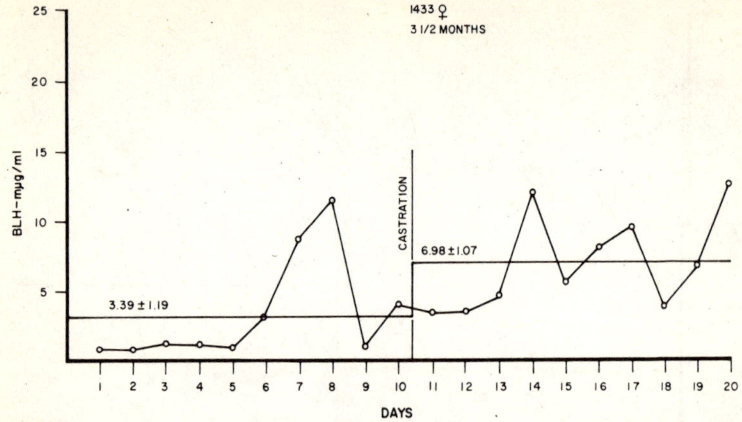

Figure 8. Effects of castration on serum LH in a 3½-month-old heifer. Mean LH is significantly greater after castration than prior to castration ($P < 0.05$).

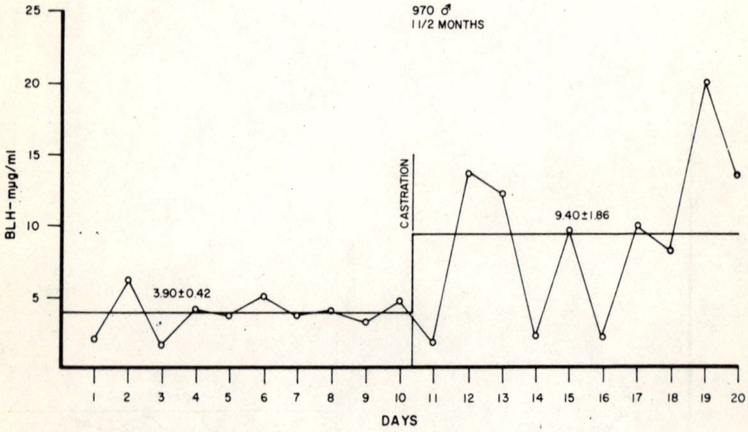

Figure 9. Effects of castration on serum LH in a 1½-month-old male. Mean LH is greater after castration than prior to castration ($P < 0.05$).

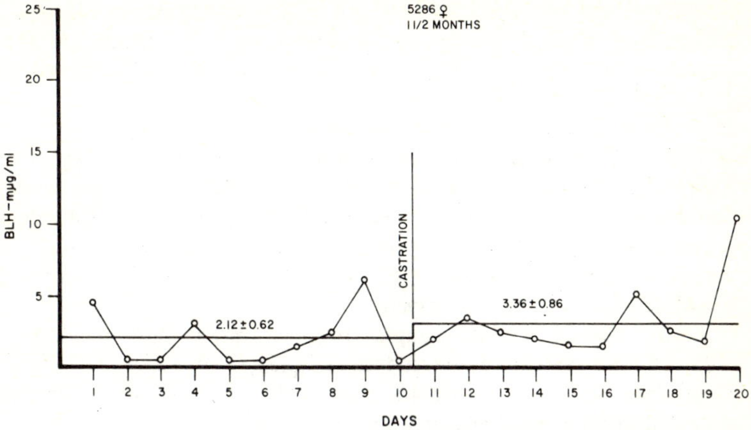

Figure 10. Effects of castration on serum LH in a 1½-month-old heifer. The mean LH before and after castration was not significantly different (P > 0.05).

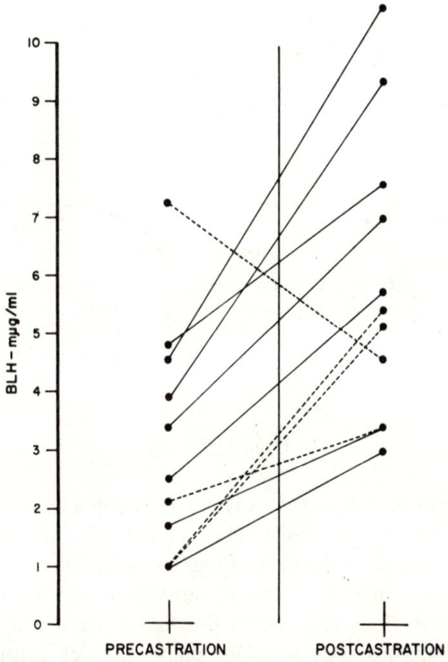

Figure 11. Mean serum LH concentrations calculated from 10 or more consecutive daily samples in 11 cattle before and after castration performed at 1½ to 3½ months of age. In all but one study, LH concentrations increased after castration. Using a paired 't' test, the increase in LH for the group of 11 animals was significant (P <0.05). (.———. = males, .-------. = females).

samples were assayed in one assay and the data are shown in Figure 12. Although a small tendency to show increased concentrations with increasing age may be suggested, the changes are not significant and we are unable to state that there are differences between BLH concentrations in adult and prepubertal cattle.

Four of the animals that were castrated at 3 to 4 months of age were followed until 13 to 15 months of age by obtaining weekly serum samples to see if a second clearly-definable rise of gonadotrophin might occur at the age when sexual maturity would normally occur. The data for three are shown in Figures 13 to 15. A second rise in gonadotrophin was not observed.

Lastly, several random samples from castrated adult males were obtained and compared to random samples from eugonadal adults. These data, shown in Figure 16, show a similar magnitude of change after castration in adult and prepubertal animals.

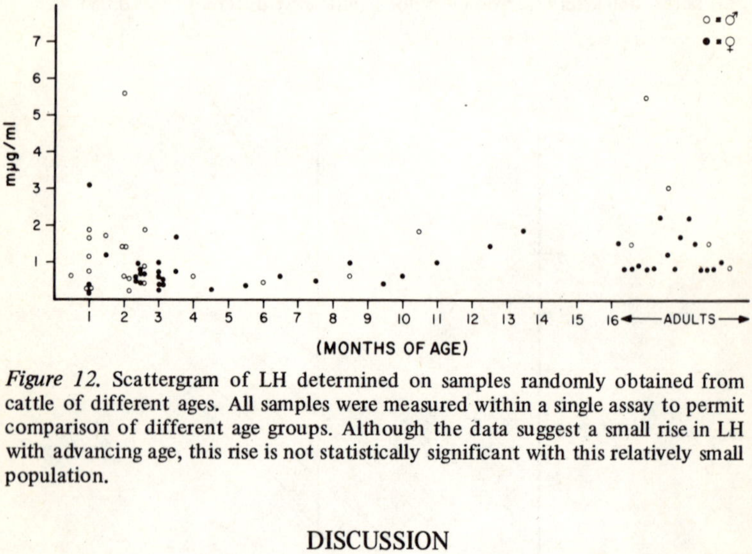

Figure 12. Scattergram of LH determined on samples randomly obtained from cattle of different ages. All samples were measured within a single assay to permit comparison of different age groups. Although the data suggest a small rise in LH with advancing age, this rise is not statistically significant with this relatively small population.

DISCUSSION

The radioimmunoassay for bovine LH is capable of quantifying LH in serum from cattle of both sexes of all ages. Specificity has been shown by correlation between bioassay and immunoassay potency estimates, by immunological studies, and by demonstrating appropriate changes in LH in the adult female during the oestrus cycle. Using this assay and measuring 10 or more daily samples prior to and after castration, we have shown that castration of cattle of both sexes at ages of one to five months results in a significant elevation in serum LH. Considerable variation in day-to-day samples occurred, making a comparison of gonadal status based on single determinations difficult.

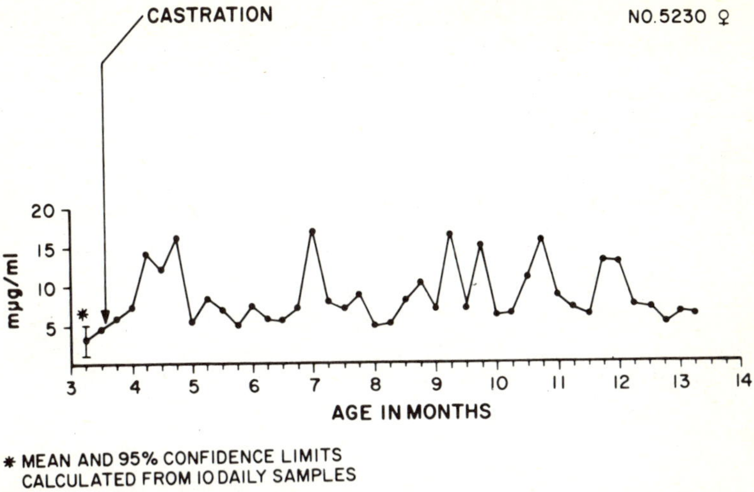

Figure 13. Weekly measurements of LH in a cow castrated at 3-4 months of age and studied through the age of sexual maturation. The pre-castration mean LH and 95 per cent confidence limits are shown.

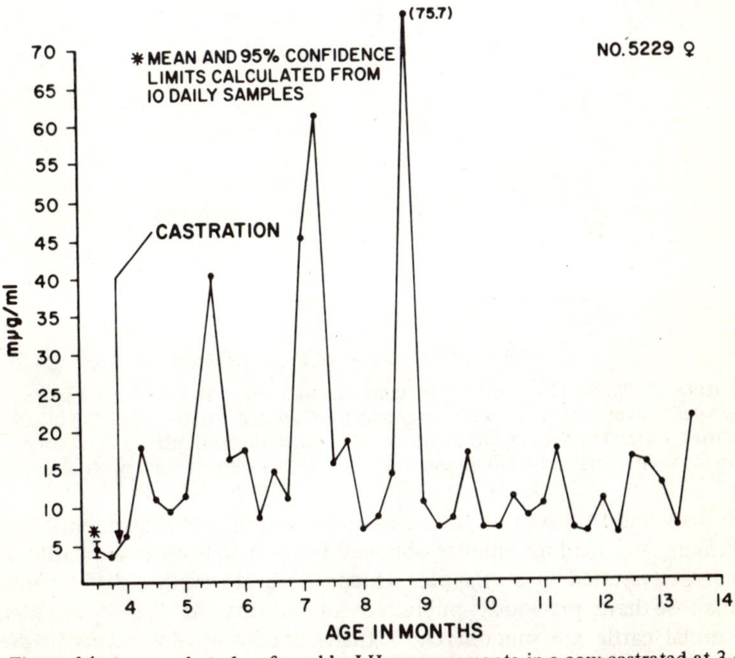

Figure 14. A second study of weekly LH measurements in a cow castrated at 3-4 months of age and studies through the age of sexual maturation as in Figure 13.

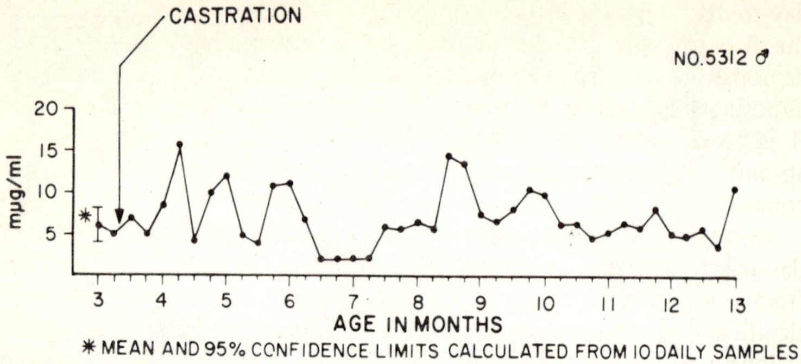

Figure 15. A third example of LH levels in a 3-4 month-old castrate followed through sexual maturation with weekly bleedings as in Figures 13 and 14.

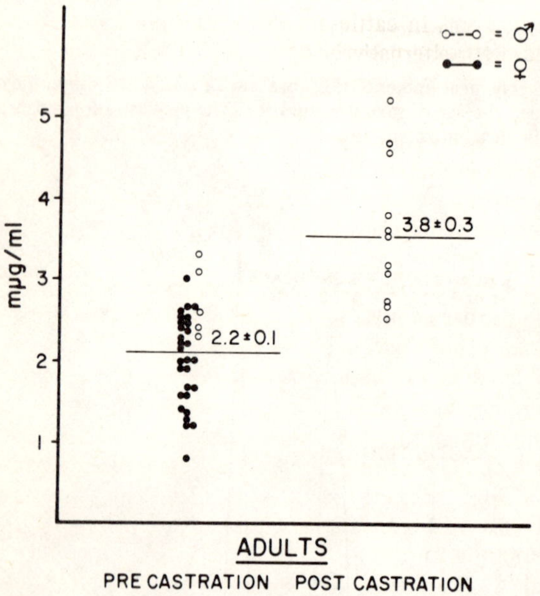

Figure 16. Scattergram of serum samples randomly obtained from eugonadal sexually mature cattle of both sexes and from a small number of castrated adult males. Note that the magnitude of rise after castration is similar to that observed in the prepubertal cattle after castration. All samples were run in a single assay.

In fact, the LH variation from day-to-day within individual animals was striking and random samples obtained from castrate or intact cattle at all ages resulted in overlapping groups. This observation differs from that we have previously published for humans. At 1 to 5 months, normal cattle are prepubertal; sexual maturity usually occurs in the cattle studied at 12 to 13 months of age. After castration, considerable day-to-day variation in LH occurred in all animals. These studies are the first utilizing direct measurements of serum gonadotrophin which demonstrate an active positive feedback between gonadal secretions and hypothalamic-pituitary-gonadotrophin secretion in prepubertal animals. It is to be noted that we obtained blood from the jugular vein; the absolute concentrations may be higher than other peripheral vein concentrations and the fluctuations may be greater.

However, several puzzling observations result from these studies: (1) day-to-day variation in LH was great, but taking random serum samples from cattle 2 weeks of age through adulthood, we were unable to show significant increases; (2) four animals castrated at 5 months of age and followed to beyond the normal age of sexual maturity failed to show a secondary rise in LH; (3) after castration, adult males showed approximately a two-fold rise in serum LH concentrations; prepubertal animals also showed a similar increase. These observations suggest that pubertal mechanisms in cattle are different from those we believe to exist in children or alternatively, that some problem with our LH assay at lower concentrations prevents us from 'seeing' small gradations. The fact that we did observe increases in LH concentrations after prepubertal castration and that these elevated values were in the same range as in adult castrates makes the latter unlikely. Furthermore, any secondary rise from postulated 'prepubertal castrate' concentrations to adult castrate concentrations should have been easily discernible, but no changes were observed.

We have recently reviewed pubertal mechanisms in children, relating available studies in animals and children (Odell and Moyer, 1969). Johanson *et al.* (1969) have recently measured LH in serum from large numbers of children and have demonstrated that LH remains low and relatively constant until about age 8 when it begins to increase gradually, reaching adult concentrations at approximately ages 15 to 16. It is to be emphasized that these were population changes and may not portray changes in a single subject. Guyda *et al.*, (1969) have measured LH in prepubertal girls with gonadal agenesis and in eugonadal prepubertal girls and found the former to be statistically higher than the latter, 4.5 $\pm$ 0.3 (mean $\pm$ SE) versus 2.7 $\pm$ 0.3. Even the elevated prepubertal castrate concentrations were not as high as adult women (10.2 $\pm$ 0.8). Thus, the studies described herein indicate that in cattle the mechanisms of pubertal control may be different from those in humans. These differences may permit important insights into the control mechanisms of puberty in both species and additional studies are in progress designed to clarify them.

CONCLUSIONS

A radioimmunoassay for bovine LH has been utilized to study

changes in circulating serum LH prior to and after castration of 1- to 5-month-old cattle of both sexes. Puberty normally occurs at about 12 to 13 months of age. Castration resulted in elevation in serum LH; this observation demonstrates the existence of a dynamic hypothalamic - pituitary - gonadal interrelationship prior to sexual maturation.

Concentrations of LH measured daily in individual prepubertal animals both before and after castration showed considerable day-to-day variation. No differences in mean LH concentrations calculated from single random samples obtained from animals of varying ages ranging from 1 to 15 months could be observed. Four animals castrated at 3 to 4 months of age had serum LH measured weekly until 14 months of age and no secondary rise was observed at 12 to 14 months. After castration, adult animals appeared to exhibit a rise in serum LH similar in magnitude to that observed in castrated prepubertal animals. These studies indicate that the dynamic control system which exists in cattle prior to puberty may differ from that in children.

GENERAL DISCUSSION

PARTICIPANTS: LLOYD, GREEP, DONOVAN, PRESL, LUNENFELD, LOUWERENS, ESHKOL, WEISZ AND STEGNER.

1. LH levels in the adult cow are in general very low except at the mid-cycle peak. Mean concentrations are slightly higher before than after the peak. Maximal oestrogen output occurs 12 to 24 hours before the LH peak while progesterone reaches its maximum on Day 13 of the 20-day cycle (Presl). The marked peak and the large day to day fluctuations in LH levels resemble closely the variations found in the human. LH levels in women rise very sharply immediately after ovariectomy, then fall rapidly. Subsequently, as in post-menopausal women, they show greater fluctuations than in the eugonadal subject, perhaps because the ovarian steroid feedback, necessary to 'damp down' the system, is lacking. Little information is available from cows immediately after ovariectomy but there does not seem in all cases to be an immediate large rise. Data on seasonal variations in cattle, which breed less well in hot conditions, are not available.

2. In the 14-day-old prepubertal rat a very transient LH peak coincides with a 'maturation crisis' in the ovary when certain steroidogenic enzymes first appear in the interstitial gland tissue and oestrogenic activity is first evident (Weisz). No data on the cow ovary at a comparable age is available, but specialized interstitial tissue is already present in the calf foetus.

3. Measurement of urinary LH in groups of girls aged 6 to 14 years, by the ovarian ascorbic acid depletion assay, suggested no significant change in spite of the onset of menstruation (probably anovulatory bleeding – Lunenfeld). However, if boys or girls are grouped according to

the stage of sexual maturity, immunoassays indicate a definite upward trend in both LH and FSH.

4. The agonadal human produces more gonadotrophin after the normal age of puberty than before. If maturity is estimated by bone development maturation appears to be delayed in such subjects. Possibly maturation of the hypothalamus is also delayed, and shifts in the level of LH in animals castrated while very young should be sought later than the age of normal puberty.

REFERENCES

Bagshawe, K.D., Wilde, C.E. & Orr, A. Hilary (1966). *Lancet,* **1,** 1118.

Byrnes, W.W. & Meyer, R.K. (1951). *Endocrinology,* **48,** 133.

Greep, R.O., Van Dyke, H.B. & Chow, B.F. (1941). *Proc. Soc. exp. Biol. Med.* **46,** 644.

Guyda, H.J., Johanson, A.J., Migeon, C.J. & Blizzard, R.M. (1969). *Pediat. Res.* **3,** 538.

Johanson, A.J., Guyda, H., Light, C., Migeon, C.J. & Blizzard, R.M. (1969). *J. Pediat.* **74,** 416.

Kallas, H.C. (1929). *C. r. Seanc. Soc. Biol.* **100,** 979.

Kiddy, C.A. & Odell, W.D. (1969). *J. Anim. Sci.* **29,** 192.

Kulin, H.E., Kaplan, S.L. & Grumbach, M.M. (1969). *Prog. Endocr. Soc.,* **87.**

Kulin, H.E., Rifkind, A.B., Ross, G.T. & Odell, W.D. (1967). *J. clin. Endocr. Metab.* **27,** 1123.

Odell, W.D., Abraham, G., Raud, H.R., Swerdloff, R.S. & Fisher, D.A. (1969). *Acta Endocr. Copenh.* Supp. **142,** 54.

Odell, W.D. & Kiddy, C.A. (1969). *Clin. Res.* **27,** 145.

Odell, W.D. & Moyer, D. (1969). *Reproductive Physiology.* St. Louis: Mosby.

Odell, W.D., Parlow, A.F., Cargille, C.M. & Ross, G.T. (1968). *J. clin. Invest.* **47,** 2551.

Odell, W.D. & Ross, G.T. (1966). *J. clin. Invest.* **45,** 1052.

Odell, W.D., Ross, G.T. & Rayford, P.L. (1966). *Metabolism* **15,** 287.

Odell, W.D., Ross, G.T. & Rayford, P.L. (1967). *J. clin. Invest.* **46,** 248.

Raiti, S., Johanson, A., Light, C., Migeon, C.J. & Blizzard, R.M. (1969). *Metabolism,* **18,** 234.

Rifkind, A.B., Kulin, H.E. & Ross, G.T. (1967). *J. clin. Invest.* **46,** 1925.

Footnote: This manuscript was presented in Birmingham, England, in September 1969. Since that time, while this book was in press, a great number of additional publications have appeared. We have recently published a detailed review of prepubertal pituitary – gonadal relations (Odell *et al*, Pfizer Symposium Proceedings 1970) in which these later publications are reviewed.

Index Part I

(NOTE.–Where the same subject is mentioned on more than one consecutive page of the same paper, only the first page number is given.)

A

N-Acetylheptulosaminic acid, in assay of NANA, 92
N-Acetylneuraminic acid, *see* Neuraminic acid
Actinomycin D, as inhibitor of protein synthesis, 207
Adsorption, of gonadotrophins to glass surfaces, etc., 109, 205
Alanine, tritiated, use of, 219
Albumin, bovine serum, gel filtration of, 48
  in preparation of gonadotrophins, 16, 18, 39, 52
Alkali, use of, in structural studies on gonadotrophins, 77, 81, 93
Amberlite IRC-50, use of, in purification of gonadotrophins, 131, 138, 177
Amidoblack 10B, use of, as protein stain, 18, 72, 122, 132
α-Amino-butyric acid, in analysis of amino acids, 83, 94
Ampholytes, use of, in electrofocusing, 70, 200, 203
Asparagine, in structural studies on gonadotrophins, 208

B

Benzoic acid, use of, in preparation of HCG, 167
Biogel, use of, in purification of gonadotrophins, 77, 79, 85, 90, 167, 171
Borohydride, use of, in structural studies on gonadotrophins, 77, 81, 93, 171, 173
n-Butanol, use of, in purification of HMG, 57, 200

C

CM-cellulose (and Sephadex), use of, in purification of gonadotrophins, 3, 7, 9, 15, 79, 117, 119, 138, 146, 203
Carbohydrates, in gonadotrophins, *see* FSH, HCG, LH, etc.
Carboxypeptidase, use of, in structural studies, 142, 156
Choriocarcinoma, male, source of HCG, 166, 207
Chorionic gonadotrophin, *see* Human chorionic gonadotrophins
Chymotrypsin, effect of, on gonadotrophins, 90, 107, 172, 215
Chymotrypsinogen A, gel filtration of, 48
Coccobacillus enzymes, use of, in structural studies, 79, 92, 172
Countercurrent distributions, use of, in structural studies, 143, 149, 152, 207
Criteria of purity, 2, 11, 70, 205
Cyanogen bromide, in structural analysis, 146
Cysteic acid, in oxidised gonadotrophins, 83, 88, 91, 95, 126, 134, 146, 174
Cysteine (and cystine), and activity of gonadotrophins, 53, 95, 171, 174, 176
  in gonadotrophins, 50, 53, 78, 83, 88, 91, 94, 96, 113, 125, 127, 129, 134, 142, 147, 161, 165, 168, 171, 173, 176
  use of, in papain digestions, 107
Cytochrome, gel filtration of, 48

D

DEAE-cellulose, use of, in purification of gonadotrophins, 23, 35, 39, 58, 61, 75, 117, 119, 131, 200, 203
DEAE-Sephadex, use of, in purification of gonadotrophins, 3, 7, 9, 75, 78, 159, 167, 169, 170

DNA, ovarian, and influence of FSH *in vitro*, 98, 100, 105, 210
Density gradient electrophoresis, use of, in studies on gonadotrophins, 1, 4, 8
Dextran blue, gel filtration of, 47
Di-isopropyl fluorophosphate (DFP), use of, in extraction of gonadotrophins, 1, 6, 8, 13
in treatment of enzymes, 156
5-Dimethylamino-l-naphthalene sulphonyl chloride (DNS-Cl 'dansyl' chloride), use of, in structural studies, 107, 110, 112, 132
Disc electrophoresis, use of, in studies on gonadotrophins, 1, 4, 11, 15, 18, 39, 46, 53, 123, 131, 136, 139, 143, 150, 161, 167, 186, 188, 205
Dithiothreitol (Cleland's Reagent), use of, in structural studies of HCG, 155, 161
Dowex 50X2, use of, in structural studies of FSH, 108, 112

E

Edman reaction, use of, in end group analysis, 141
Electrofocusing, for purification of gonadotrophins, 21, 70, 200
Eutectic points, of phosphates, 218

F

Fluorodinitrobenzene (FDNB), use of, in structural studies, 141
Follicle stimulating hormone (FSH)
Adsorption of, to surfaces, 109
Bioassay of, *in vivo,* 2, 13, 15, 24, 41, 49, 51, 53, 58, 98, 102, 105, 109, 172, 189, 195, 201
in vitro, 98, 175, 178, 210
Disappearance rate (half-life) of, 179, 185, 188
Diurnal variation of, 190
Inhibitors of, 100, 102, 210
Ovine, 57, 218
Pituitary, alkaline degradation of, 209
amino acids in, 78, 81, 83, 88, 94, 110, 113, 115
antisera to, 109, 114, 177, 186, 188, 193, 196, 210
biological activity of, 15, 21, 24, 29, 33, 35, 57, 77, 98, 102, 105, 107, 110, 114, 171, 189, 195, 205, 209, 214, 219
carbohydrates in, 77, 90, 110, 113, 115, 171, 175, 209
effect on, of acid, 209
of chloramine T, 77, 81, 84, 95, 171, 174, 209
of chymotrypsin, 90, 107, 172
of diazo-sulphanilic acid, 77, 85, 90, 96, 171, 175
of photo-oxidation, 85, 89, 95, 171, 173, 175
electrofocusing of, 21
electrophoresis, use of, in purification of, 18, 23, 30, 34, 114, 215
end group analysis of, 209
homogeneity of, tests for, 37, 77, 209, 215
immunoelectrophoresis of, 109, 114
immunological activity of, 26, 29, 33, 36, 57, 98, 105, 108, 112, 114, 171, 177, 209
LH activity, in preparations of, 2, 15, 21, 24, 29, 33, 35, 100, 177, 178, 196
LH, separation from, 3, 15, 21
molecular weight of, 77, 81, 85, 90, 92, 96, 173, 175, 209, 214
neuraminic acid (N-acetylneuraminic acid, NANA, sialic acid) in, 19, 78, 91, 98, 101, 105, 110, 113, 172, 175, 206, 209
preparation and purification of, 77, 107, 114, 171
radioimmunoassay of, 24, 108, 112, 114, 171, 177, 181, 185, 188, 193, 209, 218
radioiodination of, 110, 114, 177, 185, 205, 211, 215
sedimentation coefficient of, 77
stability of, 54, 78, 105, 172, 175, 205, 209, 215, 218
standard preparations of, 185, 189, 195, 219
structural studies on, 77, 105, 171, 209
sub-units of, 175
ultraviolet absorption of, 54, 79,

85, 87, 90, 108
Presence of, in preparation of HCG, 203
Production rate of, 179, 185
Renal clearance of, 185, 190, 194
Urinary, amino acids in, 47, 50, 53, 206
antisera to, 52, 216
biological activity of, 22, 24, 29, 33, 35, 39, 57, 200, 204, 206, 216
carbohydrates in, 39, 48, 51, 53, 204
diffusion constant of, 39, 46, 53
electrofocusing of, 200
electrophoresis, use of, in purification of, 23, 30, 34, 39, 43, 46
excretion of, during menstrual cycle, 216
extraction and purification of, 39, 57, 191, 200
homogeneity of, tests for, 37, 39, 46, 53, 205
immunoelectrophoresis of, 39, 52, 54, 109, 114, 205
immunological activity of, 26, 29, 33, 36, 39, 57, 216
levels of, in menstrual cycle, 216
LH activity, in preparations of, 22, 24, 29, 33, 35, 39, 65, 67, 117, 200, 204
LH, separation from, 22, 39, 117, 128, 200
molecular weight of, 39, 46, 50, 53
neuraminic acid in, 39, 48, 51, 53, 206
radioimmunoassay of, 24, 216
sedimentation coefficient of, 39, 46, 53
stability of, 49, 51, 54, 200, 204, 217
standard preparations of, 24, 40, 55, 218
ultraviolet absorption of, 39, 40, 46, 49, 53
β-Formyl pyruvic acid, in assay of NANA, 92
Freeze-drying (lyophilization), of gonadotrophins, 54, 109, 114, 119, 172, 200, 204, 217
L-Fucopyranose, formula of, 94
Fucose, in gonadotrophins, 80, 91, 94, 113, 134, 140. 145, 172, 208
Fucosidase, use of, in structural studies, 80, 92

G

D-Galactopyranose, formula of, 94
Galactosamine (2-amino-2-deoxy-D-galactose), and acetyl derivative, in gonadotrophins, 78, 91, 94, 134, 140, 208
Galactose, in gonadotrophins, 78, 80, 91, 93, 134, 140, 172, 209
Galactose oxidase, use of, in structural studies, 77, 81, 171, 175
Galactosidase, use of, in structural studies, 79, 172
Gas liquid chromatography, use of, in carbohydrate studies, 80, 209
Gastrin, adsorption of, to glass, 205
Gel filtration, *see* Biogel, Sephadex
Genetic polymorphism, and gonadotrophins, 12, 166, 207
Glucosamine (2-amino-2-deoxy-D-glucose), and acetyl derivative, 78, 80, 83, 91, 94, 113, 115, 134, 140, 208
Gonadotrophins, *see also* Follicle stimulating hormone, Human chorionic gonadotrophin, Luteinizing hormone
Standard preparations, use of, 217, 219
HCG (2nd international standard), 71, 218
HMG (2nd IRP-HMG), 24, 41 58, 118, 181, 190, 196, 217
NIH preparations, 2, 111, 131, 133, 145, 149, 150, 181, 217
pituitary (LER preparations, MRC preparations), 149, 185, 189, 195, 219
Growth hormone, chemical studies on, 2, 13, 205, 216
labelling of, 219
standard preparation of, 217

H

Haemagglutination inhibition assay, use of, 201, 215
Heptalosaminic acid, in assay of NANA, 92
Hexosamines, in gonadotrophins, 39,

48, 51, 53, 108, 110, 112, 134, 140, 145
Hexoses, in gonadotrophins, 39, 48, 51, 53, 110, 112, 115, 134, 140, 145, 209
Histidine, and activity of FSH, 171, 173, 175
and structures of FSH, 95
Human chorionic gonadotrophin (HCG), amino acids in, 134, 156, 161, 206, 219
antisera to, 6, 9, 52, 54, 67, 72, 109, 118, 124, 181, 196, 205, 210, 218
biological activity of, 12, 70, 74, 134, 159, 181, 205, 210, 214
biological assay of, 72, 74, 166, 218
C-terminal amino acids of, 155, 161, 208
carbohydrates in, 12, 155, 159, 207, 208
carboxymethylation of, 155, 208
choriocarcinoma, presence in, 166, 207
criteria for purity of, 70, 155, 161, 205
effect on, of acid, 156, 215
of acetylation, 210
of guanidine, 160, 208
of succinylation, 210
of tetranitromethane, 210
of urea, 159, 161, 208
electrofocusing of, 70
electrophoresis of, 70, 72, 75, 161, 167, 181 205
homogeneity of, tests for, 12, 40, 70, 72, 75, 155, 161, 167, 170, 205, 209
immunological properties of, activity, 134, 205, 210
complement fixation, 71, 74, 215
double diffusion, 72, 181
haemagglutination, 70, 74, 181, 215
immunoelectrophoresis, 40, 72, 109, 181
radioimmunoassay, 167, 181, 210, 218
use in absorption of antisera to FSH, 52, 186, 188, 196, 212
isoelectric point of, 72, 74
molecular weight of, 155, 161, 164, 214
N-terminal amino acids of, 155
neuraminic acid (N-acetylneuraminic acid, NANA, sialic acid) in, 12, 71, 74, 156, 159, 162, 206, 218
polymorphism of, 12, 166, 207
'pro'-HCG, possible existence of, 166
purification of, 40, 70, 155, 160, 166, 203, 205, 218
radioactive labelling of, 6, 181, 211, 219
reduction of, 155, 161
standard preparations of, 71, 218
structure studies of, 77, 155, 161, 208, 210, 219
sub-units of, 155, 161, 207, 219
testicular tumours, presence in, 166, 207
trophoblastic tumours, presence in, 70, 207
Human menopausal gonadotrophin (HMG), *see* Follicle stimulating hormone, urinary and Luteinizing hormone, urinary
Hydrazinolysis, use of, in structural studies, 156, 159

I

Immunoelectrophoresis, use of, in studies on gonadotrophins, 39, 52, 54, 72, 109, 114, 118, 123, 128, 132, 181, 205
Immunological properties and assays, *see* separate hormones, Follicle stimulating hormone, Human chorionic gonadotrophin, Luteinizing hormone
Insulin, radioimmunoassay of, 179
structure of, 166, 206
Iodination, damage, 215
See also radioiodinated hormones
Iodoacetamide, use of, in structural studies, 155, 161, 165
Ion-exchange chromatography, *see* CM-cellulose, DEAE-cellulose and Sephadex, etc.
Iso-electric focusing, *see* Electrofocusing

J

Jack bean mannosidase, use of, in structural studies on FSH, 79, 172

K

Kaolin, use of, in extraction of HCG, 166

L

Leucine aminopeptidase, use of, in structural studies on HCG, 156, 159

Luteinizing hormone, (LH, interstitial cell stimulating hormone, ICSH), bioassay of, 2, 10, 15, 24, 40, 58, 100, 118, 120, 145, 149, 152, 196, 203, 207
  bovine, structural, studies on, 133, 139, 147, 149, 152
  inhibitor, removal of during purification, 133
  ovine, studies on, 12, 57, 131, 133, 138, 149, 207, 219
  pituitary, amino acid sequence of, 143, 147, 215
    amino acids in, 117, 127, 129, 131, 139, 152, 163, 166, 206
    and HCG, comparison of, 181
    antisera to, 178, 205, 210
    biological activity of, 1, 10, 13, 15, 21, 24, 29, 30, 33, 35, 98, 100, 131, 140, 144, 147, 149, 166, 177, 196, 203, 205, 210, 214, 216
    C-terminal amino acids of, 140, 142, 149
    carbohydrates in, 134, 139, 144
    diffusion coefficient of, 153
    dissociation of, *see* sub-units
    distribution coefficients of, 6
    effect on, of acid, 131, 136, 139, 143, 149, 207
      of guanidine, 143
      of urea, 143, 149, 151, 208
    electrophoresis of, 1, 4, 11, 15, 18, 23, 30, 34, 131, 136, 138, 182
    frictional ratio of, 153
    FSH activity in preparation of, 2, 15, 21, 24, 29, 33, 35, 219
    homogeneity of, 2, 5, 8, 11, 21, 37, 131, 136, 139, 150, 177
    immunological properties of
      activity, 1, 13, 26, 29, 33, 36, 132, 147, 149, 178, 216
      immunoelectrophoresis, 132, 181
      radioimmunoassay, 1, 6, 13, 24, 109, 114, 149, 177, 181, 203, 218
    isoelectric point of, 139
    molecular weight of, 133, 139, 153, 207, 214
    N-terminal amino acids in, 132, 140, 143, 146, 149
    neuraminic acid (N-acetylneuraminic acid, NANA) in, 19, 140, 208, 210
    polymorphism of, 12
    'pro'-LH, possible existence of, 166
    purification of, 117, 131, 138, 150, 203, 205, 208, 216
    radioiodination of, 1, 6, 9, 178, 181, 211
    sedimentation coefficient of, 139, 152
    separation of, from FSH, 3, 15, 21
    stability of, 131, 205
    standard preparations of, 219
    Stokes' radius of, 152
    sub-units of, 131, 136, 141, 143, 149, 163, 166, 204, 207, 219
    TSH activity, in preparations of, 1, 6, 11
  porcine, amino acids in, 133, 149
  urinary, amino acids in, 117, 206
    and HCG, comparison of, 215
    biological activity of, 22, 24, 29, 33, 35, 40, 119, 128, 133, 204, 216
    carbohydrates in, 117
    effect on, of urea, 67
    electrophoresis of, 23, 30, 34, 114, 118, 122, 128, 211
    FSH activity, in preparations of, 22, 24, 29, 33, 35
    homogeneity of, 37, 123, 128
    immunoelectrophoresis of, 118, 123, 125, 128, 205
    immunological activity of, 26, 29, 33, 36, 204, 215
    levels of, in menstrual cycle, 216
    molecular weight of, 54, 121, 126, 129
    purification of, 40, 117, 128, 200, 204
    radioimmunoassay of, 24, 40, 45, 53, 118, 202, 204, 210
    separation of, from FSH, 22, 117, 200

stability of, 119, 128, 204
standard preparations of, 24, 40, 55, 118, 218
Lysosyme, crystal structure of, 219

M

D-Mannopyranose, formula of, 94
Mannose, in gonadotrophins, 78, 91, 94, 134, 140, 209
Metaphosphoric acid, use of, in purification of gonadotrophins, 57, 138, 200
Methionine, and activity of gonadotrophins, 95, 171, 174, 176
in gonadotrophins, 50, 78, 83, 88, 91, 94, 113, 125, 135, 142, 144, 152, 158, 164, 168, 171, 173, 176, 207
Methionine sulphoxide, presence of, in hydrolysates, 126
Methylene blue, use of, in photo-oxidation of FSH, 85, 173

N

Neuramanic acid, (N-acetylneuraminic acid, NANA, sialic acid), and activity of gonadotrophins, 53, 75, 98, 140, 175, 206, 210
determination of, 16, 19, 49, 71, 79, 91, 110
N-glycoll derivatve, 79
presence of, in gonadotrophins, 12, 19, 39, 48, 51, 53, 71, 74, 77, 98, 110, 113, 140, 156, 159, 160, 162, 167, 172, 175, 206, 218
Neuraminidase, action of, on gonadotrophins, 79, 92, 96, 98, 140, 171

O

Ovarian ascorbic acid depletion test (OAAD), use of, 2, 10, 15, 24, 40, 58, 100, 118, 145, 149, 152, 196, 207
Ovarian augmentation test, use of, 2, 15, 24, 41, 51, 58, 99, 105, 109, 172, 189, 195, 201

P

PAS, stain for glycoproteins, 19
Papain, use of, in structural studies, 107, 209
Periodate oxidation, use of, in structural studies, 77, 81, 90, 171, 173
Photo-oxidation, effect of, on FSH, 85, 89, 95, 171, 173, 175
Pituitary glands, source of gonadotrophin, 2, 6, 13, 15, 22, 54, 107, 131, 137, 203
Polyacrylamide gel, use of, in preparative electrophoresis, 15, 23, 30, 34, 39, 43, 49, 52, 114, 118, 122, 128, 136, 162, 203
See also, Disc electrophoresis
Polyalanine, attachment of, to proteins, 219
Polymorphism, in preparations of gonadotrophin, 12, 207
Progesterone, production rate of, 185
'Pro'-HCG, LH, possible existence of, 166
Proline, estimation of, 125, 141
in structural studies, 83, 141, 146
oxidation of, in FSH, 84, 88, 91
presence of, in FSH, 78, 83, 113
in HCG, 163, 167, 206
in HMG, 50, 126, 129
in LH, 126, 129, 135, 206
Proteins, determination of, by Lowry colorimetric method, 2, 22, 24, 108, 204
by U.V. absorption, 23, 44, 53, 108, 150, 162, 167, 178, 204
precipitation of, 187, 191, 196
Proteolytic enzymes, action of, 6, 8, 13, 77, 90, 96, 105, 107, 165, 171, 204, 207, 209

Q

QAE-Sephadex, use of, in purification of gonadotrophins, 15, 200

R

Radiation,–inactivation, 92, 214
Radioimmunoassay, of gonadotrophins, *see* individual hormones
statistical procedures for, 216
Radioiodination, of FSH, 110, 114, 177, 185, 205, 211, 215
of HCG, 6, 181, 211, 219
of LH, 1, 6, 9, 178, 181, 211

of TSH, 211
Ram, testicular enzymes of, 79, 92, 172
Riboflavin, 23, 37
Ribonuclease, 12

S

SE-Sephadex, use of, in purification of gonadotrophins, 58, 65, 117, 120, 124, 128, 138, 167, 200
Sedimentation studies, *see* Ultracentrifugal studies
Seminal vesicles, assay method for gonadotrophin, 72, 203, 218
Sephadex gels, use of, in purification of gonadotrophin G25, 44, 51, 108, 110, 112, 114, 146, 209
G50, 200
G75, 186, 189, 191
G100, 1, 3, 6, 23, 27, 37, 39, 43, 46, 54, 58, 70, 72, 75, 117, 119, 121, 128, 132, 136, 138, 143, 151, 203, 216
G200, 58, 65, 117, 119, 124, 200
See also CM-Sephadex, DEAE-Sephadex, QAE-Sephadex, SE-Sephadex
Serine, and carbohydrate-peptide linkage in gonadotrophins, 93, 209
Sialic acid, *see* Neuraminic acid
Sodium dodecyl sulphate, 67, 200
Starch gel electrophoresis, use of, in studies on gonadotrophins, 12, 70, 72, 139, 181, 205
Steroids, radiommunoassay of, 206
Sucrose density gradient, *see* Density gradient electrophoresis
Sulphosalicylic acid, use of, in purification of HMG, 57, 200

T

Tertiary structure, of gonadotrophins, 105, 166, 175, 205, 210, 214, 219
Thin layer chromatography, use of, 48, 107, 141
Threonine, and carbohydrate-peptide linkage in gonadotrophins, 93
Thymidine, in assay for FSH, 98, 100, 210
Thyroid stimulating hormone (thyrotrophin, TSH), bioassay of, 2, 13
in preparations of LH, 1, 219
polymorphism of, 12
radioimmunoassay of, 177, 205, 211
standard preparation of, 2
Trophoblastic tumours, source of HCG, 70, 207
Trypsin, 90, 96, 107, 141, 165, 172, 208, 219
Tryptophan, and activity of gonadotrophins, 171, 173
estimation of, 48, 78, 90, 134, 163
in gonadotrophins, 50, 84, 89, 112, 126, 135, 141, 164, 169
oxidation of, 89, 95, 171, 173
Tyrosine, and activity of gonadotrophins, 171, 173
estimation of, 47, 125, 134, 163
in gonadotrophins, 50, 78, 83, 88, 91, 113, 126, 135, 141, 156, 164, 169
in structural studies, 143, 146, 157, 208, 210, 219
oxidation of, 88, 91, 95, 171, 173

U

Ultracentrifugal studies, on gonadotrophins, 39, 44, 46, 53, 75, 77, 92, 139, 140, 149, 155, 159, 205, 208
Ultraviolet absorption, and determination of protein, 23, 44, 53, 108, 150, 162, 167, 178, 204
of gonadotrophin, 39, 46, 79, 85, 87, 140, 204
Uterine weight assay, use of, 118

X

X-ray crystallography, use of, 210

Y

Yeast, source of galactosidase, 79

Z

Zinc acetate, use of, in preparation of HMG, 58
Zone electrophoresis, use of, in preparation of gonadotrophins, 12, 138

Index Part II

(Note.–Where the same subject is mentioned on more than one consecutive page of the same paper, only the first page number is given.)

A

Accelerated development, periods of, 326, 330, 342, 349, 369
Actinomycin D, effects on follicle histology, 290
effects on protein synthesis, 287
interaction with PMSG, 288
Adrenal gland, and gonad development, 244
ultrastructure of cortical cells, 237, 364
Age, and gonadotrophin production, 262, 342, 345, 369
and sensitivity to gonadotrophins, 263, 266, 322, 325, 329, 330, 332, 334, 349
and steroid synthesis, 224, 237, 244, 263, 267, 270, 309, 333, 345, 349, 384
of mouse, *foetus,* 232, 233, 237; *1 day,* 233, 237, 264, 267; *2 days,* 264; *3 days,* 268; *4 days,* 229, 237; *5 days,* 234, 237, 260, 352; *6 days,* 266, 355; *7 days,* 223, 229, 231, 250, 266, 307, 352, 354, 359, 364; *8 days,* 260; *10 days,* 231, 354; *11 days,* 229, 326, 362; *12 days,* 249; *13 days,* 326, 349, 362; *14 days,* 223, 229, 250, 253, 314, 320, 322, 349, 352, 354; *15 days,* 266, 272, 326; *16 days,* 266, 272; *17 days,* 272; *18 days,* 229, 268, 283, 324, 349; *21 days,* 223, 230, 292, 314, 320, 322, 352, 355; *22 days,* 283, 292; *23 days,* 292; *28 days,* 223, 316, 320, 322, 352, 355; *30 days,* 266; *35 days,* 223, 352, 355
of rat, *1 day,* 333; *2 days,* 333; *4 days,* 249, 347; *5 days,* 311, 342, 348, 368; *7 days,* 224, 342, 345; *8 days,* 237; *9 days,* 249, 345; *10 days,* 249, 344, 345, 349, 368; *11 days,* 333, 345; *12 days,* 346; *14 days,* 344, 345, 384; *15 days,* 349, 369; *18 days,* 224, 344, 345, 348; *20 days,* 369; *21 days,* 225, 346; *24 days,* 344, 349; *26 days,* 225; *28 days,* 346.
Aging ovary, similarities with ovary of androgenized animal, 366
Agranular reticulum, see Ultrastructure
Amino acids, uptake and incorporation into protein, effects of gonadotrophins, 280, 282, 292, 325
Androgens, direct effect on ovary, 351, 354
lack of production by grafted ovaries, 333
See also Androgenized mouse and Androgenized rat
Androgenized mouse (treated neonatally with androgen), 351, 362
abnormal enzyme production and activity, 359, 360
optimum age for androgen injection, 360
potential normality of ovary, 366
ultrastructure of ovary, 363
Androgenized rat, 332, 339
precocious vaginal opening, 347, 349
response to gonadotrophins, 351
serum LH, 344 346
Anterior pituitary, basophils (gonadotrophs) in rats, 345, 349
gonadotrophin content, normal and androgenized rats, 340
uptake of ^{3}H-oestradiol, 368
See also Feed-back between gonad and hypothalamus-pituitary
Antibodies, to rabbit serum, failure to develop in immature mice, 258
Antisera, against HCG, absence of cross reaction with mouse LH, 324
against rat pituitary gonadotrophins, 250, 259

for eliminating endogenous gonadotrophins, 250, 259
for production of biologically pure FSH, 250, 313
Antra, development in response to gonadotrophins, 255, 263, 272, 278
Atresia of follicles after gonadotrophin deprivation, 261
of follicles after hypophysectomy, 280
of follicles in FSH-stimulated ovaries, 322
of follicles in grafted infantile ovaries, 333
of oocytes in androgenized mice, 351, 363
of oocytes, transformation of associated granulosa cells, 231
Autoradiography, for studying development of stroma, 229
for studying kinetics of follicle development, 233, 314, 352

B

Basal lamina (basement membrane of electron microscopy) of follicles in androgenized mouse, 364
of young normal follicle, 233
Basement membrane, of granulosa, dependence on gonadotrophins, 252, 255
formátion by pregranulosa cells, 231
in androgenized mouse, 354, 359
of 'bizarre' follicles, 230, 231
Basophils, of rat anterior pituitary, 345, 349
Blood vessels, development in theca of rat, 349
effects of gonadotrophins, 255, 263
in foetal ovary, 233
in infantile ovary, 229, 234
Brain, uptake of labelled oestrogen, 368

C

Castration, effects on gonadotrophin release in normal and androgenized rat, 334
effects on serum LH, 347, 373
Cat, transformation of oogonia, 227
Cattle, age of puberty, 372
effects of castration on serum LH, 373
LH fluctuations during oestrus cycle, 373
Cell population kinetics, *see* Follicles
Cell proliferation, *see* Mitosis
Cerebellum, uptake of labelled oestrogen, 368
Cerebral cortex, uptake of labelled oestrogen, 368
Collagen, formation in ovarian stroma of androgenized mouse, 364
Connective tissue, in neonatal ovary, 234
in ovary of androgenized animal, 351, 362
Constant oestrus syndrome, *see* Androgenized mouse and Androgenized rat
Corpora lutea, in androgenized animals, 351, 355, 362
in transplanted ovaries, 333
Cytology, of infantile stroma, 229

D

Degeneration, of interstitial and stromal cells in androgenized mouse, 364
Dehydroepiandrosterone, as substrate for 3β-hydroxysteroid dehydrogenase, 352
Deoxyribonucleic acid (DNA) synthesis, depression by HCG and LH, 279, 303
in absence of FSH (*in vitro*), 330
in germ cells of rabbit ovary, 240
procedure for measuring rate of synthesis, 293
stimulation by FSH and PMSG, 279, 292, 294, 325
synthesis phase, *see* S-phase
Differential counts of Follicles, *see* Follicles
Doubling time, of granulosa cell populations, *see* Follicles, developmental kinetics

E

Early androgen syndrome, *see* Andro-

genized mouse and Androgenized rat
Early oestrogen syndrome, and oestrogen uptake by brain tissues, 368
Electron microscopy, preparation of ovarian tissue, 232, 260, 362
See also Ultrastructure
Endoplasmic lamellae, *recticulum, vesicles, see* Ultrastructure
Enzymes, (of) citric acid cycle, 308
deficiencies in androgenized animals, 346, 354, 359, 365
glucose-6-phosphate dehydrogenase, 308
glyceraldehyde phosphate dehydrogenase, 308
intracellular location, 235
lactic dehydrogenase, 308
NADP diaphorase, 308
NADP-specific isocitrate dehydrogenase, 308
RNA polymerase, after PMSG stimulation, 301
steroid enzymes, abnormality in androgenized animals, 360, 365
first appearance in rat, 349, 384
in foetal ovary, 244, 309
stimulation of activity by gonadotrophins, 308, 349
3β-hydroxysteroid dehydrogenase, 224, 236, 237, 263, 309, 345, 349, 354
Ergastoplasmic lamellae, *see* Ultrastructure

F

Feed-back, between gonad and hypothalamus-pituitary, 227, 334, 347, 369, 371, 383
in androgenized animals, 347
Ferret, ovarian development, 226
Follicle cells, origin, 333
transformations, 231
ultrastructure, effects of gonadotrophins, 216
See also Granulosa and Theca
Follicle cysts, 355, 362
Follicle stimulating hormone (FSH)
duration of stimulating effect, 295, 305, 327
effects in androgenized rats, 351
effects on DNA, RNA and protein synthesis, 292, 303
effects of follicle cell population kinetics, 305, 312
effects of follicular growth, 252, 257, 272, 274, 308, 320
effects on ovarian histology and histochemistry, 252, 257, 308
effects on ultrastructure, 260
effects on uptake of thymidine, 325
exclusive stimulation of granulosa cells in mouse, 305
in normal and androgenized rats, 340
production of biologically pure FSH from HMG, 293
release in castrated animals, 334
Follicles, abnormal histology in presence of Actinomycin D, 290
'bizarre' follicles, 230, 309, 354
development in response to gonadotrophins, 252, 260, 266, 272, 274, 308
developmental kinetics, 223, 312, 352, 355, 359
differential counts of size classes, 223, 250, 253, 274, 308, 320, 322, 352, 355
effects of androgenization on kinetics and histology, 352, 354
effects of gonadotrophin deprivation, 251, 261
effects of grafting infantile ovaries into adult hosts, 267, 333
Graafian follicles in grafted ovaries, 268
inhibition of development, 225, 309
interaction with interstitial tissue, 270
isolated follicles, arrested development, 269
loss during *in vitro* culture, 277
luteinization without rupture, 335
normal histology 251, 253
procedures for isolating primary follicles, 269
ultrastructure in androgenized mice, 363

G

Germ cells, connections with granulosa cells, 240, 243
maturation, 226 239
syncytia, 240

ultrastructure, 240
See also Oocytes
Golden hamster, bridges between oocytes, 240
transformation of oogonia, 227
Golgi apparatus, *see* Ultrastructure
Gonadotrophs, *see* Basophils
Gonadotrophins, deprivation, effects on infantile ovary, 251, 259, 365
effects on granulosa ultrastructure, 244, 261, 262
effects on protein and RNA synthesis, 282
in pituitaries of androgenized rats, 340
onset of secretion in mouse, 262
release, effects of steroids in intact and castrated animals, 366
stimulation of interstitial cell maturation, 234, 236, 264
stimulation of 3β-hydroxysteroid dehydrogenase activity, 311
See also Follicle stimulating hormone, Human chorionic gonadotrophin, Luteinizing hormone, and Pregnant mare serum gonadotrophin
Grafts, into eye, 231, 269, 335
into kidney, 334
into spleen, effect on hypothalamus-pituitary feed-back, 334
of infantile ovary into adult host, 267, 333
of isolated follicles, 269
of juvenile ovary into infant host, 268
Granulosa, content of 3β-hydroxysteroid dehydrogenase, 224, 309
development from infantile stroma cells, 233
enzymes, 308
in androgenized animals, 354, 363
intercellular spaces, dependence on gonadotrophins, 251, 261
stimulation by gonadotrophins, 244, 252, 261, 305, 307
transformations of structure and activity, 231
ultrastructure, 233, 244, 260
Growth hormone, stimulation of RNA and protein synthesis, 301

H

Histochemistry, of ovary, in normal and androgenized animals, 345, 354
responses to gonadotrophins, 308
Histology, of normal and androgenized animal, 351, 308
Human chorionic gonadotrophin (HCG), effects on histology, enzyme activity and formation of theca interna, 308
Human menopausal gonadotrophin (HMG)
preparation and purification, 292
stimulation of DNA synthesis, 294
stimulation of ovarian growth, 252
Hypophysectomy, reduction in number of atretic follicles, 280
Hypothalamus, age of susceptibility to androgenization, 366
and early androgen syndrome, 332, 339, 366
change in sensitivity to gonadal hormones, 348, 371
development of vascular system, 345
lesions with effects similar to androgenization, 360
maturation, 345, 385
uptake of labelled oestrogen, 369
See also Feed-back between gonad and hypothalamus-pituitary

I

Immature animals, possible differences in steroid metabolism, 370
Immune response, to ovarian grafts, 268, 335
Intercellular bridges, between germ cells, 240, 245
Intercellular connections, in granulosa of androgenized mice, 354, 363
Intercellular spaces in granulosa, effects of gonadotrophins, 251, **255, 261**
Interstitial tissue, acquisition of sensitivity to gonadotrophins, 267
dependence on gonadotrophins, 261, 308
development, 226, 232, 345
development of steroidogenic enzymes, 345, 349
histochemical responses to gonadotrophins, 308
hypertrophy in response to PMSG, 270

in androgenized animals, 346, 364
in foetal calf, 384
in neonatal rat ovaries grafted into adults, 333
interaction with follicle cells, 270
juvenile, function, 271
maintaining LH release in rabbit, 227
origin, 226, 233, 309
possible production of follicle growth inhibitor, 271
ultrastructure, 232, 237, 261, 364
In vitro culture, effects of gonadotrophins, 272, 284, 325
loss of small follicles, 278
procedures, 99, 283, 326

L

Labelled hormones, for locating target sites, limitations, 369
Labelling index, of granulosa cells, *see* Follicles, developmental kinetics
Leydig cells, compared to adrenal cortical cells and ovarian interstitial cells, 237
Lipid droplets, in androgenized mouse ovary, 354
in infantile rat ovary, 345
possible significance, 236
See also Ultrastructure
Luteinizing hormone (LH) cyclic discharge, absence in androgenized animals, 332
depression of thymidine uptake and DNA synthesis, 303, 305
effects on histochemistry, 307, 310
effects on ovarian development, 257, 272
fluctuations during oestrus cycle, cattle, 373
improvement of tissue condition *in vitro,* 279
in a highly purified FSH preparation, 326
in normal and androgenized rats, 340, 348
pituitary concentrations, 340, 347
release, following castration, 334
release, positive feedback, 227, 366
removal from HMG, 250, 313
serum levels, 341, 348, 366, 373, 383
stimulation of follicle growth, 272, 305

M

Man, bridges between germ cells, 240
gonadotrophins in agonadal subject at age of puberty, 371
serum LH levels, 371, 383
steroid enzymes in foetal ovary, 244
Maturation of oocytes, synchronized, 239
Median eminence, maturation, 345
uptake of labelled oestrogen, 369
Microvilli, of oocyte surface, 262
Mink, transformation of oogonia, 227
Mitochondria, transformation in follicles with atretic oocytes, 231. *See also* Ultrastructure.
Mitosis, in absence of gonadotrophins (*in vitro*), 258
increased rate in response to gonadotrophins, 252, 270
Monoaminergic nerve terminals in median eminence, 346
Mouse, bridges between germ cells, 240
3β-hydroxysteroid dehydrogenase in foetal ovary, 309
See also Age, of mouse
Muscle, uptake of labelled oestrogen, 368

N

Nucleic acids, *see* Deoxyribonucleic acid and Ribonucleic acid

O

Oestradiol, for oestrogenization of male rat, 333
rate of metabolism in immature rat, 347
synthesis in ovaries of androgenized rats, 346
uptake by target tissues, 368
Oestrogenized male rat, 333
Oestrogens, chronic stimulation of LH release, 366
first appearance in rat, 384
in rat ovary, 345
in rat plasma, 263, 349

negative feed-back, 369
positive feed-back, 366, 369
production by ovarian grafts, 267, 270, 333
retention in uterus, 369
secretion in cow, 384
stimulation of RNA synthesis in uterus, 303
synthesis by ovary of androgenized rat, 346
target tissues, 368
uptake by hypothalamus and median eminence, 368

Oestrus cycle, in cattle, LH fluctuations, 373

Oocytes, differentiation and gonadotrophins, 260, 262
growth, independence of gonadotrophins and granulosa, 252, 264
loss in androgenized mice, 354, 359
microvilli, 262
See also Germ cells

Oogenesis, synchronization in rabbit, 239

Organelles, transfer between germ cells, 242. *See also* Ultrastructure

Ovary, foetal, presence of 3β-hydroxysteroid dehydrogenase, 244, 309
histology and size, effects of gonadotrophins, 251, 320
histology and size, in androgenized animals, 351
ultrastructure, effects of gonadotrophins, 244, 260
ultrastructure in androgenized animals, 363
volume, determination of, 313
weight in androgenized animals, 346, 351

Ovulation, age at onset, 267, 372
in androgenized mice and rats, 332, 355
interval after LH peak, cattle, 373
reflex, 227

P

Parabiosis, for studying gonad-hypothalamus-pituitary interrelations, 371

Pinocytosis, by germ cells, 244, 262

Polycystic ovary syndrome, 339. *See also* Androgenized mouse, Androgenized rat

Pregnancy, in androgenized animals, 360

Pregnant mare serum gonadotrophin (PMSG)
abnormal response in 7-15 day old mouse, 266
abnormal response in transplanted infantile ovary, 267
absence of response in neonate, 249, 266
half-life in circulation after injection, 271, 290
ratio of FSH- to LH-like activities, 271
response of grafted 18-day ovary, 268
stimulation of protein and RNA synthesis, 282, 301

Pregnenolone, as substrate for histochemical location of 3β-hydroxysteroid dehydrogenase, 310

Pregranulosa cells, definition, 245
in mice deprived of gonadotrophins, 260
origin, 229
ultrastructure, 233, 260

Progesterone, secretion in cow, 384

Protein synthesis, estimation, 283, 293
in presence of Actinomycin D, 287,
in response to gonadotrophins, 282, 284, 292, 295
in response to various hormones, 301

Puberty, age of onset, various species, 348, 372, 383
fluctuations of LH before onset, 349
pituitary and serum gonadotrophins, 341
precocious, in androgenized rat, 347, 349

Pulse-labelling, with ^{3}H-thymidine, 223, 229, 312, 352

R

Rabbit, germ cell relationships, 239
maturation of germ cells, 227, 239
ultrastructure of germ cells, 240

Radioimmunoassay, for rat pituitary LH,348
for serum LH, 341, 372

Rat, bridges between germ cells, 240

serum LH, 341
transplantation of neonatal ovaries into adults, 332
See also Age, of rat
Recruitment, of cells into proliferating pool by FSH, 305
Ribonucleic acid (RNA), determination of profiles, 293
determination of rate of synthesis, 293
effects on synthesis, of gonadotrophins, 282, 292, 295
effects on synthesis, of various hormones, 301, 303
polymerase, stimulation of activity by PMSG, 301
stability of messenger RNA, 290
Ribosomes, procedure for determining profiles, 293
profiles after FSH stimulation, 301, 303
Rough membranes, *see* Ultrastructure

S

S-phase of mitosis, duration after FSH treatment, 317
duration in granulosa cells of mouse, 223
in androgenized mice, 357
Seminal tubules, in oestrogenized male rat, 333
Seminal vesicles, absence of stimulation by grafted ovaries, 333
Smooth vesicles, *see* Ultrastructure
Sterility, of androgenized mice and rats, 332, 339, 351, 362
Steroids, altered synthesis in androgenized mouse, 359
metabolism in immature animals, 370
transport within cell, 235
See also Oestradiol, Oestrogens etc.
Steroid secreting cells, histochemistry, 310. *See also* Ultrastructure
Strain differences, in age-related responses to gonadotrophins, 323, 349
Stroma, *adult,* development from follicle cells, 230
in androgenized animals, 355, 364
juvenile, dependence on gonadotrophins, 253
development, 226, 228, 253
in androgenized animals, 355, 364
in grafted infantile rat ovaries, 333
ultrastructure, 233, 364

T

Target tissues, for oestrogens, 368
identification with labelled hormones, 369
Testis, ultrastructure of interstitial cells, 364
Testosterone propionate, direct effect on ovary, 354
for induction of early androgen syndrome, 333, 339, 351
Theca interna, dependence on gonadotrophins, 252, 255, 261
derivation from primary interstitial cells, 309
first appearance, 233, 349
histochemical development, 349
importance for normal response to PMSG, 270
in grafted infantile rat ovary, 333
initiation of formation by HCG, 308
vascularization, 349
Thymidine, labelled, for study of cell population kinetics, 223, 294, 312, 352
uptake and incorporation into DNA in response to gonadotrophins, 279, 294, 325
Thyroxine, stimulation of RNA synthesis, 301
Time relations, of RNA, DNA and protein synthesis after hormone stimulation, 282, 284, 289, 295, 301, 303, 314
Transit time, of growing follicles, *see* Follicles developmental kinetics
Transplantation, *see* Grafts

U

Ultrastructure, dependence on gonadotrophins, 260
differences between species, 237
effects of androgenization, 362
of capillaries, 234
connective tissue, 234
germ cells, 240, 260, 363

granulosa cells, 233, 243, 261, 310, 363
interstitial cells, 234, 237, 261, 364
pregranulosa cells, 233, 260
steroid-producing cells, 234, 364
stromal cells, 233, 364
thecal cells, 233, 261
zona pellucida, 233, 261
Uridine, incorporation into RNA, effects of gonadotrophins, 282, 292, 295
Uterus, oedema in androgenized mouse, 355
relative weight increase in 14–18 day old rat, 345
stimulation by grafted infantile ovaries, 333
uptake of labelled oestrogen, 368

V

Vagina, constant cornification in androgenized rats, 339
stimulation by infantile ovaries grafted into adults, 333
Vaginal opening, as index of sexual maturity, 346
precocious, in androgenized rat, 347
serum LH, at time of, (rats), 342
Vole, transformation of oogonia, 227

Z

Zona pellucida, effect of gonadotrophins, 258, 262
formation, 233, 244, 261
ultrastructure, 261

Printed in Great Britain by
Robert Cunningham and Sons Ltd., Alva